# Calculus of Several Variables

Third Edition

# Student Solutions Manual

ROBERT A. ADAMS
Department of Mathematics
University of British Columbia

Addison-Wesley Publishers Limited
Don Mills, Ontario • Reading, Massachusetts • Menlo Park, California • New York • Wokingham, England • Amsterdam • Bonn • Sydney • Singapore • Tokyo • Madrid • San Juan • Paris • Seoul • Milan • Mexico City • Taipei

SPONSORING EDITOR: Ron Doleman, Publisher
MANAGING EDITOR: Linda Scott
PRODUCTION COORDINATOR: Melanie van Rensburg
MANUFACTURING COORDINATOR: Sharon Latta Paterson
COVER DESIGN: Pronk & Associates



ISBN 0-201-88196-9

Printed and bound in Canada

A B C D E - MP - 99 98 97 96 95

# FOREWORD

These solutions are provided for the benefit of students using the textbook *Calculus Of Several Variables (Third Edition)* by R. A. Adams, published by Addison-Wesley Publishers Ltd. For the most part, the solutions are detailed, especially in exercises on core material and techniques. Occasionally some details are omitted — for example, in exercises on applications of multiple integration, the evaluation of the definite integrals encountered is not always given with the same degree of detail as they would be if those exercises were dealing specifically with techniques of integration. Only the even-numbered exercises in the text are solved in this Manual.

As a student of Calculus, you should use this Manual with caution. It is always more beneficial to attempt exercises and problems on your own, before you look at solutions prepared by others. If you use these solutions as "study material" prior to attempting the exercises, you can lose much of the benefit that can follow from diligent attempts to develop your own analytical powers. When you have tried unsuccessfully to solve a problem, then look at these solutions to try to get a "hint" for a second attempt.

The author is grateful to Ken MacKenzie of McGill University who did a very thorough checking of the solutions for this and earlier editions, thereby exposing for correction several errors that would otherwise have made it into this printed edition. Of course, the author accepts full responsibility for any errors that remain and would be grateful to readers who call them to his attention.

July 1995.

R. A. Adams
Department of Mathematics
The University of British Columbia
Vancouver, B.C., Canada. V6T 1Z2
adms@math.ubc.ca

# Contents

# CHAPTER 1. SEQUENCES AND SERIES

## Section 1.1 Sequences and Convergence (page 35)

**2.** $\left\{\dfrac{2n}{n^2+1}\right\} = \left\{1, \dfrac{4}{5}, \dfrac{3}{5}, \dfrac{8}{17}, \ldots\right\}$ is bounded, positive, decreasing, and converges to 0.

**4.** $\left\{\sin\dfrac{1}{n}\right\} = \left\{\sin 1, \sin\left(\dfrac{1}{2}\right), \sin\left(\dfrac{1}{3}\right), \ldots\right\}$ is bounded, positive, decreasing, and converges to 0.

**6.** $\left\{\dfrac{e^n}{\pi^n}\right\} = \left\{\dfrac{e}{\pi}, \left(\dfrac{e}{\pi}\right)^2, \left(\dfrac{e}{\pi}\right)^3, \ldots\right\}$ is bounded, positive, decreasing, and converges to 0, since $e < \pi$.

**8.** $\left\{\dfrac{(-1)^n n}{e^n}\right\} = \left\{\dfrac{-1}{e}, \dfrac{2}{e^2}, \dfrac{-3}{e^3}, \ldots\right\}$ is bounded, alternating, and converges to 0.

**10.** $\dfrac{(n!)^2}{(2n)!} = \dfrac{1}{n+1}\,\dfrac{2}{n+2}\,\dfrac{3}{n+3}\cdots\dfrac{n}{2n} \le \left(\dfrac{1}{2}\right)^n.$

Also, $\dfrac{a_{n+1}}{a_n} = \dfrac{(n+1)^2}{(2n+2)(2n+1)} < \dfrac{1}{2}$. Thus the sequence $\left\{\dfrac{(n!)^2}{(2n)!}\right\}$ is positive, decreasing, bounded, and convergent to 0.

**12.** $\left\{\dfrac{\sin n}{n}\right\} = \left\{\sin 1, \dfrac{\sin 2}{2}, \dfrac{\sin 3}{3}, \ldots\right\}$ is bounded and converges to 0.

**14.** $\lim \dfrac{5-2n}{3n-7} = \lim \dfrac{\dfrac{5}{n}-2}{3-\dfrac{7}{n}} = -\dfrac{2}{3}.$

**16.** $\lim \dfrac{n^2}{n^3+1} = \lim \dfrac{\dfrac{1}{n}}{1+\dfrac{1}{n^3}} = 0.$

**20.** $\lim n \sin\dfrac{1}{n} = \lim\limits_{x\to 0+} \dfrac{\sin x}{x} = \lim\limits_{x\to 0+} \dfrac{\cos x}{1} = 1.$

**22.**
$$\lim \frac{n}{\ln(n+1)} = \lim_{x\to\infty} \frac{x}{\ln(x+1)}$$
$$= \lim_{x\to\infty} \frac{1}{\left(\dfrac{1}{x+1}\right)} = \lim_{x\to\infty} x+1 = \infty.$$

**24.**
$$\lim\left(n - \sqrt{n^2-4n}\right) = \lim \frac{n^2-(n^2-4n)}{n+\sqrt{n^2-4n}}$$
$$= \lim \frac{4n}{n+\sqrt{n^2-4n}} = \lim \frac{4}{1+\sqrt{1-\dfrac{4}{n}}} = 2.$$

**26.** If $a_n = \left(\dfrac{n-1}{n+1}\right)^n$, then
$$\begin{aligned}
\lim a_n &= \lim \left(\frac{n-1}{n}\right)^n \left(\frac{n}{n+1}\right)^n \\
&= \lim\left(1-\frac{1}{n}\right)^n \Big/ \lim\left(1+\frac{1}{n}\right)^n \\
&= \frac{e^{-1}}{e} = e^{-2}.
\end{aligned}$$

**28.** We have $\lim \dfrac{n^2}{2^n} = 0$ since $2^n$ grows much faster than $n^2$ and $\lim \dfrac{4^n}{n!} = 0$ by Theorem 3(b). Hence,
$$\lim \frac{n^2\,2^n}{n!} = \lim \frac{n^2}{2^n}\cdot\frac{2^{2n}}{n!} = \left(\lim \frac{n^2}{2^n}\right)\left(\lim \frac{4^n}{n!}\right) = 0.$$

**30.** Let $a_1 = 1$ and $a_{n+1} = \sqrt{1+2a_n}$ for $n = 1, 2, 3, \ldots$. Then we have $a_2 = \sqrt{3} > a_1$. If $a_{k+1} > a_k$ for some $k$, then
$$a_{k+2} = \sqrt{1+2a_{k+1}} > \sqrt{1+2a_k} = a_{k+1}.$$
Thus, $\{a_n\}$ is increasing by induction. Observe that $a_1 < 3$ and $a_2 < 3$. If $a_k < 3$ then
$$a_{k+1} = \sqrt{1+2a_k} < \sqrt{1+2(3)} = \sqrt{7} < \sqrt{9} = 3.$$
Therefore, $a_n < 3$ for all $n$, by induction. Since $\{a_n\}$ is increasing and bounded above, it converges. Let $\lim a_n = a$. Then
$$a = \sqrt{1+2a} \Rightarrow a^2 - 2a - 1 = 0 \Rightarrow a = 1 \pm \sqrt{2}.$$
Since $a = 1 - \sqrt{2} < 0$, it is not appropriate. Hence, we must have $\lim a_n = 1 + \sqrt{2}$.

**32.** Let $a_n = \left(1+\dfrac{1}{n}\right)^n$ so $\ln a_n = n\ln\left(1+\dfrac{1}{n}\right)$.

a) If $f(x) = x\ln\left(1+\dfrac{1}{x}\right) = x\ln(x+1) - x\ln x$, then
$$\begin{aligned}
f'(x) &= \ln(x+1) + \frac{x}{x+1} - \ln x - 1 \\
&= \ln\left(\frac{x+1}{x}\right) - \frac{1}{x+1} \\
&= \int_x^{x+1} \frac{dt}{t} - \frac{1}{x+1} \\
&> \frac{1}{x+1}\int_x^{x+1} dt - \frac{1}{x+1} \\
&= \frac{1}{x+1} - \frac{1}{x+1} = 0.
\end{aligned}$$

Since $f'(x) > 0$, $f(x)$ must be an increasing function. Thus, $\{a_n\} = \{e^{f(x_n)}\}$ is increasing.

b) Since $\ln x \le x - 1$,

$$\ln a_k = k \ln\Big(1 + \frac{1}{k}\Big) \le k\Big(1 + \frac{1}{k} - 1\Big) = 1$$

which implies that $a_k \le e$ for all $k$. Since $\{a_n\}$ is increasing, $e$ is an upper bound for $\{a_n\}$.

**34.** If $\{|a_n|\}$ is bounded then it is bounded above, and there exists a constant $K$ such that $|a_n| \le K$ for all $n$. Therefore, $-K \le a_n \le K$ for all $n$, and so $\{a_n\}$ is bounded above and below, and is therefore bounded.

**36.** a) "If $\lim a_n = \infty$ and $\lim b_n = L > 0$, then $\lim a_n b_n = \infty$" is TRUE. Let $R$ be an arbitrary, large positive number. Since $\lim a_n = \infty$, and $L > 0$, it must be true that $a_n \ge \frac{2R}{L}$ for $n$ sufficiently large. Since $\lim b_n = L$, it must also be that $b_n \ge \frac{L}{2}$ for $n$ sufficiently large. Therefore $a_n b_n \ge \frac{2R}{L}\frac{L}{2} = R$ for $n$ sufficiently large. Since $R$ is arbitrary, $\lim a_n b_n = \infty$.

b) "If $\lim a_n = \infty$ and $\lim b_n = -\infty$, then $\lim(a_n + b_n) = 0$" is FALSE. Let $a_n = 1 + n$ and $b_n = -n$; then $\lim a_n = \infty$ and $\lim b_n = -\infty$ but $\lim(a_n + b_n) = 1$.

c) "If $\lim a_n = \infty$ and $\lim b_n = -\infty$, then $\lim a_n b_n = -\infty$" is TRUE. Let $R$ be an arbitrary, large positive number. Since $\lim a_n = \infty$ and $\lim b_n = -\infty$, we must have $a_n \ge \sqrt{R}$ and $b_n \le -\sqrt{R}$, for all sufficiently large $n$. Thus $a_n b_n \le -R$, and $\lim a_n b_n = -\infty$.

d) "If neither $\{a_n\}$ nor $\{b_n\}$ converges, then $\{a_n b_n\}$ does not converge" is FALSE. Let $a_n = b_n = (-1)^n$; then $\lim a_n$ and $\lim b_n$ both diverge. But $a_n b_n = (-1)^{2n} = 1$ and $\{a_n b_n\}$ does converge (to 1).

e) "If $\{|a_n|\}$ converges, then $\{a_n\}$ converges" is FALSE. Let $a_n = (-1)^n$. Then $\lim_{n\to\infty} |a_n| = \lim_{n\to\infty} 1 = 1$, but $\lim_{n\to\infty} a_n$ does not exist.

## Section 1.2 Infinite Series (page 42)

**2.** $$3 - \frac{3}{4} + \frac{3}{16} - \frac{3}{64} + \cdots = \sum_{n=1}^{\infty} 3\left(-\frac{1}{4}\right)^{n-1} = \frac{3}{1 + \frac{1}{4}} = \frac{12}{5}.$$

**4.** $$\sum_{n=0}^{\infty} \frac{5}{10^{3n}} = 5\left[1 + \frac{1}{1000} + \left(\frac{1}{1000}\right)^2 + \cdots\right] = \frac{5}{1 - \frac{1}{1000}} = \frac{5000}{999}.$$

**6.** $$\sum_{n=0}^{\infty} \frac{1}{e^n} = 1 + \frac{1}{e} + \left(\frac{1}{e}\right)^2 + \cdots = \frac{1}{1 - \frac{1}{e}} = \frac{e}{e-1}.$$

**8.** $\sum_{j=1}^{\infty} \pi^{j/2}\cos(j\pi) = \sum_{j=2}^{\infty}(-1)^j \pi^{j/2}$ diverges because $\lim_{j\to\infty}(-1)^j\pi^{j/2}$ does not exist.

**10.** $$\sum_{n=0}^{\infty} \frac{3 + 2^n}{3^{n+2}} = \frac{1}{3}\sum_{n=0}^{\infty}\left(\frac{1}{3}\right)^n + \frac{1}{9}\sum_{n=0}^{\infty}\left(\frac{2}{3}\right)^n = \frac{1}{3}\cdot\frac{1}{1 - \frac{1}{3}} + \frac{1}{9}\cdot\frac{1}{1 - \frac{2}{3}} = \frac{1}{2} + \frac{1}{3} = \frac{5}{6}.$$

**12.** Let

$$\sum_{n=1}^{\infty} \frac{1}{(2n-1)(2n+1)} = \frac{1}{1\times 3} + \frac{1}{3\times 5} + \frac{1}{5\times 7} + \cdots.$$

Since $\frac{1}{(2n-1)(2n+1)} = \frac{1}{2}\left(\frac{1}{2n-1} - \frac{1}{2n+1}\right)$, the partial sum is

$$s_n = \frac{1}{2}\Big(1 - \frac{1}{3}\Big) + \frac{1}{2}\Big(\frac{1}{3} - \frac{1}{5}\Big) + \cdots + \frac{1}{2}\Big(\frac{1}{2n-3} - \frac{1}{2n-1}\Big) + \frac{1}{2}\Big(\frac{1}{2n-1} - \frac{1}{2n+1}\Big) = \frac{1}{2}\Big(1 - \frac{1}{2n+1}\Big).$$

Hence,

$$\sum_{n=1}^{\infty} \frac{1}{(2n-1)(2n+1)} = \lim s_n = \frac{1}{2}.$$

**14.** Since

$$\frac{1}{n(n+1)(n+2)} = \frac{1}{2}\left[\frac{1}{n} - \frac{2}{n+1} + \frac{1}{n+2}\right],$$

the partial sum is

$$s_n = \frac{1}{2}\Big(1 - \frac{2}{2} + \frac{1}{3}\Big) + \frac{1}{2}\Big(\frac{1}{2} - \frac{2}{3} + \frac{1}{4}\Big) + \cdots + \frac{1}{2}\Big(\frac{1}{n-1} - \frac{2}{n} + \frac{1}{n+1}\Big) + \frac{1}{2}\Big(\frac{1}{n} - \frac{2}{n+1} + \frac{1}{n+2}\Big) = \frac{1}{2}\Big(\frac{1}{2} - \frac{1}{n+1} + \frac{1}{n+2}\Big).$$

Hence,

$$\sum_{n=1}^{\infty} \frac{1}{n(n+1)(n+2)} = \lim s_n = \frac{1}{4}.$$

**16.** $\sum_{n=1}^{\infty} \frac{n}{n+2}$ diverges to infinity since $\lim \frac{n}{n+2} = 1 > 0$.

**18.** $\displaystyle\sum_{n=1}^{\infty} \frac{2}{n+1} = 2\left(\frac{1}{2} + \frac{1}{3} + \frac{1}{4} + \cdots\right)$ diverges to infinity since it is just twice the harmonic series with the first term omitted.

**20.** If $s = \displaystyle\sum_{n=2}^{\infty}\left(3e^{-n} - \frac{2}{n^2-1}\right)$, then

$$s_n = \sum_{k=2}^{n} 3\left(\frac{1}{e}\right)^k - \sum_{k=2}^{n} \frac{2}{k^2-1} = A_n - B_n$$

where

$$A_n = \frac{3}{e^2}\left(1 + \frac{1}{e} + \frac{1}{e^2} \cdots + \frac{1}{e^{n-2}}\right) = \frac{3}{e^2}\frac{1 - \frac{1}{e^{n-1}}}{1 - \frac{1}{e}} = \frac{3}{e(e-1)}\left(1 - \frac{1}{e^{n-1}}\right),$$

and

$$\begin{aligned} B_n &= \sum_{k=2}^{n} \frac{2}{k^2-1} = \sum_{k=2}^{n}\left(\frac{1}{k-1} - \frac{1}{k+1}\right) \\ &= \left(1 - \frac{1}{3}\right) + \left(\frac{1}{2} - \frac{1}{4}\right) + \left(\frac{1}{3} - \frac{1}{5}\right) + \cdots + \\ &\quad + \left(\frac{1}{n-2} - \frac{1}{n}\right) + \left(\frac{1}{n-1} - \frac{1}{n+1}\right) \\ &= 1 + \frac{1}{2} - \frac{1}{n} - \frac{1}{n+1} \\ &= \frac{3}{2} - \frac{1}{n} - \frac{1}{n+1}. \end{aligned}$$

Hence,

$$s = \lim s_n = \lim(A_n - B_n) = \frac{3}{e(e-1)} - \frac{3}{2}.$$

**22.** The balance at the end of 8 years is

$$\begin{aligned} s_n &= 1000\big[(1.1)^8 + (1.1)^7 + \cdots + (1.1)^2 + (1.1)\big] \\ &= 1000(1.1)\left(\frac{(1.1)^8 - 1}{1.1 - 1}\right) \approx \$12,579.48. \end{aligned}$$

**24.** If $\{a_n\}$ is ultimately positive, then the sequence $\{s_n\}$ of partial sums of the series must be ultimately increasing. By Theorem 2, if $\{s_n\}$ is ultimately increasing, then either it is bounded above, and therefore convergent, or else it is not bounded above and diverges to infinity. Since $\sum a_n = \lim s_n$, $\sum a_n$ must either converge when $\{s_n\}$ converges and $\lim s_n = s$ exists, or diverge to infinity when $\{s_n\}$ diverges to infinity.

**26.** "If $a_n = 0$ for every $n$, then $\sum a_n$ converge" is TRUE because $s_n = \sum_{k=0}^{n} 0 = 0$, for every $n$, and so $\sum a_n = \lim s_n = 0$.

**28.** "If $\sum a_n$ and $\sum b_n$ both diverge, then so does $\sum(a_n + b_n)$" is FALSE. Let $a_n = \dfrac{1}{n}$ and $b_n = -\dfrac{1}{n}$, then $\sum a_n = \infty$ and $\sum b_n = -\infty$ but $\sum(a_n + b_n) = \sum(0) = 0$.

**30.** "If $\sum a_n$ diverges and $\{b_n\}$ is bounded, then $\sum a_n b_n$ diverges" is FALSE. Let $a_n = \dfrac{1}{n}$ and $b_n = \dfrac{1}{n+1}$. Then $\sum a_n = \infty$ and $0 \le b_n \le 1/2$. But $\sum a_n b_n = \sum \dfrac{1}{n(n+1)}$ which converges by Example 3.

## Section 1.3 Convergence Tests for Positive Series (page 51)

**2.** $\displaystyle\sum_{n=1}^{\infty} \frac{n}{n^4 - 2}$ converges by comparison with $\displaystyle\sum_{n=1}^{\infty} \frac{1}{n^3}$ since

$$\lim \frac{\left(\frac{n}{n^4-2}\right)}{\left(\frac{1}{n^3}\right)} = 1, \quad \text{and} \quad 0 < 1 < \infty.$$

**4.** $\displaystyle\sum_{n=1}^{\infty} \frac{\sqrt{n}}{n^2+n+1}$ converges by comparison with $\displaystyle\sum_{n=1}^{\infty} \frac{1}{n^{3/2}}$ since

$$\lim \frac{\left(\frac{\sqrt{n}}{n^2+n+1}\right)}{\left(\frac{1}{n^{3/2}}\right)} = 1, \quad \text{and} \quad 0 < 1 < \infty.$$

**6.** $\displaystyle\sum_{n=8}^{\infty} \frac{1}{\pi^n + 5}$ converges by comparison with the geometric series $\displaystyle\sum_{n=8}^{\infty}\left(\frac{1}{\pi}\right)^n$ since $0 < \dfrac{1}{\pi^n + 5} < \dfrac{1}{\pi^n}$.

**8.** $\displaystyle\sum_{n=1}^{\infty} \frac{1}{\ln(3n)}$ diverges to infinity by comparison with the harmonic series $\displaystyle\sum_{n=1}^{\infty} \frac{1}{3n}$ since $\dfrac{1}{\ln(3n)} > \dfrac{1}{3n}$ for $n \ge 1$.

**10.** $\displaystyle\sum_{n=0}^{\infty} \frac{1+n}{2+n}$ diverges to infinity since $\lim \dfrac{1+n}{2+n} = 1 > 0$.

**12.** $\displaystyle\sum_{n=1}^{\infty} \frac{n^2}{1 + n\sqrt{n}}$ diverges to infinity since $\lim \dfrac{n^2}{1 + n\sqrt{n}} = \infty$.

**14.** $\displaystyle\sum_{n=2}^{\infty} \frac{1}{n \ln n (\ln \ln n)^2}$ converges by the integral test:

$$\int_a^{\infty} \frac{dt}{t \ln t (\ln \ln t)^2} = \int_{\ln \ln a}^{\infty} \frac{du}{u^2} < \infty \quad \text{if } \ln \ln a > 0.$$

**16.** The series

$$\sum_{n=1}^{\infty} \frac{1 + (-1)^n}{\sqrt{n}} = 0 + \frac{2}{\sqrt{2}} + 0 + \frac{2}{\sqrt{4}} + 0 + \frac{2}{\sqrt{6}} + \cdots$$
$$= 2 \sum_{k=1}^{\infty} \frac{1}{\sqrt{2k}} = \sqrt{2} \sum_{k=1}^{\infty} \frac{1}{\sqrt{k}}$$

diverges to infinity.

**18.** $\displaystyle\sum_{n=1}^{\infty} \frac{e^n \cos^2 n}{1 + \pi^n}$ converges by comparison with the convergent geometric series $\displaystyle\sum_{n=1}^{\infty} \left(\frac{e}{\pi}\right)^n$ since

$$0 < \frac{e^n \cos^2 n}{1 + \pi^n} < \left(\frac{e}{\pi}\right)^n \quad \text{and} \quad \left(\frac{e}{\pi}\right) < 1.$$

**20.** $\displaystyle\sum_{n=1}^{\infty} \frac{n^4}{n!}$ converges by the ratio test since

$$\lim \frac{\dfrac{(n+1)^4}{(n+1)!}}{\dfrac{n^4}{n!}} = \lim \left(\frac{n+1}{n}\right)^4 \frac{1}{n+1} = 0.$$

**22.** $\displaystyle\sum_{n=1}^{\infty} \frac{(2n)! 6^n}{(3n)!}$ converges by the ratio test since

$$\lim \frac{(2n+2)! 6^{n+1}}{(3n+3)!} \bigg/ \frac{(2n)! 6^n}{(3n)!}$$
$$= \lim \frac{(2n+2)(2n+1)6}{(3n+3)(3n+2)(3n+1)} = 0.$$

**24.** $\displaystyle\sum_{n=0}^{\infty} \frac{n^{100} 2^n}{\sqrt{n!}}$ converges by the ratio test since

$$\lim \frac{(n+1)^{100} 2^{n+1}}{\sqrt{(n+1)!}} \bigg/ \frac{n^{100} 2^n}{\sqrt{n!}}$$
$$= \lim 2 \left(\frac{n+1}{n}\right)^{100} \frac{1}{\sqrt{n+1}} = 0.$$

**26.** $\displaystyle\sum_{n=1}^{\infty} \frac{1 + n!}{(1 + n)!}$ diverges by comparison with the harmonic series $\displaystyle\sum_{n=1}^{\infty} \frac{1}{n+1}$ since $\dfrac{1 + n!}{(1 + n)!} > \dfrac{n!}{(1 + n)!} = \dfrac{1}{n+1}$.

**28.** $\displaystyle\sum_{n=1}^{\infty} \frac{n^n}{\pi^n n!}$ converges by the ratio test since

$$\lim \frac{(n+1)^{n+1}}{\pi^{(n+1)}(n+1)!} \bigg/ \frac{n^n}{\pi^n n!} = \frac{1}{\pi} \lim \left(1 + \frac{1}{n}\right)^n = \frac{e}{\pi} < 1.$$

**30.** Let $u = \ln \ln t$, $du = \dfrac{dt}{t \ln t}$ and $\ln \ln a > 0$; then

$$\int_a^{\infty} \frac{dt}{t \ln t (\ln \ln t)^p} = \int_{\ln \ln a}^{\infty} \frac{du}{u^p}$$

will converge if and only if $p > 1$. Thus, $\displaystyle\sum_{n=3}^{\infty} \frac{1}{n \ln n (\ln \ln n)^p}$ will converge if and only if $p > 1$.
Similarly,

$$\sum_{n=N}^{\infty} \frac{1}{n (\ln n)(\ln \ln n) \cdots (\ln_j n)(\ln_{j+1} n)^p}$$

converges if and only if $p > 1$, where $N$ is large enough that $\ln_j N > 1$.

**32.** Let $a_n = 2^{n+1}/n^n$. Then

$$\lim_{n \to \infty} \sqrt[n]{a_n} = \lim_{n \to \infty} \frac{2 \times 2^{1/n}}{n} = 0.$$

Since this limit is less than 1, $\sum_{n=1}^{\infty} a_n$ converges by the root test.

**34.** Let $a_n = \dfrac{2^{n+1}}{n^n}$. Then

$$\frac{a_{n+1}}{a_n} = \frac{2^{n+2}}{(n+1)^{n+1}} \cdot \frac{n^n}{2^{n+1}}$$
$$= \frac{2}{(n+1)\left(\dfrac{n}{n+1}\right)^n} = \frac{2}{n+1} \cdot \frac{1}{\left(1 + \dfrac{1}{n}\right)^n}$$
$$\to 0 \times \frac{1}{e} = 0 \text{ as } n \to \infty.$$

Thus $\sum_{n=1}^{\infty} a_n$ converges by the ratio test.
(Remark: the question contained a typo. It was intended to to ask that #33 be repeated, using the Ratio test. That is a little harder.)

**36.** We have

$$\begin{aligned} a_n &= \frac{(2n)!}{2^{2n}(n!)^2} = \frac{1\times 2\times 3\times 4\times\cdots\times 2n}{(2\times 4\times 6\times 8\times\cdots\times 2n)^2} \\ &= \frac{1\times 3\times 5\times\cdots\times(2n-1)}{2\times 4\times 6\times\cdots\times(2n-2)\times 2n} \\ &= 1\times\frac{3}{2}\times\frac{5}{4}\times\frac{7}{6}\times\cdots\times\frac{2n-1}{2n-2}\times\frac{1}{2n} > \frac{1}{2n}. \end{aligned}$$

Therefore $\displaystyle\sum_{n=1}^{\infty}\frac{(2n)!}{2^{2n}(n!)^2}$ diverges to infinity by comparison with the harmonic series $\displaystyle\sum_{n=1}^{\infty}\frac{1}{2n}$.

## Section 1.4 Absolute and Conditional Convergence (page 57)

**2.** $\displaystyle\sum_{n=1}^{\infty}\frac{(-1)^n}{n^2+\ln n}$ converges absolutely since $\displaystyle\left|\frac{(-1)^n}{n^2+\ln n}\right| \le \frac{1}{n^2}$ and $\displaystyle\sum_{n=1}^{\infty}\frac{1}{n^2}$ converges.

**4.** $\displaystyle\sum_{n=1}^{\infty}\frac{(-1)^{2n}}{2^n} = \sum_{n=1}^{\infty}\frac{1}{2^n}$ is a positive, convergent geometric series so must converge absolutely.

**6.** $\displaystyle\sum_{n=1}^{\infty}\frac{(-2)^n}{n!}$ converges absolutely by the ratio test since

$$\lim\left|\frac{(-2)^{n+1}}{(n+1)!}\cdot\frac{n!}{(-2)^n}\right| = 2\lim\frac{1}{n+1} = 0.$$

**8.** $\displaystyle\sum_{n=0}^{\infty}\frac{-n}{n^2+1}$ diverges to $-\infty$ since all terms are negative and $\displaystyle\sum_{n=0}^{\infty}\frac{n}{n^2+1}$ diverges to infinity by comparison with $\displaystyle\sum_{n=0}^{\infty}\frac{1}{n}$.

**10.** $\displaystyle\sum_{n=1}^{\infty}\frac{100\cos(n\pi)}{2n+3} = \sum_{n=1}^{\infty}\frac{100(-1)^n}{2n+3}$ converges by the alternating series test but only conditionally since

$$\left|\frac{100(-1)^n}{2n+3}\right| = \frac{100}{2n+3}$$

and $\displaystyle\sum_{n=1}^{\infty}\frac{100}{2n+3}$ diverges to infinity.

**12.** $\displaystyle\sum_{n=10}^{\infty}\frac{\sin(n+\frac{1}{2})\pi}{\ln\ln n} = \sum_{n=10}^{\infty}\frac{(-1)^n}{\ln\ln n}$ converges by the alternating series test but only conditionally since $\displaystyle\sum_{n=10}^{\infty}\frac{1}{\ln\ln n}$ diverges to infinity by comparison with $\displaystyle\sum_{n=10}^{\infty}\frac{1}{n}$. ($\ln\ln n < n$ for $n \ge 10$.)

**14.** Since the terms of the series $s = \sum_{n=0}^{\infty}\dfrac{(-1)^n}{(2n)!}$ are alternating in sign and decreasing in size, the size of the error in the approximation $s \approx s_n$ does not exceed that of the first omitted term:

$$|s - s_n| \le \frac{1}{(2n+2)!} < 0.001$$

if $n = 3$. Hence $s \approx 1 - \dfrac{1}{2!} + \dfrac{1}{4!} - \dfrac{1}{6!}$; four terms will approximate $s$ with error less than 0.001 in absolute value.

**16.** Since the terms of the series $s = \sum_{n=0}^{\infty}(-1)^n\dfrac{3^n}{n!}$ are alternating in sign and ultimately decreasing in size (they decrease after the third term), the size of the error in the approximation $s \approx s_n$ does not exceed that of the first omitted term (provided $n \ge 3$): $|s - s_n| \le \dfrac{3^{n+1}}{(n+1)!} < 0.001$ if $n = 12$. Thus twelve terms will suffice to approximate $s$ with error less than 0.001 in absolute value.

**18.** Let $a_n = \dfrac{(x-2)^n}{n^2 2^{2n}}$. Apply the ratio test

$$\rho = \lim\left|\frac{(x-2)^{n+1}}{(n+1)^2 2^{2n+2}}\times\frac{n^2 2^{2n}}{(x-2)^n}\right| = \frac{|x-2|}{4} < 1$$

if and only if $|x-2| < 4$, that is $-2 < x < 6$. If $x = -2$, then $\displaystyle\sum_{n=1}^{\infty}a_n = \sum_{n=1}^{\infty}\frac{(-1)^n}{n^2}$, which converges absolutely. If $x = 6$, then $\displaystyle\sum_{n=1}^{\infty}a_n = \sum_{n=1}^{\infty}\frac{1}{n^2}$, which also converges absolutely. Thus, the series converges absolutely if $-2 \le x \le 6$ and diverges elsewhere.

**20.** Let $a_n = \dfrac{1}{2n-1}\left(\dfrac{3x+2}{-5}\right)^n$. Apply the ratio test

$$\begin{aligned} \rho &= \lim\left|\frac{1}{2n+1}\left(\frac{3x+2}{-5}\right)^{n+1}\times\frac{2n-1}{1}\left(\frac{3x+2}{-5}\right)^{-n}\right| \\ &= \left|\frac{3x+2}{5}\right| < 1 \end{aligned}$$

if and only if $\left|x+\frac{2}{3}\right| < \frac{5}{3}$, that is $-\frac{7}{3} < x < 1$. If $x = -\frac{7}{3}$, then $\sum_{n=1}^{\infty} a_n = \sum_{n=1}^{\infty} \frac{1}{2n-1}$, which diverges. If $x = 1$, then $\sum_{n=1}^{\infty} a_n = \sum_{n=1}^{\infty} \frac{(-1)^n}{2n-1}$, which converges conditionally. Thus, the series converges absolutely if $-\frac{7}{3} < x < 1$, converges conditionally if $x = 1$ and diverges elsewhere.

**22.** Let $a_n = \dfrac{(4x+1)^n}{n^3}$. Apply the ratio test

$$\rho = \lim \left| \frac{(4x+1)^{n+1}}{(n+1)^3} \times \frac{n^3}{(4x+1)^n} \right| = |4x+1| < 1$$

if and only if $-\frac{1}{2} < x < 0$. If $x = -\frac{1}{2}$, then $\sum_{n=1}^{\infty} a_n = \sum_{n=1}^{\infty} \frac{(-1)^n}{n^3}$, which converges absolutely. If $x = 0$, then $\sum_{n=1}^{\infty} a_n = \sum_{n=1}^{\infty} \frac{1}{n^3}$, which also converges absolutely. Thus, the series converges absolutely if $-\frac{1}{2} \le x \le 0$ and diverges elsewhere.

**24.** Let $a_n = \dfrac{1}{n}\left(1+\dfrac{1}{x}\right)^n$. Apply the ratio test

$$\rho = \lim \left| \frac{1}{n+1}\left(1+\frac{1}{x}\right)^{n+1} \times \frac{n}{1}\left(1+\frac{1}{x}\right)^{-n} \right| = \left|1+\frac{1}{x}\right| < 1$$

if and only if $|x+1| < |x|$, that is, $-2 < \frac{1}{x} < 0 \Rightarrow x < -\frac{1}{2}$. If $x = -\frac{1}{2}$, then $\sum_{n=1}^{\infty} a_n = \sum_{n=1}^{\infty} \frac{(-1)^n}{n}$, which converges conditionally. Thus, the series converges absolutely if $x < -\frac{1}{2}$, converges conditionally if $x = -\frac{1}{2}$ and diverges elsewhere. It is undefined at $x = 0$.

**26.** Let $a_n = \dfrac{(x^2-1)^n}{\sqrt{n}}$. Apply the ratio test

$$\rho = \lim \left| \frac{(x^2-1)^{n+1}}{\sqrt{n+1}} \times \frac{\sqrt{n}}{(x^2-1)^n} \right| = |x^2-1| < 1$$

if and only if $-\sqrt{2} < x < 0$ or $0 < x < \sqrt{2}$. If $x = \pm\sqrt{2}$, then $\sum_{n=1}^{\infty} a_n = \sum_{n=1}^{\infty} \frac{1}{\sqrt{n}}$, which diverges. If $x = 0$, then $\sum_{n=1}^{\infty} a_n = \sum_{n=1}^{\infty} \frac{(-1)^n}{\sqrt{n}}$, which converges conditionally. Thus, the series converges absolutely if $-\sqrt{2} < x < 0$ or $0 < x < \sqrt{2}$, converges conditionally if $x = 0$ and diverges elsewhere.

**28.** Let $a_n = \dfrac{\sin^n x}{n}$. Apply the ratio test

$$\rho = \lim \left| \frac{\sin^{n+1} x}{n+1} \times \frac{n}{\sin^n x} \right| = |\sin x| < 1$$

for $x \ne (k+\frac{1}{2})\pi$ where $k = 0, \pm1, \pm2, \ldots$. If $x = (2k+\frac{1}{2})\pi$, then $\sin x = 1$ and $\sum_{n=1}^{\infty} a_n = \sum_{n=1}^{\infty} \frac{1}{n}$, which diverges. If $x = (2k-\frac{1}{2})\pi$, then $\sin x = -1$ and $\sum_{n=1}^{\infty} a_n = \sum_{n=1}^{\infty} \frac{(-1)^n}{n}$, which converges conditionally. Thus, the series converges absolutely for all $x \ne (k+\frac{1}{2})\pi$, converges conditionally if $x = (2k-\frac{1}{2})\pi$ and diverges elsewhere.

**30.** If

$$a_n = \begin{cases} \dfrac{10}{n^2}, & \text{if } n \text{ is even;} \\ \dfrac{-1}{10n^3}, & \text{if } n \text{ is odd;} \end{cases}$$

then $|a_n| \le \frac{10}{n^2}$ for every $n \ge 1$. Hence, $\sum_{n=1}^{\infty} a_n$ converges absolutely by comparison with $\sum_{n=1}^{\infty} \frac{10}{n^2}$.

**32.**

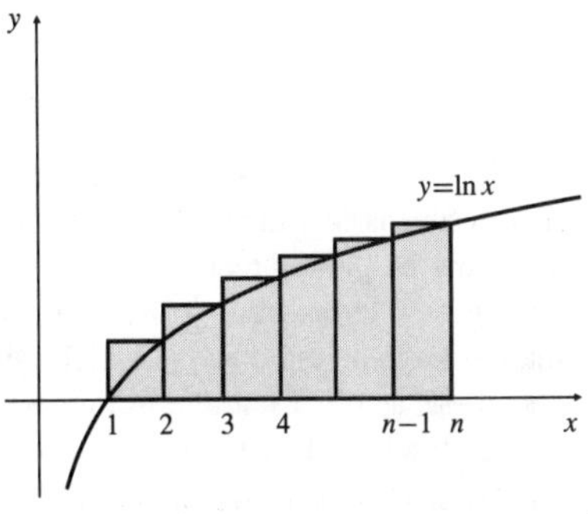

Fig. 1.4.32

a) We have

$$\begin{aligned}\ln(n!) &= \ln 1 + \ln 2 + \ln 3 + \cdots + \ln n \\ &= \text{sum of area of the shaded rectangles} \\ &> \int_1^n \ln t\,dt = (t\ln t - t)\Big|_1^n \\ &= n\ln n - n + 1.\end{aligned}$$

b) Let $a_n = \dfrac{n!x^n}{n^n}$. Apply the ratio test

$$\begin{aligned}\rho &= \lim \left|\frac{(n+1)!x^{n+1}}{(n+1)^{n+1}} \times \frac{n^n}{n!x^n}\right| \\ &= \lim \frac{|x|}{\left(1+\frac{1}{n}\right)^n} = \frac{|x|}{e} < 1\end{aligned}$$

if and only if $-e < x < e$. If $x = \pm e$, then, by (a),

$$\begin{aligned}\ln\left|\frac{n!e^n}{n^n}\right| &= \ln(n!) + \ln e^n - \ln n^n \\ &> (n\ln n - n + 1) + n - n\ln n = 1. \\ \Rightarrow \left|\frac{n!e^n}{n^n}\right| &> e.\end{aligned}$$

Hence, $\sum_{n=1}^{\infty} a_n$ converges absolutely if $-e < x < e$ and diverges elsewhere.

**34.** Let $p_n = \dfrac{1}{2n-1}$ and $q_n = -\dfrac{1}{2n}$. Then $\sum p_n$ diverges to $\infty$ and $\sum q_n$ diverges to $-\infty$. Also, the alternating harmonic series is the sum of all the $p_n$s and $q_n$s in a specific order:

$$\sum_{n=1}^{\infty} \frac{(-1)^{n-1}}{n} = \sum_{n=1}^{\infty} (p_n + q_n).$$

a) Rearrange the terms as follows: first add terms of $\sum p_n$ until the sum exceeds 2. Then add $q_1$. Then add more terms of $\sum p_n$ until the sum exceeds 3. Then add $q_2$. Continue in this way; at the $n$th stage, add new terms from $\sum p_n$ until the sum exceeds $n+1$, and then add $q_n$. All partial sums after the $n$th stage exceed $n$, so the rearranged series diverges to infinity.

b) Rearrange the terms of the original alternating harmonic series as follows: first add terms of $\sum q_n$ until the sum is less than $-2$. Then add $p_1$. The sum will now be greater than $-2$. (Why?) Then resume adding new terms from $\sum q_n$ until the sum is less than $-2$ again, and add $p_2$, which will raise the sum above $-2$ again. Continue in this way. After the $n$th stage, all succeeding partial sums will differ from $-2$ by less than $1/n$, so the rearranged series will converge to $-2$.

## Section 1.5 Estimating the Sum of a Series (page 63)

**2.** Since $f(x) = \dfrac{1}{x^3}$ is positive, continuous and decreasing on $[1,\infty)$, for any $n = 1, 2, 3, \ldots$, we have

$$s_n + A_{n+1} \le s \le s_n + A_n$$

where $s_n = \sum_{k=1}^{n} \dfrac{1}{k^3}$ and $A_n = \displaystyle\int_n^\infty \frac{dx}{x^3} = \frac{1}{2n^2}$. If $s_n^* = s_n + \frac{1}{2}(A_{n+1} + A_n)$, then

$$\begin{aligned}|s_n - s_n^*| &\le \frac{A_n - A_{n+1}}{2} = \frac{1}{4}\left[\frac{1}{n^2} - \frac{1}{(n+1)^2}\right] \\ &= \frac{1}{4}\frac{2n+1}{n^2(n+1)^2} < 0.001\end{aligned}$$

if $n = 8$. Thus, the error in the approximation $s \approx s_8^*$ is less than 0.001.

**4.** As in Exercise 2, now we have

$$s_n = \sum_{k=2}^{n} \frac{1}{k(\ln k)^2} \quad \text{and} \quad A_n = \int_n^\infty \frac{dx}{x(\ln x)^2} = \frac{1}{\ln(n)}.$$

If $s_n^* = s_n + \frac{1}{2}(A_{n+1} + A_n)$, then

$$\begin{aligned}|s_n - s_n^*| &\le \frac{A_n - A_{n+1}}{2} = \frac{1}{2}\left[\frac{1}{\ln(n)} - \frac{1}{\ln(n+1)}\right] \\ &= \frac{1}{2}\frac{\ln\left(\frac{n+1}{n}\right)}{\ln n\,\ln(n+1)} < \frac{1}{2n(\ln n)^2} < 0.001\end{aligned}$$

if $n(\ln n)^2 > 500$, i.e., if $n \ge 38$. Hence, the approximation $s \approx s_{38}^*$ has error less than 0.001.

**6.** Again, we have $s_n + A_{n+1} \le s \le s_n + A_n$ where $s_n = \sum_{k=1}^{n} \dfrac{1}{k^2+4}$ and

$$A_n = \int_n^\infty \frac{dx}{x^2+4} = \frac{1}{2}\tan^{-1}\left(\frac{x}{2}\right)\Big|_n^\infty = \frac{\pi}{4} - \frac{1}{2}\tan^{-1}\left(\frac{n}{2}\right).$$

If $s_n^* = s_n + \frac{1}{2}(A_{n+1} + A_n)$, then

$$\begin{aligned}
|s_n - s_n^*| &\le \frac{A_n - A_{n+1}}{2} \\
&= \frac{1}{2}\left[\frac{\pi}{4} - \frac{1}{2}\tan^{-1}\left(\frac{n}{2}\right) - \frac{\pi}{4} + \frac{1}{2}\tan^{-1}\left(\frac{n+1}{2}\right)\right] \\
&= \frac{1}{4}\left[\tan^{-1}\left(\frac{n+1}{2}\right) - \tan^{-1}\left(\frac{n}{2}\right)\right] = \frac{1}{4}(a-b),
\end{aligned}$$

where $a = \tan^{-1}\left(\frac{n+1}{2}\right)$ and $b = \tan^{-1}\left(\frac{n}{2}\right)$. Now

$$\begin{aligned}
\tan(a-b) &= \frac{\tan a - \tan b}{1 + \tan a \tan b} \\
&= \frac{\left(\frac{n+1}{2}\right) - \left(\frac{n}{2}\right)}{1 + \left(\frac{n+1}{2}\right)\left(\frac{n}{2}\right)} \\
&= \frac{2}{n^2 + n + 4} \\
\Leftrightarrow \quad a - b &= \tan^{-1}\left(\frac{2}{n^2 + n + 4}\right).
\end{aligned}$$

We want error less than 0.001:

$$\begin{aligned}
&\frac{1}{4}(a-b) = \frac{1}{4}\tan^{-1}\left(\frac{2}{n^2+n+4}\right) < 0.001 \\
\Leftrightarrow \quad &\frac{2}{n^2+n+4} < \tan 0.004 \\
\Leftrightarrow \quad &n^2 + n > 2\cot(0.004) - 4 \approx 496.
\end{aligned}$$

$n = 22$ will do. The approximation $s \approx s_{22}^*$ has error less than 0.001.

**8.** We have $s = \sum_{k=1}^{\infty} \frac{1}{(2k-1)!}$ and

$$s_n = \sum_{k=1}^{n} \frac{1}{(2k-1)!} = \frac{1}{1!} + \frac{1}{3!} + \frac{1}{5!} + \cdots + \frac{1}{(2n-1)!}.$$

Then

$$\begin{aligned}
0 < s - s_n &= \frac{1}{(2n+1)!} + \frac{1}{(2n+3)!} + \frac{1}{(2n+5)!} + \cdots \\
&= \frac{1}{(2n+1)!}\left[1 + \frac{1}{(2n+2)(2n+3)} + \right. \\
&\qquad \left. \frac{1}{(2n+2)(2n+3)(2n+4)(2n+5)} + \cdots\right] \\
&< \frac{1}{(2n+1)!}\left[1 + \frac{1}{(2n+2)(2n+3)} + \right. \\
&\qquad \left. \frac{1}{[(2n+2)(2n+3)]^2} + \cdots\right] \\
&= \frac{1}{(2n+1)!}\left[\frac{1}{1 - \frac{1}{(2n+2)(2n+3)}}\right] \\
&= \frac{1}{(2n+1)!}\frac{4n^2+10n+6}{4n^2+10n+5} < 0.001
\end{aligned}$$

if $n = 3$. Thus, $s \approx s_3 = 1 + \frac{1}{3!} + \frac{1}{5!} = 1.175$ with error less than 0.001.

**10.** We have $s = \sum_{k=1}^{\infty} \frac{1}{k^k}$ and

$$s_n = \sum_{k=1}^{n} \frac{1}{k^k} = \frac{1}{1} + \frac{1}{2^2} + \frac{1}{3^3} + \cdots + \frac{1}{n^n}.$$

Then

$$\begin{aligned}
0 < s - s_n &= \frac{1}{(n+1)^{n+1}} + \frac{1}{(n+2)^{n+2}} + \frac{1}{(n+3)^{n+3}} + \cdots \\
&< \frac{1}{(n+1)^{n+1}}\left[1 + \frac{1}{n+1} + \frac{1}{(n+1)^2} + \cdots\right] \\
&= \frac{1}{(n+1)^{n+1}}\left[\frac{1}{1 - \frac{1}{n+1}}\right] \\
&= \frac{1}{n(n+1)^n} < 0.001
\end{aligned}$$

if $n = 4$. Thus, $s \approx s_4 = 1 + \frac{1}{2^2} + \frac{1}{3^3} + \frac{1}{4^4} = 1.291$ with error less than 0.001.

**12.** If $s = \sum_{k=1}^{\infty} c_k = \sum_{k=1}^{\infty} \frac{1}{k^2(k+1)}$, then we have

$$s_n + A_{n+1} \le s \le s_n + A_n$$

where $s_n = \sum_{k=1}^{n} \frac{1}{k^2(k+1)}$ and

$$\begin{aligned} A_n &= \int_n^\infty \frac{dx}{x^2(x+1)} = \int_n^\infty \left(\frac{-1}{x} + \frac{1}{x^2} + \frac{1}{x+1}\right) dx \\ &= -\ln x - \frac{1}{x} + \ln(x+1)\Big|_n^\infty \\ &= \ln\left(1+\frac{1}{x}\right) - \frac{1}{x}\Big|_n^\infty \\ &= \frac{1}{n} - \ln\left(1+\frac{1}{n}\right). \end{aligned}$$

If $s_n^* = s_n + \frac{1}{2}(A_{n+1} + A_n)$, then

$$\begin{aligned} |s_n - s_n^*| &\le \frac{A_n - A_{n+1}}{2} \\ &= \frac{1}{2}\left[\frac{1}{n} - \ln\left(1+\frac{1}{n}\right) - \frac{1}{n+1} + \ln\left(1+\frac{1}{n+1}\right)\right] \\ &= \frac{1}{2}\left[\frac{1}{n(n+1)} + \ln\left(\frac{n^2+2n}{n^2+2n+1}\right)\right] \\ &\le \frac{1}{2}\left[\frac{1}{n(n+1)} + \left(\frac{n^2+2n}{n^2+2n+1} - 1\right)\right] \\ &= \frac{1}{2n(n+1)^2} < 0.001 \end{aligned}$$

if $n = 8$. Thus,

$$\begin{aligned} \sum_{n=1}^{\infty} \frac{1}{n^2} &= 1 + s_8^* = 1 + s_8 + \frac{1}{2}(A_9 + A_8) \\ &= 1 + \left[\frac{1}{2} + \frac{1}{2^2(3)} + \frac{1}{3^2(4)} + \cdots + \frac{1}{8^2(9)}\right] + \\ &\quad \frac{1}{2}\left[\left(\frac{1}{9} - \ln\frac{10}{9}\right) + \left(\frac{1}{8} - \ln\frac{9}{8}\right)\right] \\ &= 1.6450 \end{aligned}$$

with error less than 0.001.

## Review Exercises 1 (page 64)

**2.** $\lim_{n\to\infty} \frac{n^{100} + 2^n\pi}{2^n} = \lim_{n\to\infty} \left(\pi + \frac{n^{100}}{2^n}\right) = \pi.$
The sequence converges.

**4.** $$\begin{aligned} \lim_{n\to\infty} \frac{(2n)!}{(n!)^2} &= \lim_{n\to\infty} \frac{2n(2n-1)\cdots(n+2)(n+1)}{n(n-1)\cdots 2\cdot 1} \\ &\ge \lim_{n\to\infty} (2\times 2\times\cdots\times 2) = \lim_{n\to\infty} 2^n = \infty. \end{aligned}$$
The sequence diverges to infinity.

**6.** $\lim_{n\to\infty} \frac{(-1)^n n^2}{\pi n(n-\pi)} = \lim_{n\to\infty} \frac{(-1)^n}{1-(\pi/n)}$ does not exist.
The sequence diverges (oscillates).

**8.** By l'Hôpital's Rule,

$$\lim_{x\to\infty} \frac{\ln(x+1)}{\ln x} = \lim_{x\to\infty} \frac{1/(x+1)}{1/x} = \lim_{x\to\infty} \frac{x}{x+1} = 1.$$

Thus

$$\lim_{n\to\infty} \big(\ln\ln(n+1) - \ln\ln n\big) = \lim_{n\to\infty} \ln\frac{\ln(n+1)}{\ln n} = \ln 1 = 0.$$

**10.** $$\begin{aligned} \sum_{n=0}^{\infty} \frac{4^{n-1}}{(\pi-1)^{2n}} &= \frac{1}{4}\sum_{n=0}^{\infty} \left(\frac{4}{(\pi-1)^2}\right)^n \\ &= \frac{1}{4}\cdot\frac{1}{1 - \dfrac{4}{(\pi-1)^2}} = \frac{(\pi-1)^2}{4(\pi-1)^2 - 16}, \end{aligned}$$
since $(\pi-1)^2 > 4$.

**12.** $$\begin{aligned} \sum_{n=1}^{\infty} \frac{1}{n^2 - \frac{9}{4}} &= \sum_{n=1}^{\infty} \frac{1}{3}\left(\frac{1}{n-\frac{3}{2}} - \frac{1}{n+\frac{3}{2}}\right) \quad \text{(telescoping)} \\ &= \frac{1}{3}\left[\frac{1}{-1/2} - \frac{1}{5/2} + \frac{1}{1/2} - \frac{1}{7/2}\right. \\ &\quad \left. + \frac{1}{3/2} - \frac{1}{9/2} + \frac{1}{5/2} - \frac{1}{11/2} + \cdots\right] \\ &= \frac{1}{3}\left[-2 + 2 + \frac{2}{3}\right] = \frac{2}{9}. \end{aligned}$$

**14.** $\sum_{n=1}^{\infty} \cos\frac{1}{n}$ diverges to $\infty$ because $\lim_{n\to\infty} \cos\frac{1}{n} = 1 > 0$.

**16.** $\sum_{n=1}^{\infty} \frac{n + 2^n}{1 + 3^n}$ converges by comparison with the convergent geometric series $\sum_{n=1}^{\infty} \left(\frac{2}{3}\right)^n$ because

$$\lim_{n\to\infty} \frac{\dfrac{n+2^n}{1+3^n}}{(2/3)^n} = \lim_{n\to\infty} \frac{(n/2^n) + 1}{(1/3^n) + 1} = 1.$$

**18.** $\sum_{n=1}^{\infty} \frac{n^2}{(1+n)(1+n\sqrt{n})}$ diverges by comparison with the divergent $p$-series $\sum_{n=1}^{\infty} \frac{1}{\sqrt{n}}$ because

$$\lim_{n\to\infty} \frac{\dfrac{n^2}{(1+n)(1+n\sqrt{n})}}{\dfrac{1}{\sqrt{n}}} = \lim_{n\to\infty} \frac{1}{\left(\dfrac{1}{n}+1\right)\left(\dfrac{1}{n^{3/2}}+1\right)} = 1.$$

**20.** $\sum_{n=1}^{\infty} \frac{n^2}{(1+2^n)(1+n\sqrt{n})}$ converges by comparison with the convergent series $\sum_{n=1}^{\infty} \frac{\sqrt{n}}{2^n}$ (which converges by the ratio test) because

$$\lim_{n\to\infty} \frac{\frac{n^2}{(1+2^n)(1+n\sqrt{n})}}{\frac{\sqrt{n}}{2^n}} = \lim_{n\to\infty} \frac{1}{\left(\frac{1}{2^n}+1\right)\left(\frac{1}{n^{3/2}+1}\right)} = 1.$$

**22.** $\sum_{n=1}^{\infty} \frac{n!}{(n+2)!+1}$ converges by comparison with the convergent $p$-series $\sum_{n=1}^{\infty} \frac{1}{n^2}$, because

$$0 \le \frac{n!}{(n+2)!+1} < \frac{n!}{(n+2)!} = \frac{1}{(n+2)(n+1)} < \frac{1}{n^2}.$$

**24.** $\sum_{n=1}^{\infty} \frac{(-1)^n}{2^n - n}$ converges absolutely by comparison with the convergent geometric series $\sum_{n=1}^{\infty} \frac{1}{2^n}$, because

$$\lim_{n\to\infty} \frac{\left|\frac{(-1)^n}{2^n-n}\right|}{\frac{1}{2^n}} = \lim_{n\to\infty} \frac{1}{1-\frac{n}{2^n}} = 1.$$

**26.** $\sum_{n=1}^{\infty} \frac{n^2\cos(n\pi)}{1+n^3}$ converges by the alternating series test (note that $\cos(n\pi) = (-1)^n$), but the convergence is only conditional because

$$\left|\frac{n^2\cos(n\pi)}{1+n^3}\right| = \frac{n^2}{1+n^3} \ge \frac{1}{2n}$$

for $n \ge 1$, and $\sum_{n=1}^{\infty} \frac{1}{2n}$ is a divergent harmonic series.

**28.** $$\lim_{n\to\infty} \left|\frac{\frac{(5-2x)^{n+1}}{n+1}}{\frac{(5-2x)^n}{n}}\right| = \lim_{n\to\infty} |5-2x|\frac{n}{n+1} = |5-2x|.$$

$\sum_{n=1}^{\infty} \frac{(5-2x)^n}{n}$ converges absolutely if $|5-2x| < 1$, that is, if $2 < x < 3$, and diverges if $x < 2$ or $x > 3$.

If $x = 2$ the series is $\sum \frac{1}{n}$, which diverges.

If $x = 3$ the series is $\sum \frac{(-1)^n}{n}$, which converges conditionally.

**30.** $$\lim_{n\to\infty} \left|\frac{\frac{1}{\sqrt{n+2}}\left(1-\frac{1}{x}\right)^{n+1}}{\frac{1}{\sqrt{n+1}}\left(1-\frac{1}{x}\right)^n}\right| = \lim_{n\to\infty} \left|\frac{x-1}{x}\right| \sqrt{\frac{n+1}{n+2}} = \left|\frac{x-1}{x}\right|.$$

$\sum_{n=0}^{\infty} \frac{1}{\sqrt{n+1}}\left(1-\frac{1}{x}\right)^n$ converges absolutely if $\left|\frac{x-1}{x}\right| < 1$, that is, if $x > 1/2$ (the distance of $x$ from 1 is less than the distance of $x$ from 0), and diverges if $x < 1/2$.

If $x = 1/2$ the series is $\sum \frac{(-1)^n}{\sqrt{n+1}}$, which converges conditionally.

**32.** $s = \sum_{n=1}^{\infty} \frac{(-1)^{n-1}}{n^2}$ satisfies the conditions of the alternating series test, so the absolute value of the error if we use $s_n = \sum_{k=1}^{n} \frac{(-1)^{k-1}}{k^2}$ to approximate $s$ is less than $1/(n+1)^2$, which is less than 0.001 if $n \ge 31$. Thus

$$s \approx \sum_{k=1}^{31} \frac{(-1)^{k-1}}{k^2} \approx 0.8230$$

with error less than 0.001.

**34.** Let $s = \sum_{k=1}^{\infty} \frac{1}{4+k^2}$ and $s_n = \sum_{k=1}^{n} \frac{1}{4+k^2}$. Then

$$\int_{n+1}^{\infty} \frac{dt}{4+t^2} < s - s_n < \int_n^{\infty} \frac{dt}{4+t^2}$$
$$s_n + \frac{\pi}{4} - \frac{1}{2}\tan^{-1}\frac{n+1}{2} < s < s_n + \frac{\pi}{4} - \frac{1}{2}\tan^{-1}\frac{n}{2}.$$

Let

$$s_n^* = s_n + \frac{\pi}{4} - \frac{1}{4}\left[\tan^{-1}\frac{n+1}{2} + \tan^{-1}\frac{n}{2}\right].$$

Then $s \approx s_n^*$ with error satisfying

$$|s - s_n^*| < \frac{1}{4}\left[\tan^{-1}\frac{n+1}{2} - \tan^{-1}\frac{n}{2}\right].$$

This error is less than 0.001 if $n \ge 22$. Hence

$$s \approx \sum_{k=1}^{22} \frac{1}{4+k^2} + \frac{\pi}{4} - \frac{1}{4}\left[\tan^{-1}\frac{23}{2} + \tan^{-1}(11)\right] \approx 0.6605$$

with error less than 0.001.

## Challenging Problems 1 (page 64)

**2.** a) If $s_n = \sum_{k=1}^n v_k$ for $n \ge 1$, and $s_0 = 0$, then $v_k = s_k - s_{k-1}$ for $k \ge 1$, and

$$\sum_{k=1}^n u_k v_k = \sum_{k=1}^n u_k s_k - \sum_{k=1}^n u_k s_{k-1}.$$

In the second sum on the right replace $k$ with $k+1$:

$$\begin{aligned}\sum_{k=1}^n u_k v_k &= \sum_{k=1}^n u_k s_k - \sum_{k=0}^{n-1} u_{k+1} s_k \\ &= \sum_{k=1}^n (u_k - u_{k+1}) s_k - u_1 s_0 + u_{n+1} s_n \\ &= u_{n+1} s_n + \sum_{k=1}^n (u_k - u_{k+1}) s_k.\end{aligned}$$

b) If $\{u_n\}$ is positive and decreasing, and $\lim_{n\to\infty} u_n = 0$, then

$$\begin{aligned}\sum_{k=1}^n (u_k - u_{k+1}) &= u_1 - u_2 + u_2 - u_3 + \cdots + u_n - u_{n+1} \\ &= u_1 - u_{n+1} \to u_1 \text{ as } n \to \infty.\end{aligned}$$

Thus $\sum_{k=1}^n (u_k - u_{k+1})$ is a convergent, positive, telescoping series.

If the partial sums $s_n$ of $\{v_n\}$ are bounded, say $|s_n| \le K$ for all $n$, then

$$|(u_n - u_{n+1})s_n| \le K(u_n - u_{n+1}),$$

so $\sum_{n=1}^\infty (u_n - u_{n+1}) s_n$ is absolutely convergent (and therefore convergent) by the comparison test. Therefore, by part (a),

$$\begin{aligned}\sum_{k=1}^\infty u_k v_k &= \lim_{n\to\infty}\left(u_{n+1}s_n + \sum_{k=1}^n (u_k - u_{k+1}) s_k\right) \\ &= \sum_{k=1}^\infty (u_k - u_{k+1}) s_k\end{aligned}$$

converges.

**4.** Let $a_n$ be the $n$th integer that has no zeros in its decimal representation. The number of such integers that have $m$ digits is $9^m$. (There are nine possible choices for each of the $m$ digits.) Also, each such $m$-digit number is greater than $10^{m-1}$ (the smallest $m$-digit number). Therefore the sum of all the terms $1/a_n$ for which $a_n$ has $m$ digits is less than $9^m/(10^{m-1})$. Therefore,

$$\sum_{n=1}^\infty \frac{1}{a_n} < 9 \sum_{m=1}^\infty \left(\frac{9}{10}\right)^{m-1} = 90.$$

**6.**

$$\begin{aligned}\frac{2^{2n}(n!)^2}{(2n)!\sqrt{n}} &= \frac{\big(2\cdot4\cdot6\cdots(2n-2)(2n)\big)^2}{1\cdot2\cdot3\cdots(2n-1)(2n)\sqrt{n}} \\ &= \frac{2\cdot4\cdot6\cdots(2n-2)(2n)}{1\cdot3\cdot5\cdots(2n-1)\sqrt{n}} \\ \left(\frac{2^{2n}(n!)^2}{(2n)!\sqrt{n}}\right)^2 &= \frac{2^2\cdot4^2\cdot6^2\cdots(2n-2)^2(2n)^2}{1\cdot3^2\cdot5^2\cdots(2n-1)^2\cdot n} \\ &= \left[\frac{2}{1}\cdot\frac{2}{3}\cdot\frac{4}{3}\cdot\frac{4}{5}\cdots\frac{2n}{2n-1}\cdot\frac{2n}{2n+1}\right]\frac{2n+1}{n}\end{aligned}$$

The expression in the square bracket above is the Wallis product, whose limit as $n \to \infty$ is $\pi/2$. Therefore

$$\lim_{n\to\infty} \frac{2^{2n}(n!)^2}{(2n)!\sqrt{n}} = \sqrt{\frac{\pi}{2}\cdot 2} = \sqrt{\pi}.$$

**8.** a) Since $e = \sum_{j=0}^\infty \frac{1}{j!}$, we have

$$\begin{aligned}0 < e - \sum_{j=0}^n \frac{1}{j!} &= \sum_{j=n+1}^\infty \frac{1}{j!} \\ &= \frac{1}{(n+1)!}\left(1 + \frac{1}{n+2} + \frac{1}{(n+2)(n+3)} + \cdots\right) \\ &\le \frac{1}{(n+1)!}\left(1 + \frac{1}{n+2} + \frac{1}{(n+2)^2} + \cdots\right) \\ &= \frac{1}{(n+1)!}\cdot\frac{1}{1-\frac{1}{n+2}} = \frac{n+2}{(n+1)!(n+1)} < \frac{1}{n!n}.\end{aligned}$$

The last inequality follows from $\frac{n+2}{(n+1)^2} < \frac{1}{n}$, that is, $n^2 + 2n < n^2 + 2n + 1$.

b) Suppose $e$ is rational, say $e = M/N$ where $M$ and $N$ are positive integers. Then $N!e$ is an integer and $N!\sum_{j=0}^{N}(1/j!)$ is an integer (since each $j!$ is a factor of $N!$). Therefore the number

$$Q = N!\left(e - \sum_{j=0}^{N} \frac{1}{j!}\right)$$

is a difference of two integers and so is an integer.

c) By part (a), $0 < Q < \dfrac{1}{N} \leq 1$. By part (b), $Q$ is an integer. This is not possible; there are no integers between 0 and 1. Therefore $e$ cannot be rational.

# CHAPTER 2. POWER SERIES

## Section 2.1 Power Series (page 76)

**2.** We have $\sum_{n=0}^{\infty} 3n(x+1)^n$. The centre of convergence is $x = -1$. The radius of convergence is

$$R = \lim \frac{3n}{3(n+1)} = 1.$$

The series converges absolutely on $(-2, 0)$ and diverges on $(-\infty, -2)$ and $(0, \infty)$. At $x = -2$, the series is $\sum_{n=0}^{\infty} 3n(-1)^n$, which diverges. At $x = 0$, the series is $\sum_{n=0}^{\infty} 3n$, which diverges to infinity. Hence, the interval of convergence is $(-2, 0)$.

**4.** We have $\sum_{n=1}^{\infty} \frac{(-1)^n}{n^4 2^{2n}} x^n$. The centre of convergence is $x = 0$. The radius of convergence is

$$\begin{aligned} R &= \lim \left| \frac{(-1)^n}{n^4 2^{2n}} \cdot \frac{(n+1)^4 2^{2n+2}}{(-1)^{n+1}} \right| \\ &= \lim \left| \left( \frac{n+1}{n} \right)^4 \cdot 4 \right| = 4. \end{aligned}$$

At $x = 4$, the series is $\sum_{n=1}^{\infty} \frac{(-1)^n}{n^4}$, which converges.
At $x = -4$, the series is $\sum_{n=1}^{\infty} \frac{1}{n^4}$, which also converges.
Hence, the interval of convergence is $[-4, 4]$.

**6.** We have $\sum_{n=1}^{\infty} \frac{e^n}{n^3} (4-x)^n$. The centre of convergence is $x = 4$. The radius of convergence is

$$R = \lim \frac{e^n}{n^3} \cdot \frac{(n+1)^3}{e^{n+1}} = \frac{1}{e}.$$

At $x = 4 + \frac{1}{e}$, the series is $\sum_{n=1}^{\infty} \frac{(-1)^n}{n^3}$, which converges.
At $x = 4 - \frac{1}{e}$, the series is $\sum_{n=1}^{\infty} \frac{1}{n^3}$, which also converges.
Hence, the interval of convergence is $\left[4 - \frac{1}{e}, 4 + \frac{1}{e}\right]$.

**8.** We have $\sum_{n=1}^{\infty} \frac{(4x-1)^n}{n^n} = \sum_{n=1}^{\infty} \left(\frac{4}{n}\right)^n \left(x - \frac{1}{4}\right)^n$. The centre of convergence is $x = \frac{1}{4}$. The radius of convergence is

$$\begin{aligned} R &= \lim \frac{4^n}{n^n} \cdot \frac{(n+1)^{n+1}}{4^{n+1}} \\ &= \frac{1}{4} \lim \left( \frac{n+1}{n} \right)^n (n+1) = \infty. \end{aligned}$$

Hence, the interval of convergence is $(-\infty, \infty)$.

**10.** We have

$$1 + x + x^2 + x^3 + \cdots = \frac{1}{1-x} = \sum_{n=0}^{\infty} x^n$$

and

$$1 - x + x^2 - x^3 + \cdots = \frac{1}{1+x} = \sum_{n=0}^{\infty} (-1)^n x^n$$

holds for $-1 < x < 1$. Since $a_n = 1$ and $b_n = (-1)^n$ for $n = 0, 1, 2, \ldots$, we have

$$C_n = \sum_{j=0}^{n} (-1)^{n-j} = \begin{cases} 0, & \text{if } n \text{ is odd;} \\ 1, & \text{if } n \text{ is even.} \end{cases}$$

Then the Cauchy product is

$$1 + x^2 + x^4 + \cdots = \sum_{n=0}^{\infty} x^{2n} = \frac{1}{1-x} \cdot \frac{1}{1+x} = \frac{1}{1-x^2}$$

for $-1 < x < 1$.

**12.**
$$\begin{aligned} \frac{1}{2-x} &= \frac{1}{2} \frac{1}{\left(1 - \frac{x}{2}\right)} = \frac{1}{2} \sum_{n=0}^{\infty} \left(\frac{x}{2}\right)^n \\ &= \frac{1}{2} + \frac{x}{2^2} + \frac{x^2}{2^3} + \frac{x^3}{2^4} + \cdots \qquad (-2 < x < 2). \end{aligned}$$

**14.**
$$\begin{aligned} \frac{1}{1+2x} &= \sum_{n=0}^{\infty} (-2x)^n \\ &= 1 - 2x + 2^2x^2 - 2^3x^3 + \cdots \qquad (-\tfrac{1}{2} < x < \tfrac{1}{2}). \end{aligned}$$

**16.** Let $y = x - 1$. Then $x = 1 + y$ and

$$\begin{aligned} \frac{1}{x} &= \frac{1}{1+y} = \sum_{n=0}^{\infty} (-y)^n \qquad (-1 < y < 1) \\ &= \sum_{n=0}^{\infty} \left[-(x-1)\right]^n \\ &= 1 - (x-1) + (x-1)^2 - (x-1)^3 + (x-1)^4 - \cdots \\ &\quad \text{(for } 0 < x < 2\text{)}. \end{aligned}$$

**18.** 
$$\begin{aligned}\frac{1-x}{1+x} &= \frac{2}{1+x} - 1\\ &= 2(1-x+x^2-x^3+\cdots)-1\\ &= 1+2\sum_{n=1}^{\infty}(-x)^n \qquad (-1<x<1).\end{aligned}$$

**20.** Let $y = x-4$. Then $x = 4+y$ and

$$\begin{aligned}\frac{1}{x} &= \frac{1}{4+y} = \frac{1}{4}\frac{1}{\left(1+\frac{y}{4}\right)} = \frac{1}{4}\sum_{n=0}^{\infty}\left(-\frac{y}{4}\right)^n\\ &= \frac{1}{4}\sum_{n=0}^{\infty}\left[-\frac{(x-4)}{4}\right]^n\\ &= \frac{1}{4} - \frac{(x-4)}{4^2} + \frac{(x-4)^2}{4^3} - \frac{(x-4)^3}{4^4} + \cdots\end{aligned}$$

for $0<x<8$. Therefore,

$$\begin{aligned}\ln x &= \int_1^x \frac{dt}{t} = \int_1^4 \frac{dt}{t} + \int_4^x \frac{dt}{t}\\ &= \ln 4 + \int_4^x\left[\frac{1}{4} - \frac{(t-4)}{4^2} + \frac{(t-4)^2}{4^3} - \frac{(t-4)^3}{4^4} + \cdots\right]dt\\ &= \ln 4 + \frac{x-4}{4} - \frac{(x-4)^2}{2\cdot 4^2} + \frac{(x-4)^3}{3\cdot 4^3} - \frac{(x-4)^4}{4\cdot 4^4} + \cdots\\ &\quad (\text{for } 0<x\le 8).\end{aligned}$$

**22.** We differentiate the series

$$\sum_{n=0}^{\infty} x^n = 1+x+x^2+x^3+\cdots = \frac{1}{1-x}$$

and multiply by $x$ to get

$$\sum_{n=0}^{\infty} nx^n = x+2x^2+3x^3+\cdots = \frac{x}{(1-x)^2}$$

for $-1<x<1$. Therefore,

$$\begin{aligned}\sum_{n=0}^{\infty}(n+3)x^n &= \sum_{n=0}^{\infty} nx^n + 3\sum_{n=0}^{\infty} x^n\\ &= \frac{x}{(1-x)^2} + \frac{3}{1-x}\\ &= \frac{3-2x}{(1-x)^2} \qquad (-1<x<1).\end{aligned}$$

**24.** We start with

$$1-x+x^2-x^3+x^4-\cdots = \frac{1}{1+x}$$

and differentiate to get

$$-1+2x-3x^3+4x^3-\cdots = -\frac{1}{(1+x)^2}.$$

Now we multiply by $-x^3$:

$$x^3-2x^4+3x^5-4x^6+\cdots = \frac{x^3}{(1+x)^2}.$$

Differentiating again we get

$$3x^2 - 2\times 4x^3 + 3\times 5x^4 - 4\times 6x^5 + \cdots = \frac{x^3+3x^2}{(1+x)^3}.$$

Finally, we remove the factor $x^2$:

$$3 - 2\times 4x + 3\times 5x^2 - 4\times 6x^3 + \cdots = \frac{x+3}{(1+x)^3}.$$

All steps are valid for $-1<x<1$.

**26.** Since $x - \dfrac{x^2}{2} + \dfrac{x^3}{3} - \dfrac{x^4}{4} + \cdots = \ln(1+x)$ for $-1<x\le 1$, therefore

$$x^2 - \frac{x^4}{2} + \frac{x^6}{3} - \frac{x^8}{4} + \cdots = \ln(1+x^2)$$

for $-1\le x\le 1$, and, dividing by $x^2$,

$$1-\frac{x^2}{2}+\frac{x^4}{3}-\frac{x^6}{4}+\cdots = \begin{cases}\dfrac{\ln(1+x^2)}{x^2} & \text{if } -1\le x\le 1,\ x\ne 0\\ 1 & \text{if } x=0.\end{cases}$$

**28.** Using the result of Example 5a,

$$\sum_{n=0}^{\infty}\frac{n+1}{2^n} = \sum_{k=1}^{\infty}\frac{k}{2^{k-1}} = \frac{1}{\left(1-\frac{1}{2}\right)^2} = 4.$$

**30.** From Example 5(a),

$$\sum_{n=1}^{\infty} nx^{n-1} = \frac{1}{(1-x)^2}, \qquad (-1<x<1).$$

Differentiate with respect to $x$ and then replace $n$ by $n+1$:

$$\sum_{n=2}^{\infty} n(n-1)x^{n-2} = \frac{2}{(1-x)^3}, \qquad (-1<x<1)$$

$$\sum_{n=1}^{\infty}(n+1)nx^{n-1} = \frac{2}{(1-x)^3}, \qquad (-1<x<1).$$

Now let $x=-1/2$:

$$\sum_{n=1}^{\infty}(-1)^{n-1}\frac{n(n+1)}{2^{n-1}} = \frac{16}{27}.$$

Finally, multiply by $-1/2$:

$$\sum_{n=1}^{\infty}(-1)^n \frac{n(n+1)}{2^n} = -\frac{8}{27}.$$

**32.** Put $x = -1/2$ in the series

$$\sum_{n=1}^{\infty}(-1)^{n-1}\frac{x^n}{n} = \ln(1+x), \quad (-1 < x \le 1)$$

to obtain

$$\sum_{n=1}^{\infty}\frac{1}{n2^n} = \ln 2$$

$$\sum_{n=3}^{\infty}\frac{1}{n2^n} = \ln 2 - \frac{1}{2} - \frac{1}{8} = \ln 2 - \frac{5}{8}.$$

## Section 2.2 Taylor and Maclaurin Series (page 84)

**2.**
$$\begin{aligned}\cos(2x^3) &= 1 - \frac{(2x^3)^2}{2!} + \frac{(2x^3)^4}{4!} - \frac{(2x^3)^6}{6!} + \cdots \\ &= 1 - \frac{2^2x^6}{2!} + \frac{2^4x^{12}}{4!} - \frac{2^6x^{18}}{6!} + \cdots \\ &= \sum_{n=0}^{\infty}\frac{(-1)^n 4^n}{(2n)!}x^{6n} \quad \text{(for all } x).\end{aligned}$$

**4.**
$$\begin{aligned}\cos(2x - \pi) &= -\cos(2x) \\ &= -1 + \frac{2^2x^2}{2!} - \frac{2^4x^4}{4!} + \frac{2^6x^6}{6!} - \cdots \\ &= -\sum_{n=0}^{\infty}\frac{(-1)^n}{(2n)!}(2x)^{2n} \\ &= \sum_{n=0}^{\infty}\frac{(-1)^{n+1}}{(2n)!}4^n(x)^{2n} \quad \text{(for all } x).\end{aligned}$$

**6.**
$$\begin{aligned}\cos^2\left(\frac{x}{2}\right) &= \frac{1}{2}(1+\cos x) \\ &= \frac{1}{2}\left(1 + 1 - \frac{x^2}{2!} + \frac{x^4}{4!} - \frac{x^6}{6!} + \cdots\right) \\ &= 1 + \frac{1}{2}\sum_{n=1}^{\infty}\frac{(-1)^n}{(2n)!}x^{2n} \quad \text{(for all } x).\end{aligned}$$

**8.**
$$\begin{aligned}\tan^{-1}(5x^2) &= (5x^2) - \frac{(5x^2)^3}{3} + \frac{(5x^2)^5}{5} - \frac{(5x^2)^7}{7} + \cdots \\ &= \sum_{n=0}^{\infty}\frac{(-1)^n}{(2n+1)}(5x^2)^{2n+1} \\ &= \sum_{n=0}^{\infty}\frac{(-1)^n 5^{2n+1}}{(2n+1)}x^{4n+2} \\ &\left(\text{for } -\frac{1}{\sqrt{5}} \le x \le \frac{1}{\sqrt{5}}\right).\end{aligned}$$

**10.**
$$\begin{aligned}\ln(2+x^2) &= \ln 2\left(1 + \frac{x^2}{2}\right) \\ &= \ln 2 + \ln\left(1 + \frac{x^2}{2}\right) \\ &= \ln 2 + \left[\frac{x^2}{2} - \frac{1}{2}\left(\frac{x^2}{2}\right)^2 + \frac{1}{3}\left(\frac{x^2}{2}\right)^3 - \cdots\right] \\ &= \ln 2 + \sum_{n=1}^{\infty}\frac{(-1)^{n-1}}{n}\cdot\frac{x^{2n}}{2^n} \\ &(\text{for } -\sqrt{2} \le x \le \sqrt{2}).\end{aligned}$$

**12.**
$$\begin{aligned}\frac{e^{2x^2}-1}{x^2} &= \frac{1}{x^2}\left(e^{2x^2} - 1\right) \\ &= \frac{1}{x^2}\left(1 + 2x^2 + \frac{(2x^2)^2}{2!} + \frac{(2x^2)^3}{3!} + \cdots - 1\right) \\ &= 2 + \frac{2^2x^2}{2!} + \frac{2^3x^4}{3!} + \frac{2^4x^6}{4!} + \cdots \\ &= \sum_{n=0}^{\infty}\frac{2^{n+1}}{(n+1)!}x^{2n} \quad \text{(for all } x \ne 0).\end{aligned}$$

**14.**
$$\begin{aligned}\sinh x - \sin x &= \sum_{n=0}^{\infty}[1 - (-1)^n]\frac{x^{2n+1}}{(2n+1)!} \\ &= 2\left(\frac{x^2}{2!} + \frac{x^6}{6!} + \frac{x^{10}}{10!} + \cdots\right) \\ &= 2\sum_{n=0}^{\infty}\frac{x^{4n+3}}{(4n+3)!} \quad \text{(for all } x).\end{aligned}$$

**16.** Let $y = x - \frac{\pi}{2}$; then $x = y + \frac{\pi}{2}$. Hence,

$$\begin{aligned}\sin x = \sin\left(y + \frac{\pi}{2}\right) &= \cos y \\ &= 1 - \frac{y^2}{2!} + \frac{y^4}{4!} - \cdots \quad \text{(for all } y) \\ &= 1 - \frac{1}{2!}\left(x - \frac{\pi}{2}\right)^2 + \frac{1}{4!}\left(x - \frac{\pi}{2}\right)^4 - \cdots \\ &= \sum_{n=0}^{\infty}\frac{(-1)^n}{(2n)!}\left(x - \frac{\pi}{2}\right)^{2n} \quad \text{(for all } x).\end{aligned}$$

**18.** Let $y = x - 3$; then $x = y + 3$. Hence,

$$\begin{aligned}
\ln x &= \ln(y+3) = \ln 3 + \ln\left(1 + \frac{y}{3}\right) \\
&= \ln 3 + \frac{y}{3} - \frac{1}{2}\left(\frac{y}{3}\right)^2 + \frac{1}{3}\left(\frac{y}{3}\right)^3 - \frac{1}{4}\left(\frac{y}{3}\right)^4 + \cdots \\
&= \ln 3 + \frac{(x-3)}{3} - \frac{(x-3)^2}{2\cdot 3^2} + \frac{(x-3)^3}{3\cdot 3^3} - \frac{(x-3)^4}{4\cdot 3^4} + \cdots \\
&= \ln 3 + \sum_{n=1}^{\infty} \frac{(-1)^{n-1}}{n\cdot 3^n}(x-3)^n \qquad (0 < x \le 6).
\end{aligned}$$

**20.** Let $t = x + 1$. Then $x = t - 1$, and

$$\begin{aligned}
e^{2x+3} &= e^{2t+1} = e\, e^{2t} \\
&= e\sum_{n=0}^{\infty} \frac{2^n t^n}{n!} \qquad \text{(for all } t) \\
&= \sum_{n=0}^{\infty} \frac{e 2^n (x+1)^n}{n!} \qquad \text{(for all } x).
\end{aligned}$$

**22.** Let $y = x - \frac{\pi}{8}$; then $x = y + \frac{\pi}{8}$. Thus,

$$\begin{aligned}
\cos^2 x &= \cos^2\left(y + \frac{\pi}{8}\right) \\
&= \frac{1}{2}\left[1 + \cos\left(2y + \frac{\pi}{4}\right)\right] \\
&= \frac{1}{2}\left[1 + \frac{1}{\sqrt{2}}\cos(2y) - \frac{1}{\sqrt{2}}\sin(2y)\right] \\
&= \frac{1}{2} + \frac{1}{2\sqrt{2}}\left[1 - \frac{(2y)^2}{2!} + \frac{(2y)^4}{4!} - \cdots\right] \\
&\quad - \frac{1}{2\sqrt{2}}\left[2y - \frac{(2y)^3}{3!} + \frac{(2y)^5}{5!} - \cdots\right] \\
&= \frac{1}{2} + \frac{1}{2\sqrt{2}}\left[1 - 2y - \frac{(2y)^2}{2!} + \frac{(2y)^3}{3!}\right. \\
&\quad \left. + \frac{(2y)^4}{4!} - \frac{(2y)^5}{5!} - \cdots\right] \\
&= \frac{1}{2} + \frac{1}{2\sqrt{2}}\left[1 - 2\left(x - \frac{\pi}{8}\right) - \frac{2^2}{2!}\left(x - \frac{\pi}{8}\right)^2\right. \\
&\quad \left. + \frac{2^3}{3!}\left(x - \frac{\pi}{8}\right)^3 + \frac{2^4}{4!}\left(x - \frac{\pi}{8}\right)^4 - \frac{2^5}{5!}\left(x - \frac{\pi}{8}\right)^5 - \cdots\right] \\
&= \frac{1}{2} + \frac{1}{2\sqrt{2}} + \frac{1}{2\sqrt{2}}\sum_{n=1}^{\infty}(-1)^n\left[\frac{2^{2n-1}}{(2n-1)!}\left(x - \frac{\pi}{8}\right)^{2n-1}\right. \\
&\quad \left. + \frac{2^{2n}}{(2n)!}\left(x - \frac{\pi}{8}\right)^{2n}\right] \qquad \text{(for all } x).
\end{aligned}$$

**24.** Let $y = x - 1$; then $x = y + 1$. Thus,

$$\begin{aligned}
\frac{x}{1+x} &= \frac{1+y}{2+y} = 1 - \frac{1}{2\left(1 + \frac{y}{2}\right)} \\
&= 1 - \frac{1}{2}\left[1 - \frac{y}{2} + \left(\frac{y}{2}\right)^2 - \left(\frac{y}{2}\right)^3 + \cdots\right] \\
&= \frac{1}{2}\left[1 + \frac{y}{2} - \frac{y^2}{2^2} + \frac{y^3}{2^3} - \frac{y^4}{2^4} + \cdots\right] \qquad (-1 < y < 1) \\
&= \frac{1}{2} + \frac{1}{2^2}(x-1) - \frac{1}{2^3}(x-1)^2 + \frac{1}{2^4}(x-1)^3 - \cdots \\
&= \frac{1}{2} + \sum_{n=1}^{\infty} \frac{(-1)^{n-1}}{2^{n+1}}(x-1)^n \qquad \text{(for } 0 < x < 2).
\end{aligned}$$

**26.** Let $u = x + 2$. Then $x = u - 2$, and

$$\begin{aligned}
xe^x &= (u-2)e^{u-2} \\
&= (u-2)e^{-2}\sum_{n=0}^{\infty}\frac{u^n}{n!} \qquad \text{(for all } u) \\
&= \sum_{n=0}^{\infty}\frac{e^{-2}u^{n+1}}{n!} - \sum_{n=0}^{\infty}\frac{2e^{-2}u^n}{n!}.
\end{aligned}$$

In the first sum replace $n$ by $n - 1$.

$$\begin{aligned}
xe^x &= \sum_{n=1}^{\infty}\frac{e^{-2}u^n}{(n-1)!} - \sum_{n=0}^{\infty}\frac{2e^{-2}u^n}{n!} \\
&= -\frac{2}{e^2} + \sum_{n=1}^{\infty}\frac{1}{e^2}\left(\frac{1}{(n-1)!} - \frac{2}{n!}\right)u^n \\
&= -\frac{2}{e^2} + \sum_{n=1}^{\infty}\frac{1}{e^2}\left(\frac{1}{(n-1)!} - \frac{2}{n!}\right)(x+2)^n \qquad \text{(for all } x).
\end{aligned}$$

**28.** If we divide the first four terms of the series

$$\cos x = 1 - \frac{x^2}{2} + \frac{x^4}{24} - \frac{x^6}{720} + \cdots$$

into 1 we obtain

$$\sec x = 1 + \frac{x^2}{2} + \frac{5x^4}{24} + \frac{61x^6}{720} + \cdots.$$

Now we can differentiate and obtain

$$\sec x \tan x = x + \frac{5x^3}{6} + \frac{61x^5}{120} + \cdots.$$

(Note: the same result can be obtained by multiplying the first three nonzero terms of the series for $\sec x$ (from Exercise 25) and $\tan x$ (from Example 6(b)).)

**30.** We have

$$\begin{aligned}
e^{\tan^{-1}x}-1&=\exp\left[x-\frac{x^3}{3}+\frac{x^5}{5}-\frac{x^7}{7}+\cdots\right]-1\\
&=1+\left(x-\frac{x^3}{3}+\frac{x^5}{5}-\cdots\right)+\frac{1}{2!}\left(x-\frac{x^3}{3}+\cdots\right)^2\\
&\quad+\frac{1}{3!}(x-\cdots)^3+\cdots-1\\
&=x-\frac{x^3}{3}+\frac{x^2}{2}+\frac{x^3}{6}+\text{higher degree terms}\\
&=x+\frac{x^2}{2}-\frac{x^3}{6}+\cdots.
\end{aligned}$$

**32.** $\csc x$ does not have a Maclaurin series because $\lim_{x\to 0}\csc x$ does not exist.
Let $y=x-\dfrac{\pi}{2}$. Then $x=y+\dfrac{\pi}{2}$ and $\sin x=\cos y$.
Therefore, using the result of Exercise 25,

$$\begin{aligned}
\csc x=\sec y&=1+\frac{y^2}{2}+\frac{5y^4}{24}+\cdots\\
&=1+\frac{1}{2}\left(x-\frac{\pi}{2}\right)^2+\frac{5}{24}\left(x-\frac{\pi}{2}\right)^4+\cdots.
\end{aligned}$$

**34.**
$$\begin{aligned}
&x^3-\frac{x^9}{3!\times 4}+\frac{x^{15}}{5!\times 16}-\frac{x^{21}}{7!\times 64}+\frac{x^{27}}{9!\times 256}-\cdots\\
&=2\left[\frac{x^3}{2}-\frac{1}{3!}\left(\frac{x^3}{2}\right)^3+\frac{1}{5!}\left(\frac{x^3}{2}\right)^5-\cdots\right]\\
&=2\sin\left(\frac{x^3}{2}\right)\qquad\text{(for all }x).
\end{aligned}$$

**36.**
$$\begin{aligned}
&1+\frac{1}{2\times 2!}+\frac{1}{4\times 3!}+\frac{1}{8\times 4!}+\cdots\\
&=2\left[\frac{1}{2}+\frac{1}{2!}\left(\frac{1}{2}\right)^2+\frac{1}{3!}\left(\frac{1}{2}\right)^3+\cdots\right]\\
&=2\left(e^{1/2}-1\right).
\end{aligned}$$

**38.** If $a\neq 0$ and $|x-a|<|a|$, then

$$\begin{aligned}
\frac{1}{x}&=\frac{1}{a+(x-a)}=\frac{1}{a}\,\frac{1}{1+\dfrac{x-a}{a}}\\
&=\frac{1}{a}\left[1-\frac{x-a}{a}+\frac{(x-a)^2}{a^2}-\frac{(x-a)^3}{a^3}+\cdots\right].
\end{aligned}$$

The radius of convergence of this series is $|a|$, and the series converges to $1/x$ throughout its interval of convergence. Hence, $1/x$ is analytic at $x=a$.

**40.** If

$$f(x)=\begin{cases}e^{-1/x^2}, & \text{if } x\neq 0;\\ 0, & \text{if } x=0;\end{cases}$$

then the Maclaurin series for $f(x)$ is the identically zero series $0+0x+0x^2+\cdots$ since $f^{(k)}(0)=0$ for every $k$. The series converges for every $x$, but converges to $f(x)$ only at $x=0$, since $f(x)\neq 0$ if $x\neq 0$. Hence, $f$ cannot be analytic at $x=0$.

## Section 2.3 Applications of Taylor and Maclaurin Series (page 88)

**2.** We have

$$\frac{1}{e}=e^{-1}=1-\frac{1}{1!}+\frac{1}{2!}-\frac{1}{3!}+\frac{1}{4!}-\cdots$$

which satisfies the conditions for the alternating series test, and the error incurred in using a partial sum to approximate $e^{-1}$ is less than the first omitted term in absolute value. Now $\dfrac{1}{(n+1)!}<5\times 10^{-5}$ if $n=7$, so

$$\frac{1}{e}\approx\frac{1}{2}-\frac{1}{6}+\frac{1}{24}-\frac{1}{120}+\frac{1}{720}-\frac{1}{5040}\approx 0.36786$$

with error less than $5\times 10^{-5}$ in absolute value.

**4.** We have

$$\sin(0.1)=0.1-\frac{(0.1)^3}{3!}+\frac{(0.1)^5}{5!}-\frac{(0.1)^7}{7!}+\cdots.$$

Since $\dfrac{(0.1)^5}{5!}=8.33\times 10^{-8}<5\times 10^{-5}$, therefore

$$\sin(0.1)=0.1-\frac{(0.1)^3}{3!}\approx 0.09983$$

with error less than $5\times 10^{-5}$ in absolute value.

**6.** We have

$$\begin{aligned}
\ln\left(\frac{6}{5}\right)&=\ln\left(1+\frac{1}{5}\right)\\
&=\frac{1}{5}-\frac{1}{2}\left(\frac{1}{5}\right)^2+\frac{1}{3}\left(\frac{1}{5}\right)^3-\frac{1}{4}\left(\frac{1}{5}\right)^4+\cdots.
\end{aligned}$$

Since $\dfrac{1}{n}\left(\dfrac{1}{5}\right)^n<5\times 10^{-5}$ if $n=6$, therefore

$$\begin{aligned}
\ln\left(\frac{6}{5}\right)&\approx\frac{1}{5}-\frac{1}{2}\left(\frac{1}{5}\right)^2+\frac{1}{3}\left(\frac{1}{5}\right)^3-\frac{1}{4}\left(\frac{1}{5}\right)^4+\frac{1}{5}\left(\frac{1}{5}\right)^5\\
&\approx 0.18233
\end{aligned}$$

with error less than $5\times 10^{-5}$ in absolute value.

**8.** We have

$$\begin{aligned}
\sin 80^\circ=\cos 10^\circ&=\cos\left(\frac{\pi}{18}\right)\\
&=1-\frac{1}{2!}\left(\frac{\pi}{18}\right)^2+\frac{1}{4!}\left(\frac{\pi}{18}\right)^4-\cdots.
\end{aligned}$$

Since $\frac{1}{4!}\left(\frac{\pi}{18}\right)^4 < 5\times 10^{-5}$, therefore

$$\sin 80^\circ \approx 1 - \frac{1}{2!}\left(\frac{\pi}{18}\right)^2 \approx 0.98477$$

with error less than $5\times 10^{-5}$ in absolute value.

**10.** We have

$$\tan^{-1}(0.2) = 0.2 - \frac{(0.2)^3}{3} + \frac{(0.2)^5}{5} - \frac{(0.2)^7}{7} + \cdots.$$

Since $\frac{(0.2)^7}{7} < 5\times 10^{-5}$, therefore

$$\tan^{-1}(0.2) \approx 0.2 - \frac{(0.2)^3}{3} + \frac{(0.2)^5}{5} \approx 0.19740$$

with error less than $5\times 10^{-5}$ in absolute value.

**12.** We have

$$\begin{aligned}\ln\left(\frac{3}{2}\right) &= \ln\left(1+\frac{1}{2}\right)\\ &= \frac{1}{2} - \frac{1}{2}\left(\frac{1}{2}\right)^2 + \frac{1}{3}\left(\frac{1}{2}\right)^3 - \frac{1}{4}\left(\frac{1}{2}\right)^4 + \cdots.\end{aligned}$$

Since $\frac{1}{n}\left(\frac{1}{2}\right)^n < \frac{1}{20000}$ if $n = 11$, therefore

$$\begin{aligned}\ln\left(\frac{3}{2}\right) &\approx \frac{1}{2} - \frac{1}{2}\left(\frac{1}{2}\right)^2 + \frac{1}{3}\left(\frac{1}{2}\right)^3 - \cdots - \frac{1}{10}\left(\frac{1}{2}\right)^{10}\\ &\approx 0.40543\end{aligned}$$

with error less than $5\times 10^{-5}$ in absolute value.

**14.**
$$\begin{aligned}J(x) &= \int_0^x \frac{e^t - 1}{t}\,dt\\ &= \int_0^x \left(1 + \frac{t}{2!} + \frac{t^2}{3!} + \frac{t^3}{4!} + \cdots\right)dt\\ &= x + \frac{x^2}{2!\cdot 2} + \frac{x^3}{3!\cdot 3} + \frac{x^4}{4!\cdot 4} + \cdots\\ &= \sum_{n=1}^{\infty} \frac{x^n}{n!\cdot n}.\end{aligned}$$

**16.**
$$\begin{aligned}L(x) &= \int_0^x \cos(t^2)\,dt\\ &= \int_0^x \left(1 - \frac{t^4}{2!} + \frac{t^8}{4!} - \frac{t^{12}}{6!} + \cdots\right)dt\\ &= x - \frac{x^5}{2!\cdot 5} + \frac{x^9}{4!\cdot 9} - \frac{x^{13}}{6!\cdot 13} + \cdots\\ &= \sum_{n=0}^{\infty} (-1)^n \frac{x^{4n+1}}{(2n)!\cdot(4n+1)}.\end{aligned}$$

**18.** We have

$$L(0.5) = 0.5 - \frac{(0.5)^5}{2!\cdot 5} + \frac{(0.5)^9}{4!\cdot 9} - \frac{(0.5)^{13}}{6!\cdot 13} + \cdots.$$

Since $\frac{(0.5)^{4n+1}}{(2n)!\cdot(4n+1)} < 5\times 10^{-4}$ if $n = 2$, therefore

$$L(0.5) \approx 0.5 - \frac{(0.5)^5}{2!\cdot 5} \approx 0.497$$

rounded to three decimal places.

**20.**
$$\begin{aligned}\lim_{x\to 0}\frac{\sin(x^2)}{\sinh x} &= \lim_{x\to 0}\frac{x^2 - \frac{x^6}{3!} + \frac{x^{10}}{5!} - \cdots}{x + \frac{x^3}{3!} + \frac{x^5}{5!} + \cdots}\\ &= \lim_{x\to 0}\frac{x - \frac{x^5}{3!} + \frac{x^9}{5!} - \cdots}{1 + \frac{x^2}{3!} + \frac{x^4}{5!} + \cdots} = 0.\end{aligned}$$

**22.** We have

$$\begin{aligned}\lim_{x\to 0}\frac{(e^x - 1 - x)^2}{x^2 - \ln(1+x^2)} &= \lim_{x\to 0}\frac{\left(\frac{x^2}{2!} + \frac{x^3}{3!} + \frac{x^4}{4!} + \cdots\right)^2}{\frac{x^4}{2} - \frac{x^6}{3} + \frac{x^8}{4} - \cdots}\\ &= \lim_{x\to 0}\frac{\frac{x^4}{4}\left(1 + \frac{x}{3} + \frac{x^2}{12} + \cdots\right)^2}{\frac{x^4}{2} - \frac{x^6}{3} + \frac{x^8}{4} - \cdots} = \frac{\left(\frac{1}{4}\right)}{\left(\frac{1}{2}\right)} = \frac{1}{2}.\end{aligned}$$

**24.** We have

$$\begin{aligned}&\lim_{x\to 0}\frac{\sin(\sin x) - x}{x[\cos(\sin x) - 1]}\\ &= \lim_{x\to 0}\frac{\left(\sin x - \frac{1}{3!}\sin^3 x + \frac{1}{5!}\sin^5 x - \cdots\right) - x}{x\left[1 - \frac{1}{2!}\sin^2 x + \frac{1}{4!}\sin^4 x - \cdots - 1\right]}\\ &= \lim_{x\to 0}\frac{\left(x - \frac{x^3}{3!} + \cdots\right) - \frac{1}{3!}\left(x - \frac{x^3}{3!} + \cdots\right)^3 + \frac{1}{5!}\left(x - \cdots\right)^5 - \cdots - x}{x\left[-\frac{1}{2!}\left(x - \frac{x^3}{3!} + \cdots\right)^2 + \frac{1}{4!}\left(x - \cdots\right)^4 - \cdots\right]}\\ &= \lim_{x\to 0}\frac{-\frac{2}{3!}x^3 + \text{higher degree terms}}{-\frac{1}{2!}x^3 + \text{higher degree terms}} = \frac{\frac{2}{3!}}{\frac{1}{2!}} = \frac{2}{3}.\end{aligned}$$

**26.** If $y = \sum_{n=0}^{\infty} a_n x^n$, then $y' = \sum_{n=1}^{\infty} n a_n x^{n-1}$ and

$$y'' = \sum_{n=2}^{\infty} n(n-1)a_n x^{n-2} = \sum_{n=0}^{\infty} (n+2)(n+1)a_{n+2}x^n.$$

Thus,

$$\begin{aligned}0 &= y'' + xy' + y\\ &= \sum_{n=0}^{\infty}(n+2)(n+1)a_{n+2}x^n + x\sum_{n=1}^{\infty} na_n x^{n-1} + \sum_{n=0}^{\infty} a_n x^n\\ &= 2a_2 + a_0 + \sum_{n=1}^{\infty}\big[(n+2)(n+1)a_{n+2} + (n+1)a_n\big]x^n.\end{aligned}$$

Since coefficients of all powers of $x$ must vanish, therefore $2a_2 + a_0 = 0$ and, for $n \ge 1$,

$$(n+2)(n+1)a_{n+2} + (n+1)a_n = 0,$$

$$\text{that is,}\quad a_{n+2} = \frac{-a_n}{n+2}.$$

If $y(0) = 1$, then $a_0 = 1$, $a_2 = \frac{-1}{2}$, $a_4 = \frac{1}{2^2 \cdot 2!}$, $a_6 = \frac{-1}{2^3 \cdot 3!}$, $a_8 = \frac{1}{2^4 \cdot 4!}$, .... If $y'(0) = 0$, then $a_1 = a_3 = a_5 = \ldots = 0$. Hence,

$$y = 1 - \frac{1}{2}x^2 + \frac{1}{8}x^4 - \frac{1}{48}x^6 + \cdots = \sum_{n=0}^{\infty}\frac{(-1)^n}{2^n \cdot n!}x^{2n}.$$

## Section 2.4 Taylor Polynomials and Taylor's Formula (page 96)

**2.** If $f(x) = \sqrt{x}$ then

$$f'(x) = \frac{1}{2}x^{-1/2},\quad f''(x) = -\frac{1}{4}x^{-3/2},$$
$$f'''(x) = \frac{3}{8}x^{-5/2},\ f^{(4)}(x) = -\frac{15}{16}x^{-7/2}.$$

Hence, the Taylor polynomial of degree 4 for $f(x)$ about $x = 1$ is

$$\begin{aligned}P_4(x) &= 1 + \frac{1}{2}(x-1) + \frac{1}{2!}\left(-\frac{1}{4}\right)(x-1)^2 +\\ &\quad \frac{1}{3!}\left(\frac{3}{8}\right)(x-1)^3 + \frac{1}{4!}\left(-\frac{15}{16}\right)(x-1)^4\\ &= 1 + \frac{1}{2}(x-1) - \frac{1}{8}(x-1)^2 +\\ &\quad \frac{1}{16}(x-1)^3 - \frac{5}{128}(x-1)^4.\end{aligned}$$

**4.** If $f(x) = \sin x$, then

$$\begin{aligned}&f'(x) = \cos x && f''(x) = -\sin x\\ &f'''(x) = -\cos x && f^{(4)}(x) = \sin x\\ &f^{(5)}(x) = \cos x.\end{aligned}$$

Hence, the Maclaurin polynomial of degree 5 for $f(x)$ in powers of $x$ is

$$P_5(x) = x - \frac{x^3}{3!} + \frac{x^5}{5!} = x - \frac{x^3}{6} + \frac{x^5}{120}.$$

**6.** If $f(x) = e^{-x}$, then

$$f'(x) = -e^{-x},\ f''(x) = e^{-x}, \ldots, f^{(n)}(x) = (-1)^n e^{-x}.$$

Hence, the Maclaurin polynomial of degree $n$ for $f(x)$ in powers of $x$ is

$$P_n(x) = 1 - x + \frac{x^2}{2!} - \frac{x^3}{3!} + \cdots + (-1)^n\frac{x^n}{n!}.$$

**8.** If $f(x) = \ln x$, then

$$\begin{aligned}&f'(x) = x^{-1} && f''(x) = -x^{-2}\\ &f'''(x) = 2x^{-3} && f^{(4)}(x) = -6x^{-4}\\ &f^{(n)}(x) = (-1)^{n-1}(n-1)!x^{-n}.\end{aligned}$$

Hence, the Taylor polynomial of degree n for $f(x)$ about $x = 2$ is

$$\begin{aligned}P_n(x) &= \ln 2 + \frac{1}{2}(x-2) + \frac{1}{2!}\left(-\frac{1}{4}\right)(x-2)^2 +\\ &\quad \frac{1}{3!}\left(\frac{1}{4}\right)(x-2)^3 + \frac{1}{4!}\left(-\frac{6}{2^4}\right)(x-2)^4 + \cdots\\ &\quad + \frac{1}{n!}\left[\frac{(-1)^{n-1}(n-1)!}{2^n}\right](x-2)^n\\ &= \ln 2 + \frac{x-2}{2} - \frac{1}{2}\left(\frac{x-2}{2}\right)^2 + \frac{1}{3}\left(\frac{x-2}{2}\right)^3 + \cdots\\ &\quad + \frac{(-1)^{n-1}}{n}\left(\frac{x-2}{2}\right)^n.\end{aligned}$$

**10.** If $f(x) = \tan^{-1} x$, then

$$f'(x) = \frac{1}{x^2+1},\ f''(x) = \frac{-2x}{(x^2+1)^2},\ f'''(x) = \frac{6x^2-2}{(x^2+1)^3},$$

and $f(1) = \frac{\pi}{4}$, $f'(1) = \frac{1}{2}$, $f''(1) = -\frac{1}{2}$, $f'''(x) = \frac{1}{2}$. Hence, the Taylor polynomial of degree 3 for $f(x)$ in powers of $x - 1$ is

$$P_3(x) = \frac{\pi}{4} + \frac{1}{2}(x-1) - \frac{1}{4}(x-1)^2 + \frac{1}{12}(x-1)^3.$$

**12.** If $f(x) = x^6$, then

$$f'(x) = 6x^5,\quad f''(x) = 30x^4,\quad f'''(x) = 120x^3,$$
$$f^{(4)}(x) = 360x^2,\ f^{(5)}(x) = 720x,\ f^{(6)}(x) = 720.$$

Hence, the Taylor polynomial of degree 6 for $f(x)$ about $x = 1$ is

$$P_6(x) = 1 + 6(x-1) + 15(x-1)^2 + 20(x-1)^3 + 15(x-1)^4 + 6(x-1)^5 + (x-1)^6.$$

**14.** The Taylor polynomial of degree $n$ in powers of $x - a$ for $f(x)$ vanishes identically if $f^{(k)}(a) = 0$ for $k = 0, 1, 2, \ldots, n$, i.e., if $(x-a)^{n+1}$ is a factor of $f(x)$.

**16.**

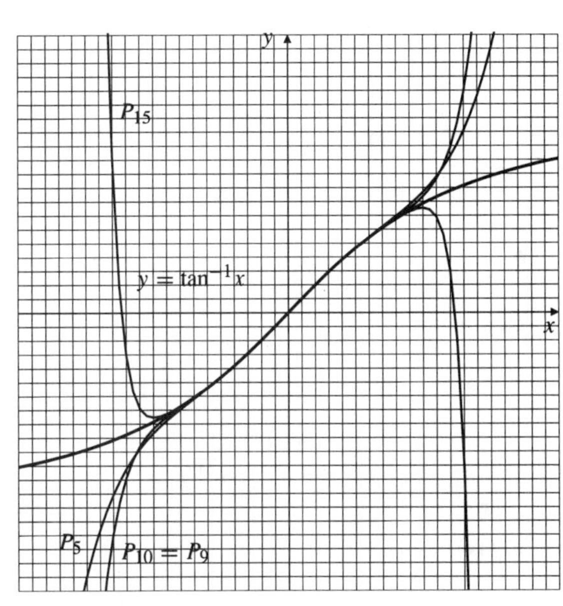


Fig. 2.4.16

The grid lines are spaced at intervals of 0.1. It appears that $P_5$, $P_{10} = P_9$ and $P_{15}$ approximate $\tan^{-1}x$ to within 0.05 to distances 0.8, 0.9, and 1.0, respectively, from $x = 0$.

**18.** Let $f(x) = \cos x$, then

$$f'(x) = -\sin x \quad f''(x) = -\cos x \quad f'''(x) = \sin x$$
$$f^{(4)}(x) = \cos x \quad f^{(5)}(x) = -\sin x \quad f^{(6)}(x) = -\cos x$$
$$f^{(7)}(x) = \sin x.$$

Since $P_6(1)$ is used to approximate $f(1)$, the error is

$$R_6(1) = \frac{f^{(7)}(X)}{7!}(1)^7 = \frac{\sin(X)}{7!}$$

for some $X$ between 0 and 1. Since $1 < \frac{\pi}{3}$, we have $\sin X < \sin\frac{\pi}{3} = \frac{\sqrt{3}}{2}$. Thus,

$$|R_6(1)| < \frac{\sqrt{3}}{2 \cdot 7!} \approx 0.000172.$$

**20.** Let $f(x) = \sec x$, then

$$f'(x) = \sec x \tan x,$$
$$f''(x) = \sec x \tan^2 x + \sec^3 x,$$
$$f'''(x) = \sec x \tan^3 x + 5\sec^3 x \tan x.$$

If $P_2(0.2)$ is used to approximate $f(0.2)$, then

$$R_2(0.2) = \frac{f'''(X)}{3!}(0.2)^3$$

for some $X$ between 0 and 0.2. Since $0.2 < \pi/6$,

$$0 \le \sec X \le \sec(0.2) < \sec\frac{\pi}{6} = \frac{2}{\sqrt{3}};$$
$$0 \le \tan X \le \tan(0.2) < \tan\frac{\pi}{6} = \frac{1}{\sqrt{3}}.$$

Thus,

$$|f'''(X)| = \sec X \tan^3 X + 5\sec^3 X \tan X < \left(\frac{2}{\sqrt{3}}\right)\left(\frac{1}{\sqrt{3}}\right)^3 + 5\left(\frac{2}{\sqrt{3}}\right)^3\left(\frac{1}{\sqrt{3}}\right) = \frac{42}{9}$$

and

$$|R_2(0.2)| < \frac{\left(\frac{42}{9}\right)}{3!}(0.2)^3 \approx 0.0062.$$

**22.** Refer to Exercise 10 above. If $f(x) = \tan^{-1}(x)$, then $f^{(4)}(x) = \frac{24(x - x^3)}{(x^2+1)^4}$. If $P_3(0.99)$ is used to approximate $f(0.99)$, the error will be

$$R_3(0.99) = \frac{f^{(4)}(X)}{4!}(0.99 - 1)^4$$

for some $X$ between 0.99 and 1. For such $X$,

$$|F^{(4)}(x)| \le \frac{24[1 - (0.99)^3]}{[(0.99)^2 + 1]^4} \approx 0.0464.$$

Hence,

$$|R_3(0.99)| \le \frac{0.0464}{4!}(0.99-1)^4 = 1.93\times 10^{-11}.$$

**24.** If $f(x) = e^{-x}$ , then $f^{(k)}(x) = (-1)^k e^{-x}$ and $f^{(k)}(0) = (-1)^k$ for $k = 0, 1, 2, \ldots$. The Lagrange remainder in Taylor's Formula is

$$R_n(x) = \frac{f^{(n+1)}(X)}{(n+1)!}x^{(n+1)} = (-1)^{n+1}e^{-X}\frac{x^{(n+1)}}{(n+1)!}$$

for some $X$ between 0 and $x$. Clearly,

$$|R_n(x)| \le e^{|x|}\frac{|x|^{(n+1)}}{(n+1)!} \to 0$$

as $n\to\infty$ for all real $x$. Thus,

$$e^{-x} = 1 - x + \frac{x^2}{2!} - \frac{x^3}{3!} + \cdots = \sum_{k=0}^{\infty}\frac{(-1)^k x^k}{k!}.$$

**26.** If $f(x) = \cos x$, then for $k = 0, 1, 2, \ldots$,

$$f^{(n)}(0) = \begin{cases} 1, & \text{if } n = 4k; \\ 0, & \text{if } n = 4k+1 \text{ or } 4k+3; \\ -1, & \text{if } n = 4k+2. \end{cases}$$

The Taylor Formula with Lagrange remainder is

$$f(x) = 1 - \frac{x^2}{2!} + \frac{x^4}{4!} + \cdots + (-1)^n\frac{x^{2n}}{2n!} + R_{2n}(x)$$

where

$$R_{2n}(x) = \frac{f^{(2n+1)}(X)}{(2n+1)!}x^{2n+1} = (-1)^{n+1}\sin X\frac{x^{2n+1}}{(2n+1)!}$$

for some $X$ between 0 and $x$. Since

$$|R_{2n}(x)| < \frac{|x|^{2n+1}}{(2n+1)!} \to 0 \text{ as } n\to\infty,$$

therefore

$$\cos x = 1 - \frac{x^2}{2!} + \frac{x^4}{4!} - \frac{x^6}{6!} + \cdots = \sum_{n=0}^{\infty}\frac{(-1)^n x^{2n}}{(2n)!}$$

for all real $x$.

**28.** If $f(x) = \sin^2 x = \dfrac{1-\cos 2x}{2}$, then, for $k = 0, 1, 2, \ldots$,

$$f^{(n)}(x) = \begin{cases} 2^{(n-1)}\sin 2x, & \text{if } n = 4k+1, \\ 2^{(n-1)}\cos 2x, & \text{if } n = 4k+2, \\ -2^{(n-1)}\sin 2x, & \text{if } n = 4k+3, \\ -2^{(n-1)}\cos 2x, & \text{if } n = 4k+4; \end{cases}$$

and

$$f^{(n)}(0) = \begin{cases} 2^{(n-1)}, & \text{if } n = 4k+2, \\ 0, & \text{if } n = 4k+1 \text{ or } 4k+3, \\ -2^{(n-1)}, & \text{if } n = 4k+4. \end{cases}$$

The Taylor Formula is

$$f(x) = \frac{2}{2!}x^2 - \frac{2^3}{4!}x^4 + \frac{2^5}{6!}x^6 - \frac{2^7}{8!}x^8 + \cdots + (-1)^{n-1}\frac{2^{2n-1}}{(2n)!}x^{2n} + R_{2n}$$

where

$$R_{2n}(x) = \frac{f^{(2n+1)}(X)}{(2n+1)!}x^{2n+1} = (-1)^n\sin 2X\frac{x^{2n+1}}{(2n+1)!}$$

for some $X$ between 0 and $x$. Since

$$|R_{2n}(x)| < \frac{|x|^{2n+1}}{(2n+1)!} \to 0 \text{ as } n\to\infty,$$

therefore

$$\begin{aligned} f(x) &= \frac{2}{2!}x^2 - \frac{2^3}{4!}x^4 + \frac{2^5}{6!}x^6 - \cdots \\ &= \sum_{n=1}^{\infty}\frac{(-1)^{n-1}2^{2n-1}}{(2n)!}x^{2n}. \end{aligned}$$

**30.** If $f(x) = \ln(1+x)$, then

$$f'(x) = \frac{1}{1+x},\ f''(x) = \frac{-1}{(1+x)^2},\ f'''(x) = \frac{2}{(1+x)^3},$$
$$f^{(4)}(x) = \frac{-3!}{(1+x)^4},\ \ldots,\ f^{(n)} = \frac{(-1)^{n-1}(n-1)!}{(1+x)^n}$$

and

$$f(0) = 0,\ f'(0) = 1,\ f''(0) = -1,\ f'''(0) = 2,$$
$$f^{(4)}(0) = -3!,\ \ldots,\ f^{(n)}(0) = (-1)^{n-1}(n-1)!.$$

Therefore, the Taylor Formula is

$$f(x) = x + \frac{-1}{2!}x^2 + \frac{2}{3!}x^3 + \frac{-3!}{4!}x^4 + \cdots + \frac{(-1)^{n-1}(n-1)!}{n!}x^n + R_n(x)$$

where

$$\begin{aligned} R_n(x) &= \frac{1}{n!}\int_0^x (x-t)^n f^{(n+1)}(t)\,dt \\ &= \frac{1}{n!}\int_0^x (x-t)^n\frac{(-1)^n n!}{(1+t)^{n+1}}\,dt \\ &= (-1)^n\int_0^x\frac{(x-t)^n}{(1+t)^{n+1}}\,dt. \end{aligned}$$

If $0 \le t \le x \le 1$, then $1+t \ge 1$ and

$$|R_n(x)| \le \int_0^x (x-t)^n\,dt = \frac{x^{n+1}}{n+1} \le \frac{1}{n+1} \to 0$$

as $n \to \infty$.
If $-1 < x \le t \le 0$, then

$$\left|\frac{x-t}{1+t}\right| = \frac{t-x}{1+t} \le |x|,$$

because $\dfrac{t-x}{1+t}$ increases from 0 to $-x = |x|$ as $t$ increases from $x$ to 0. Thus,

$$|R_n(x)| < \frac{1}{1+x}\int_0^{|x|} |x|^n\,dt = \frac{|x|^{n+1}}{1+x} \to 0$$

as $n \to \infty$ since $|x| < 1$. Therefore,

$$f(x) = x - \frac{x^2}{2} + \frac{x^3}{3} - \frac{x^4}{4} + \cdots = \sum_{n=1}^{\infty} (-1)^{n-1}\frac{x^n}{n},$$

for $-1 < x \le 1$.

**32.** If $f(x) = e^x$, then $f^{(k)}(x) = e^x$ and $f^{(k)}(a) = e^a$ for all $k$. The Lagrange remainder is

$$R_n(x) = \frac{f^{(n+1)}(X)}{(n+1)!}(x-a)^{n+1} = e^X \frac{(x-a)^{n+1}}{(n+1)!}$$

for some $X$ between $a$ and $x$. Since $e^{|X|} < e^{|x|}$, clearly,

$$|R_n(x)| < e^{|x|}\frac{|x-a|^{n+1}}{(n+1)!} \to 0 \text{ as } n \to \infty.$$

Therefore,

$$f(x) = e^a + e^a(x-a) + \frac{e^a}{2!}(x-a)^2 + \cdots = e^a \sum_{n=0}^{\infty} \frac{(x-a)^n}{n!}.$$

**34.** If $f(x) = \cos x$, then for $k = 0, 1, 2, \ldots$,

$$f^{(n)}(x) = \begin{cases} \cos x, & \text{if } n = 4k; \\ -\sin x, & \text{if } n = 4k+1; \\ -\cos x, & \text{if } n = 4k+2; \\ \sin x, & \text{if } n = 4k+3. \end{cases}$$

Therefore,

$$f^{(n)}\left(\frac{\pi}{4}\right) = \begin{cases} \dfrac{1}{\sqrt{2}}, & \text{if } n = 4k \text{ or } 4k+3; \\ -\dfrac{1}{\sqrt{2}}, & \text{if } n = 4k+1 \text{ or } 4k+2. \end{cases}$$

The Lagrange remainder is

$$|R_n(x)| = \left|\frac{f^{(n+1)}(X)}{(n+1)!}\left(x - \frac{\pi}{4}\right)^{n+1}\right| \le \frac{\left(x - \frac{\pi}{4}\right)^{n+1}}{(n+1)!},$$

since $|f^{(n+1)}(X)| \le 1$ for any $X$. Clearly, $|R_n(x)| \to 0$ as $n \to \infty$ for all real $x$. Hence,

$$\begin{aligned} f(x) &= \frac{1}{\sqrt{2}} - \frac{1}{\sqrt{2}}\left(x - \frac{\pi}{4}\right) - \frac{1}{\sqrt{2}}\left(\frac{1}{2!}\right)\left(x - \frac{\pi}{4}\right)^2 \\ &\quad + \frac{1}{\sqrt{2}}\left(\frac{1}{3!}\right)\left(x - \frac{\pi}{4}\right)^3 + \cdots \\ &= \frac{1}{\sqrt{2}}\sum_{n=0}^{\infty}(-1)^n\left[\frac{1}{(2n)!}\left(x - \frac{\pi}{4}\right)^{2n} \right. \\ &\quad \left. - \frac{1}{(2n+1)!}\left(x - \frac{\pi}{4}\right)^{2n+1}\right]. \end{aligned}$$

**36.** Let $f(x) = \ln x$, then $f^{(k)}(x) = (-1)^{k-1}(k-1)!x^{-k}$. The Taylor Formula in powers of $x-2$ is

$$f(x) = \ln 2 + \frac{x-2}{2} - \frac{1}{2}\left(\frac{x-2}{2}\right)^2 + \frac{1}{3}\left(\frac{x-2}{2}\right)^3 + \cdots + \frac{(-1)^{n-1}}{n}\left(\frac{x-2}{2}\right)^n + R_n(x)$$

where

$$\begin{aligned} R_n(x) &= \frac{1}{n!}\int_2^x (x-t)^n(-1)^n(n!)t^{-(n+1)}\,dt \\ &= (-1)^n \int_2^x \frac{(x-t)^n}{t^{n+1}}\,dt. \end{aligned}$$

If $2 \le t \le x \le 4$ then $\left|\dfrac{x-t}{t}\right|^n \le \left(\dfrac{4-t}{2}\right)^n$, so

$$|R_n| \le \frac{1}{2^{n+1}}\int_2^4 (4-t)^n\,dt = \frac{1}{n+1} \to 0$$

as $n \to \infty$.
If $0 < x \le t \le 2$ then

$$\left|\frac{x-t}{t}\right| = \frac{t-x}{t} \le \frac{2-x}{2},$$

since $\dfrac{t-x}{t}$ increases from 0 to $\dfrac{2-x}{2}$ as $t$ increases from $x$ to 2. Therefore

$$|R_n| \le \frac{1}{x}\left(\frac{2-x}{2}\right)\int_x^2 dt = \frac{(2-x)^{n+1}}{2^n x} \to 0$$

as $n \to \infty$ (since $2 - x < 2$). Therefore

$$\ln x = \ln 2 + \frac{x-2}{2} - \frac{1}{2}\left(\frac{x-2}{2}\right)^2 + \frac{1}{3}\left(\frac{x-2}{2}\right)^3 - \cdots$$
$$= \ln 2 + \sum_{n=1}^{\infty} \frac{(-1)^{n-1}}{n}\left(\frac{x-2}{2}\right)^n,$$

for $0 < x \le 4$.

## Section 2.5 The Binomial Theorem and Binomial Series (page 100)

**2.** $x\sqrt{1-x} = x(1-x)^{1/2}$

$$= x - \frac{x^2}{2} + \frac{1}{2}\left(-\frac{1}{2}\right)\frac{(-1)^2x^3}{2!} + \frac{1}{2}\left(-\frac{1}{2}\right)\left(-\frac{3}{2}\right)\frac{(-1)^3x^4}{3!} + \cdots$$
$$= x - \frac{x^2}{2} - \sum_{n=2}^{\infty} \frac{1\cdot 3\cdot 5\cdots(2n-3)}{2^n n!}x^{n+1}$$
$$= x - \frac{x^2}{2} - \sum_{n=2}^{\infty}(-1)^{n-1}\frac{(2n-2)!}{2^{2n-1}(n-1)!n!}x^{n+1} \qquad (-1 < x < 1).$$

**4.** $$\frac{1}{\sqrt{4+x^2}} = \frac{1}{2\sqrt{1+\left(\frac{x}{2}\right)^2}} = \frac{1}{2}\left[1+\left(\frac{x}{2}\right)^2\right]^{-1/2}$$
$$= \frac{1}{2}\left[1 + \left(-\frac{1}{2}\right)\left(\frac{x}{2}\right)^2 + \frac{1}{2!}\left(-\frac{1}{2}\right)\left(-\frac{3}{2}\right)\left(\frac{x}{2}\right)^4 + \frac{1}{3!}\left(-\frac{1}{2}\right)\left(-\frac{3}{2}\right)\left(-\frac{5}{2}\right)\left(\frac{x}{2}\right)^6 + \cdots\right]$$
$$= \frac{1}{2} - \frac{1}{2^4}x^2 + \frac{3}{2^7 2!}x^4 - \frac{3\times 5}{2^{10}3!}x^6 + \cdots$$
$$= \frac{1}{2} + \sum_{n=1}^{\infty}(-1)^n \frac{1\times 2\times 3\times\cdots\times(2n-1)}{2^{3n+1}n!}x^{2n}$$
$(-2 \le x \le 2)$.

**6.** $$(1+x)^{-3} = 1 - 3x + \frac{(-3)(-4)}{2!}x^2 + \frac{(-3)(-4)(-5)}{3!}x^3 + \cdots$$
$$= 1 - 3x + \frac{(3)(4)}{2}x^2 - \frac{(4)(5)}{2}x^3 + \cdots$$
$$= \sum_{n=0}^{\infty}(-1)^n\frac{(n+2)(n+1)}{2}x^n \qquad (-1 < x < 1).$$

**8.** The formula $(a+b)^n = \sum_{k=0}^{n}\binom{n}{k}a^{n-k}b^k$

holds for $n = 1$; it says $a + b = a + b$ in this case. Suppose the formula holds for $n = m$, where $m$ is some positive integer. Then

$$(a+b)^{m+1} = (a+b)\sum_{k=0}^{m}\binom{m}{k}a^{m-k}b^k$$
$$= \sum_{k=0}^{m}\binom{m}{k}a^{m+1-k}b^k + \sum_{k=0}^{m}\binom{m}{k}a^{m-k}b^{k+1}$$
(replace $k$ by $k-1$ in the latter sum)
$$= \sum_{k=0}^{m}\binom{m}{k}a^{m+1-k}b^k + \sum_{k=1}^{m+1}\binom{m}{k-1}a^{m+1-k}b^k$$
$$= a^{m+1} + \sum_{k=1}^{m}\left[\binom{m}{k} + \binom{m}{k-1}\right]a^{m+1-k}b^k + b^{m+1}$$
(by #13(i))
$$= a^{m+1} + \sum_{k=1}^{m}\binom{m+1}{k}a^{m+1-k}b^k + b^{m+1} \qquad \text{(by \#13(ii))}$$
$$= \sum_{k=0}^{m+1}\binom{m+1}{k}a^{m+1-k}b^k \qquad \text{(by \#13(i) again).}$$

Thus the formula holds for $n = m + 1$. By induction it holds for all positive integers $n$.

## Section 2.6 Fourier Series (page 106)

**2.** $g(t) = \cos(3 + \pi t)$ has fundamental period 2 since $\cos t$ has fundamental period $2\pi$:

$$g(t+2) = \cos\big(3 + \pi(t+2)\big) = \cos(3 + \pi t + 2\pi)$$
$$= \cos(3 + \pi t) = g(t).$$

**4.** Since $\sin 2t$ has periods $\pi, 2\pi, 3\pi, \ldots$, and $\cos 3t$ has periods $\frac{2\pi}{3}, \frac{4\pi}{3}, \frac{6\pi}{3} = 2\pi, \frac{8\pi}{3}, \ldots$, the sum $k(t) = \sin(2t) + \cos(3t)$ has periods $2\pi, 4\pi, \ldots$. Its fundamental period is $2\pi$.

**6.** $f(t) = \begin{cases} 0 & \text{if } 0 \le t < 1 \\ 1 & \text{if } 1 \le t < 2 \end{cases}$, $f$ has period 2.

The Fourier coefficients of $f$ are as follows:

$$\frac{a_0}{2} = \frac{1}{2}\int_0^2 f(t)\,dt = \frac{1}{2}\int_1^2 dt = \frac{1}{2}$$
$$a_n = \int_0^2 f(t)\cos(n\pi t)\,dt = \int_1^2 \cos(n\pi t)\,dt$$
$$= \frac{1}{n\pi}\sin(n\pi t)\bigg|_1^2 = 0, \qquad (n \ge 1)$$
$$b_n = \int_1^2 \sin(n\pi t)\,dt = -\frac{1}{n\pi}\cos(n\pi t)\bigg|_1^2$$
$$= -\frac{1-(-1)^n}{n\pi} = \begin{cases} -\dfrac{2}{n\pi} & \text{if } n \text{ is odd} \\ 0 & \text{if } n \text{ is even} \end{cases}$$

The Fourier series of $f$ is

$$\frac{1}{2} - \sum_{n=1}^{\infty} \frac{2}{(2n-1)\pi} \sin\big((2n-1)\pi t\big).$$

**8.** $f(t) = \begin{cases} t & \text{if } 0 \le t < 1 \\ 1 & \text{if } 1 \le t < 2, \\ 3-t & \text{if } 2 \le t < 3 \end{cases}$ $f$ has period 3.

$f$ is even, so its Fourier sine coefficients are all zero. Its cosine coefficients are

$$\frac{a_0}{2} = \frac{1}{2} \cdot \frac{2}{3} \int_0^3 f(t)\,dt = \frac{2}{3}(2) = \frac{2}{3}$$

$$\begin{aligned} a_n &= \frac{2}{3} \int_0^3 f(t) \cos \frac{2n\pi t}{3}\,dt \\ &= \frac{2}{3} \bigg[ \int_0^1 t \cos \frac{2n\pi t}{3}\,dt + \int_1^2 \cos \frac{2n\pi t}{3}\,dt \\ &\qquad + \int_2^3 (3-t) \cos \frac{2n\pi t}{3}\,dt \bigg] \\ &= \frac{3}{2n^2\pi^2} \left[ \cos \frac{2n\pi}{3} - 1 - \cos(2n\pi) + \cos \frac{4n\pi}{3} \right]. \end{aligned}$$

The latter expression was obtained using Maple to evaluate the integrals. If $n = 3k$, where $k$ is an integer, then $a_n = 0$. For other integers $n$ we have $a_n = -9/(2\pi^2 n^2)$. Thus the Fourier series of $f$ is

$$\frac{2}{3} - \frac{9}{2\pi^2} \sum_{n=1}^{\infty} \frac{1}{n^2} \cos \frac{2n\pi t}{3} + \frac{1}{2\pi^2} \sum_{n=1}^{\infty} \frac{1}{n^2} \cos(2n\pi t).$$

**10.** The Fourier sine series of $g(t) = \pi - t$ on $[0, \pi]$ has coefficients

$$b_n = \frac{2}{\pi} \int_0^{\pi} (\pi - t) \sin nt\,dt = \frac{2}{n}.$$

The required Fourier sine series is

$$\sum_{n=1}^{\infty} \frac{2}{n} \sin nt.$$

**12.** The Fourier cosine series of $f(t) = t$ on $[0, 1]$ has coefficients

$$\frac{a_0}{2} = \int_0^1 t\,dt = \frac{1}{2}$$

$$\begin{aligned} a_n &= 2 \int_0^1 t \cos(n\pi t)\,dt \\ &= \frac{2(-1)^n - 2}{n^2\pi^2} = \begin{cases} 0 & \text{if } n \text{ is even} \\ \dfrac{-4}{n^2\pi^2} & \text{if } n \text{ is odd.} \end{cases} \end{aligned}$$

The required Fourier cosine series is

$$\frac{1}{2} - \frac{4}{\pi^2} \sum_{n=1}^{\infty} \frac{\cos\big((2n-1)\pi t\big)}{(2n-1)^2}.$$

**14.** If $f$ is even and has period $T$, then

$$\begin{aligned} b_n &= \frac{2}{T} \int_{-T/2}^{T/2} f(t) \sin \frac{2n\pi t}{T}\,dt \\ &= \frac{2}{T} \left[ \int_{-T/2}^{0} f(t) \sin \frac{2n\pi t}{T}\,dt + \int_0^{T/2} f(t) \sin \frac{2n\pi t}{T}\,dt \right]. \end{aligned}$$

In the first integral in the line above replace $t$ with $-t$. Since $f(-t) = f(t)$ and sine is odd, we get

$$\begin{aligned} b_n &= \frac{2}{T} \bigg[ \int_{T/2}^{0} f(t) \left( -\sin \frac{2n\pi t}{T} \right) (-dt) \\ &\qquad + \int_0^{T/2} f(t) \sin \frac{2n\pi t}{T}\,dt \bigg] \\ &= \frac{2}{T} \left[ -\int_0^{T/2} f(t) \sin \frac{2n\pi t}{T}\,dt + \int_0^{T/2} f(t) \sin \frac{2n\pi t}{T}\,dt \right] \\ &= 0. \end{aligned}$$

Similarly,

$$\begin{aligned} a_n &= \frac{2}{T} \left[ \int_{-T/2}^{0} f(t) \cos \frac{2n\pi t}{T}\,dt + \int_0^{T/2} f(t) \cos \frac{2n\pi t}{T}\,dt \right] \\ &= \frac{2}{T} \left[ \int_{T/2}^{0} f(t) \cos \frac{2n\pi t}{T}\,(-dt) + \int_0^{T/2} f(t) \cos \frac{2n\pi t}{T}\,dt \right] \\ &= \frac{4}{T} \int_0^{T/2} f(t) \cos \frac{2n\pi t}{T}\,dt. \end{aligned}$$

The corresponding result for an odd function $f$ states that $a_n = 0$ and

$$b_n = \frac{4}{T} \int_0^{T/2} f(t) \sin \frac{2n\pi t}{T}\,dt,$$

and is proved similarly.

## Review Exercises 2 (page 107)

**2.**
$$\begin{aligned} \frac{x}{3-x^2} &= \frac{x}{3\left(1 - \dfrac{x^2}{3}\right)} = \frac{x}{3} \sum_{n=0}^{\infty} \left( \frac{x^2}{3} \right)^n \\ &= \sum_{n=0}^{\infty} \frac{x^{2n+1}}{3^{n+1}} \qquad (-\sqrt{3} < x < \sqrt{3}). \end{aligned}$$

**4.** $$\ln\frac{1}{e^2-x} = -\ln(e^2-x) = -\ln\left(e^2\left(1-\frac{x}{e^2}\right)\right)$$
$$= -2 - \sum_{n=1}^{\infty}(-1)^{n-1}\frac{(-x)^n}{e^{2n}n}$$
$$= -2 + \sum_{n=1}^{\infty}\frac{x^n}{e^{2n}n} \quad (-e^2 \le x < e^2).$$

**6.** $$\frac{1-e^{-2x}}{x} = \frac{1}{x}\left(1-1-\sum_{n=1}^{\infty}\frac{(-2x)^n}{n!}\right)$$
$$= \sum_{n=1}^{\infty}(-1)^{n-1}\frac{2^n x^{n-1}}{n!} \quad \text{(for all } x \ne 0).$$

**8.** $$\sin\left(x+\frac{\pi}{3}\right) = \sin x\cos\frac{\pi}{3} + \cos x\sin\frac{\pi}{3}$$
$$= \frac{1}{2}\sum_{n=0}^{\infty}(-1)^n\frac{x^{2n+1}}{(2n+1)!} + \frac{\sqrt{3}}{2}\sum_{n=0}^{\infty}(-1)^n\frac{x^{2n}}{(2n)!}$$
$$= \sum_{n=0}^{\infty}\frac{(-1)^n}{2}\left(\frac{\sqrt{3}x^{2n}}{(2n)!} + \frac{x^{2n+1}}{(2n+1)!}\right) \quad \text{(for all } x).$$

**10.** $$(1+x)^{1/3} = 1 + \frac{1}{3}x + \frac{\left(\frac{1}{3}\right)\left(-\frac{2}{3}\right)}{2!}x^2$$
$$+ \frac{\left(\frac{1}{3}\right)\left(-\frac{2}{3}\right)\left(-\frac{5}{3}\right)}{3!}x^3 + \cdots$$
$$= 1 + \frac{x}{3} + \sum_{n=2}^{\infty}(-1)^{n-1}\frac{2\cdot 5\cdot 8\cdots(3n-4)}{3^n n!}x^n \quad (-1 < x < 1).$$

(Remark: the series also converges at $x = 1$.)

**12.** Let $t = x-2$. Then $x = t+2$, and

$$\frac{x}{(x-4)^2} = \frac{t+2}{(t-2)^2} = \frac{1}{4}\cdot\frac{t+2}{\left(1-\frac{t}{2}\right)^2}$$
$$= \frac{t+2}{4}\sum_{n=1}^{\infty}n\left(\frac{t}{2}\right)^{n-1}$$
$$= \sum_{n=1}^{\infty}\frac{n}{2^{n+1}}t^n + \sum_{n=1}^{\infty}\frac{n}{2^n}t^{n-1} \quad (-2 < t < 2).$$

In the second sum, replace $n$ with $n+1$:

$$\frac{x}{(x-4)^2} = \sum_{n=1}^{\infty}\frac{n}{2^{n+1}}t^n + \sum_{n=0}^{\infty}\frac{n+1}{2^{n+1}}t^n$$
$$= \frac{1}{2} + \sum_{n=1}^{\infty}\frac{2n+1}{2^{n+1}}t^n$$
$$= \frac{1}{2} + \sum_{n=1}^{\infty}\frac{2n+1}{2^{n+1}}(x-2)^n \quad (0 < x < 4).$$

**14.** Let $u = x - (\pi/4)$, so $x = u + (\pi/4)$. Then

$$\sin x + \cos x = \sin\left(u+\frac{\pi}{4}\right) + \cos\left(u+\frac{\pi}{4}\right)$$
$$= \frac{1}{\sqrt{2}}\big((\sin u + \cos u) + (\cos u - \sin u)\big)$$
$$= \sqrt{2}\cos u = \sqrt{2}\sum_{n=0}^{\infty}(-1)^n\frac{u^{2n}}{(2n)!}$$
$$= \sqrt{2}\sum_{n=0}^{\infty}\frac{(-1)^n}{(2n)!}\left(x-\frac{\pi}{4}\right)^{2n} \quad \text{(for all } x).$$

**16.** $$\sin(1+x) = \sin(1)\cos x + \cos(1)\sin x$$
$$= \sin(1)\left(1-\frac{x^2}{2!}+\cdots\right) + \cos(1)\left(x-\frac{x^3}{3!}+\cdots\right)$$
$$P_3(x) = \sin(1) + \cos(1)x - \frac{\sin(1)}{2}x^2 - \frac{\cos(1)}{6}x^3.$$

**18.** $$\ln(1+xe^x) = xe^x - \frac{x^2e^{2x}}{2} + \frac{x^3e^{3x}}{3} - \frac{x^4e^{4x}}{4} + \cdots$$
$$= x\left(1+x+\frac{x^2}{2}+\frac{x^3}{6}+\cdots\right) - \frac{x^2}{2}\left(1+2x+\frac{4x^2}{2}+\cdots\right)$$
$$+ \frac{x^3}{3}(1+3x+\cdots) - \frac{x^4}{4}(1+\cdots) + \cdots$$
$$= x + \left(1-\frac{1}{2}\right)x^2 + \left(\frac{1}{2}-1+\frac{1}{3}\right)x^3$$
$$+ \left(\frac{1}{6}-1+1-\frac{1}{4}\right)x^4 + \cdots$$
$$P_4(x) = x + \frac{x^2}{2} - \frac{x^3}{6} - \frac{x^4}{12}.$$

**20.** $$\sqrt{1+\sin x} = 1 + \frac{1}{2}\sin x + \frac{\left(\frac{1}{2}\right)\left(-\frac{1}{2}\right)}{2!}(\sin x)^2$$
$$+ \frac{\left(\frac{1}{2}\right)\left(-\frac{1}{2}\right)\left(-\frac{3}{2}\right)}{3!}(\sin x)^3$$
$$+ \frac{\left(\frac{1}{2}\right)\left(-\frac{1}{2}\right)\left(-\frac{3}{2}\right)\left(-\frac{5}{2}\right)}{4!}(\sin x)^4 + \cdots$$
$$= 1 + \frac{1}{2}\left(x-\frac{x^3}{6}+\cdots\right) - \frac{1}{8}\left(x-\frac{x^3}{6}+\cdots\right)^2$$
$$+ \frac{1}{16}(x-\cdots)^3 - \frac{5}{128}(x-\cdots)^4 + \cdots$$
$$= 1 + \frac{x}{2} - \frac{x^3}{12} - \frac{x^2}{8} + \frac{x^4}{24} + \frac{x^3}{16} - \frac{5x^4}{128} + \cdots$$
$$P_4(x) = 1 + \frac{x}{2} - \frac{x^2}{8} - \frac{x^3}{48} + \frac{x^4}{384}.$$

**22.** Since

$$1 + \sum_{n=1}^{\infty}\frac{x^{2n}}{n^2} = \sum_{k=0}^{\infty}\frac{f^{(k)}(0)}{k!}x^k$$

for $x$ near 0, we have, for $n = 1, 2, 3, \ldots$

$$f^{(2n)}(0) = \frac{(2n)!}{n^2}, \quad f^{(2n-1)}(0) = 0.$$

**24.** $\displaystyle\sum_{n=0}^{\infty} nx^n = \frac{x}{(1-x)^2}$ from Section 1, Example 5a

$$\sum_{n=0}^{\infty} n^2 x^{n-1} = \frac{d}{dx}\frac{x}{(1-x)^2} = \frac{1+x}{(1-x)^3}$$

$$\sum_{n=0}^{\infty} n^2 x^n = \frac{x(1+x)}{(1-x)^3}$$

$$\sum_{n=0}^{\infty} \frac{n^2}{\pi^n} = \frac{\frac{1}{\pi}\left(1+\frac{1}{\pi}\right)}{\left(1-\frac{1}{\pi}\right)^3} = \frac{\pi(\pi+1)}{(\pi-1)^3}.$$

**26.** $\displaystyle\sum_{n=1}^{\infty} \frac{(-1)^{n-1}x^{2n-1}}{(2n-1)!} = \sin x$

$$\sum_{n=1}^{\infty} \frac{(-1)^n \pi^{2n-1}}{(2n-1)!} = -\sin\pi = 0$$

$$\sum_{n=2}^{\infty} \frac{(-1)^n \pi^{2n-4}}{(2n-1)!} = \frac{1}{\pi^3}\left(0 - \frac{(-1)\pi}{1!}\right) = \frac{1}{\pi^2}.$$

**28.** $\displaystyle\lim_{x\to 0} \frac{(x-\tan^{-1}x)(e^{2x}-1)}{2x^2 - 1 + cos(2x)}$

$$= \lim_{x\to 0} \frac{\left(x - x + \frac{x^3}{3} - \frac{x^5}{5} + \cdots\right)\left(2x + \frac{4x^2}{2!} + \cdots\right)}{2x^2 - 1 + 1 - \frac{4x^2}{2!} + \frac{16x^4}{4!} - \cdots}$$

$$= \lim_{x\to 0} \frac{x^4\left(\frac{2}{3} + \cdots\right)}{x^4\left(\frac{2}{3} + \cdots\right)} = 1.$$

**30.** If $f(x) = \ln(\sin x)$, then calculation of successive derivatives leads to

$$f^{(5)}(x) = 24\csc^4 x \cot x - 8\csc^2 \cot x.$$

Observe that $1.5 < \pi/2 \approx 1.5708$, that $\csc x \ge 1$ and $\cot x \ge 0$, and that both functions are decreasing on that interval. Thus

$$|f^{(5)}(x)| \le 24\csc^4(1.5)\cot(1.5) \le 2$$

for $1.5 \le x \le \pi/2$. Therefore, the error in the approximation

$$\ln(\sin 1.5) \approx P_4(x),$$

where $P_4$ is the 4th degree Taylor polynomial for $f(x)$ about $x = \pi/2$, satisfies

$$|\text{error}| \le \frac{2}{5!}\left|1.5 - \frac{\pi}{2}\right|^5 \le 3 \times 10^{-8}.$$

**32.** $f(t) = \begin{cases} 1 & \text{if } -\pi < t \le 0 \\ t & \text{if } 0 < t \le \pi \end{cases}$ has period $2\pi$. Its Fourier coefficients are

$$\begin{aligned}
\frac{a_0}{2} &= \frac{1}{2\pi}\int_{-\pi}^{\pi} f(t)\,dt \\
&= \frac{1}{2\pi}\left[\int_{-\pi}^{0} dt + \int_0^{\pi} t\,dt\right] = \frac{1}{2} + \frac{\pi}{4} \\
a_n &= \frac{1}{\pi}\left[\int_{-\pi}^{0} \cos(nt)\,dt + \int_0^{\pi} t\cos(nt)\,dt\right] \\
&= \frac{1}{\pi}\int_0^{\pi} (1+t)\cos(nt)\,dt \\
&= \frac{(-1)^n - 1}{\pi n^2} = \begin{cases} -2/(\pi n^2) & \text{if } n \text{ is odd} \\ 0 & \text{if } n \text{ is even} \end{cases} \\
b_n &= \frac{1}{\pi}\left[\int_{-\pi}^{0} \sin(nt)\,dt + \int_0^{\pi} t\sin(nt)\,dt\right] \\
&= \frac{1}{\pi}\int_0^{\pi} (t-1)\sin(nt)\,dt \\
&= -\frac{1 + (-1)^n(\pi - 1)}{\pi n} = \begin{cases} (\pi - 2/(\pi n) & \text{if } n \text{ is odd} \\ -(1/n) & \text{if } n \text{ is even.} \end{cases}
\end{aligned}$$

The required Fourier series is, therefore,

$$\frac{2+\pi}{4}$$

$$-\sum_{n=1}^{\infty}\left[\frac{2\cos\big((2n-1)t\big)}{\pi(2n-1)^2} + \frac{(2-\pi)\sin\big((2n-1)t\big)}{\pi(2n-1)} + \frac{\sin(2nt)}{2n}\right].$$

## Challenging Problems 2 (page 107)

**2.** Let $f$ be a polynomial and let

$$g(x) = \sum_{j=0}^{\infty} (-1)^j f^{(2j)}(x).$$

This "series" is really just a polynomial since sufficiently high derivatives of $f$ are all identically zero.

a) By replacing $j$ with $j-1$, observe that

$$\begin{aligned}
g''(x) &= \sum_{j=0}^{\infty} (-1)^j f^{(2j+2)}(x) \\
&= \sum_{j=1}^{\infty} (-1)^{j-1} f^{(2j)}(x) = -\big(g(x) - f(x)\big).
\end{aligned}$$

Also

$$\frac{d}{dx}\big(g'(x)\sin x - g(x)\cos x\big)$$
$$= g''(x)\sin x + g'(x)\cos x - g'(x)\cos x + g(x)\sin x$$
$$= \big(g''(x) + g(x)\big)\sin x = f(x)\sin x.$$

Thus

$$\int_0^\pi f(x)\sin x\,dx = \big(g'(x)\sin x - g(x)\cos x\big)\Big|_0^\pi = g(\pi) + g(0).$$

b) Suppose that $\pi = m/n$, where $m$ and $n$ are positive integers. Since $\lim_{k\to\infty} x^k/k! = 0$ for any $x$, there exists an integer $k$ such that $(\pi m)^k/k! < 1/2$. Let

$$f(x) = \frac{x^k(m-nx)^k}{k!} = \frac{1}{k!}\sum_{j=0}^{k}\binom{k}{j} m^{k-j}(-n)^j x^{j+k}.$$

The sum is just the binomial expansion.
For $0 < x < \pi = m/n$ we have

$$0 < f(x) < \frac{\pi^k m^k}{k!} < \frac{1}{2}.$$

Thus $0 < \int_0^\pi f(x)\sin x\,dx < \frac{1}{2}\int_0^\pi \sin x\,dx = 1$, and so $0 < g(\pi) + g(0) < 1$.

c) $$f^{(i)}(x) = \frac{1}{k!}\sum_{j=0}^{k}\binom{k}{j} m^{k-j}(-n)^j$$
$$\times (j+k)(j+k-1)\cdots(j+k-i+1)x^{j+k-i}$$
$$= \frac{1}{k!}\sum_{j=0}^{k}\binom{k}{j} m^{k-j}(-n)^j \frac{(j+k)!}{(j+k-i)!}x^{j+k-i}.$$

d) Evidently $f^{(i)}(0) = 0$ if $i < k$ or if $i > 2k$.
If $k \le i \le 2k$, the only term in the sum for $f^{(i)}(0)$ that is not zero is the term for which $j = i - k$. This term is the constant

$$\frac{1}{k!}\binom{k}{i-k} m^{k-j}(-n)^j \frac{i!}{0!}.$$

This constant is an integer because the binomial coefficient $\binom{k}{i-k}$ is an integer and $i!/k!$ is an integer. (The other factors are also integers.) Hence $f^{(i)}(0)$ is an integer, and so $g(0)$ is an integer.

e) Observe that $f(\pi - x) = f((m/n) - x) = f(x)$ for all $x$. Therefore $f^{(i)}(\pi)$ is an integer (for each $i$), and so $g(\pi)$ is an integer. Thus $g(\pi) + g(0)$ is an integer, which contradicts the conclusion of part (b). (There is no integer between 0 and 1.) Therefore, $\pi$ cannot be rational.

# CHAPTER 3. COORDINATE GEOMETRY AND VECTORS IN 3-SPACE

## Section 3.1 Analytic Geometry in Three and More Dimensions (page 114)

**2.** The distance between $(-1,-1,-1)$ and $(1,1,1)$ is

$$\sqrt{(1+1)^2+(1+1)^2+(1+1)^2} = 2\sqrt{3} \text{ units.}$$

**4.** The distance between $(3,8,-1)$ and $(-2,3,-6)$ is

$$\sqrt{(-2-3)^2+(3-8)^2+(-6+1)^2} = 5\sqrt{3} \text{ units.}$$

b) The shortest distance from $(x,y,z)$ to the $x$-axis is $\sqrt{y^2+z^2}$ units.

**6.** If $A=(1,2,3)$, $B=(4,0,5)$, and $C=(3,6,4)$, then

$$|AB| = \sqrt{3^2+(-2)^2+2^2} = \sqrt{17}$$
$$|AC| = \sqrt{2^2+4^2+1^2} = \sqrt{21}$$
$$|BC| = \sqrt{(-1)^2+6^2+(-1)^2} = \sqrt{38}.$$

Since $|AB|^2+|AC|^2 = 17+21 = 38 = |BC|^2$, the triangle $ABC$ has a right angle at $A$.

**8.** If $A=(1,2,3)$, $B=(1,3,4)$, and $C=(0,3,3)$, then

$$|AB| = \sqrt{(1-1)^2+(3-2)^2+(4-3)^2} = \sqrt{2}$$
$$|AC| = \sqrt{(0-1)^2+(3-2)^2+(3-3)^2} = \sqrt{2}$$
$$|BC| = \sqrt{(0-1)^2+(3-3)^2+(3-4)^2} = \sqrt{2}.$$

All three sides being equal, the triangle is equilateral.

**10.** The distance from the origin to $(1,1,1,\ldots,1)$ in $\mathbb{R}^n$ is

$$\sqrt{1^2+1^2+1^2+\cdots+1} = \sqrt{n} \text{ units.}$$

**12.** $z=2$ is a plane, perpendicular to the $z$-axis at $(0,0,2)$.

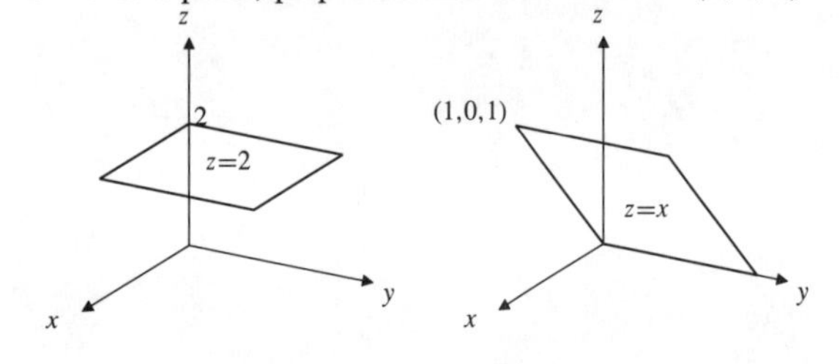


Fig. 3.1.12 Fig. 3.1.14

**14.** $z=x$ is a plane containing the $y$-axis and making $45^\circ$ angles with the positive directions of the $x$- and $z$-axes.

**16.** $x^2+y^2+z^2=4$ is a sphere centred at the origin and having radius 2 (i.e., all points at distance 2 from the origin).

**18.** $x^2+y^2+z^2=2z$ can be rewritten

$$x^2+y^2+(z-1)^2=1,$$

and so it represents a sphere with radius 1 and centre at $(0,0,1)$. It is tangent to the $xy$-plane at the origin.

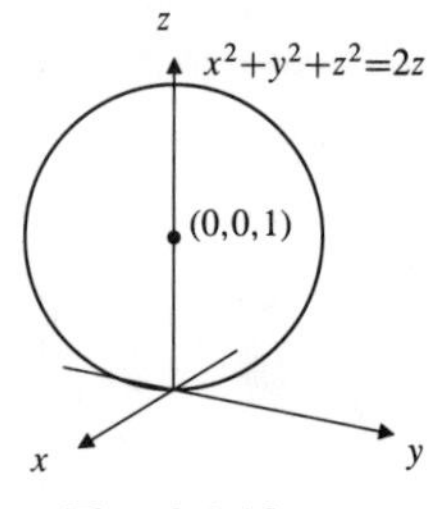


Fig. 3.1.18

**20.** $x^2+z^2=4$ is a circular cylindrical surface of radius 2 with axis along the $y$-axis.

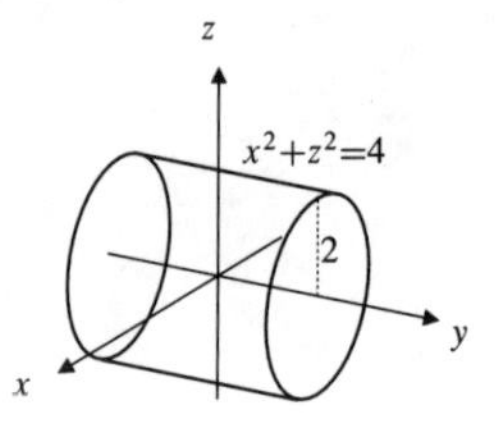


Fig. 3.1.20

**22.** $z \geq \sqrt{x^2+y^2}$ represents every point whose distance above the $xy$-plane is not less than its horizontal distance from the $z$-axis. It therefore consists of all points inside and on a circular cone with axis along the positive $z$-axis, vertex at the origin, and semi-vertical angle $45^\circ$.

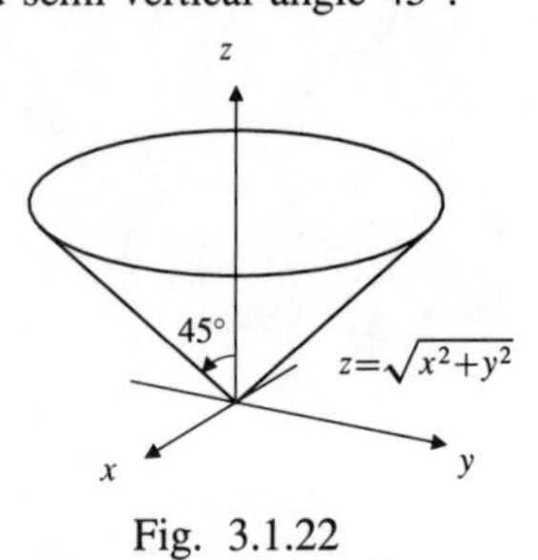


Fig. 3.1.22

**24.** $\begin{cases} x=1 \\ y=2 \end{cases}$ represents the vertical straight line in which the plane $x=1$ intersects the plane $y=2$.

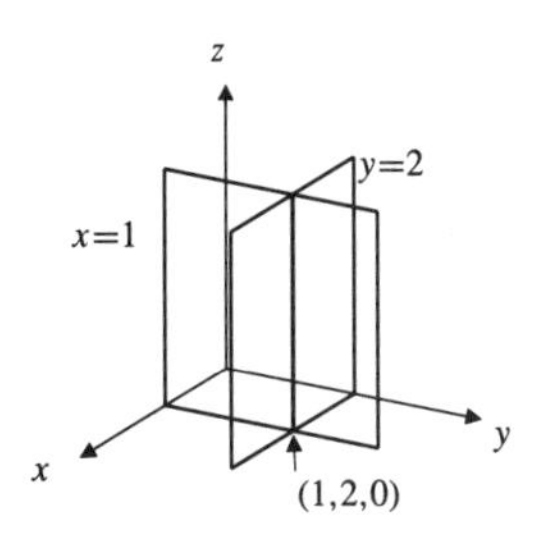


Fig. 3.1.24

26. $\begin{cases} x^2+y^2+z^2=4 \\ z=1 \end{cases}$ is the circle in which the horizontal plane $z=1$ intersects the sphere of radius 2 centred at the origin. The circle has centre $(0,0,1)$ and radius $\sqrt{4-1}=\sqrt{3}$.

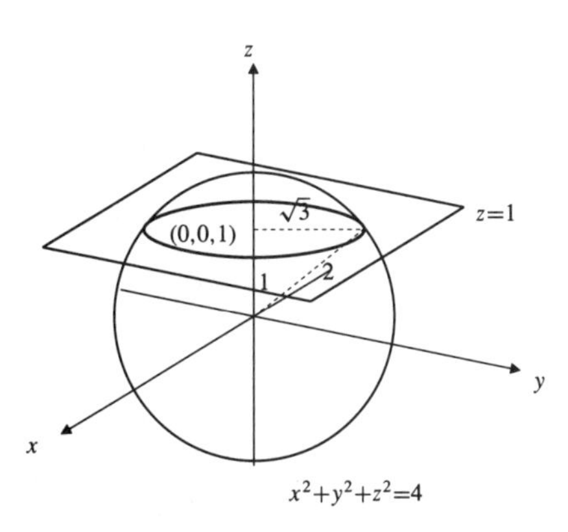


Fig. 3.1.26

28. $\begin{cases} x^2+y^2+z^2=4 \\ x^2+z^2=1 \end{cases}$ represents the two circles in which the cylinder $x^2+z^2-1$ intersects the sphere $x^2+y^2+z^2=4$. Subtracting the two equations, we get $y^2=3$. Thus, one circle lies in the plane $y=\sqrt{3}$ and has centre $(0,\sqrt{3},0)$ and the other lies in the plane $y=-\sqrt{3}$ and has centre $(0,-\sqrt{3},0)$. Both circles have radius 1.

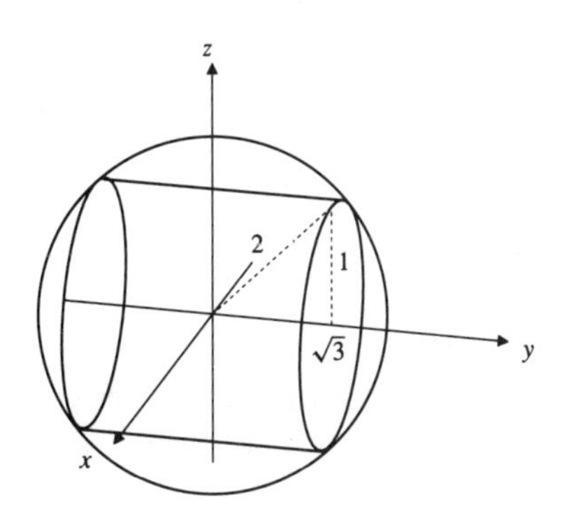


Fig. 3.1.28

30. $\begin{cases} y \ge x \\ z \le y \end{cases}$ is the quarter-space consisting of all points lying on or on the same side of the planes $y=x$ and $z=y$ as does the point $(0,1,0)$.

32. $\begin{cases} x^2+y^2+z^2 \le 1 \\ \sqrt{x^2+y^2} \le z \end{cases}$ represents all points which are inside or on the sphere of radius 1 centred at the origin and which are also inside or on the upper half of the circular cone with axis along the $z$-axis, vertex at the origin, and semi-vertical angle 45°.

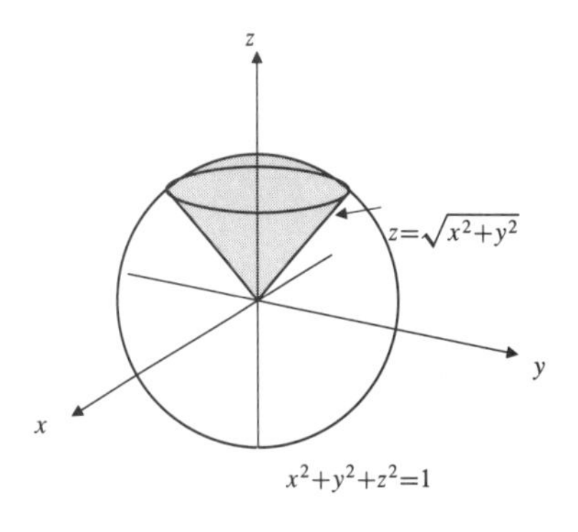


Fig. 3.1.32

## Section 3.2 Vectors in 3-Space (page 119)

2. $\mathbf{u}=\mathbf{i}-\mathbf{j}$
$\mathbf{v}=\mathbf{j}+2\mathbf{k}$

a) $\mathbf{u}+\mathbf{v}=\mathbf{i}+2\mathbf{k}$
$\mathbf{u}-\mathbf{v}=\mathbf{i}-2\mathbf{j}-2\mathbf{k}$
$2\mathbf{u}-3\mathbf{v}=2\mathbf{i}-5\mathbf{j}-6\mathbf{k}$

b) $|\mathbf{u}|=\sqrt{1+1}=\sqrt{2}$
$|\mathbf{v}|=\sqrt{1+4}=\sqrt{5}$

c) $\hat{\mathbf{u}}=\dfrac{1}{\sqrt{2}}(\mathbf{i}-\mathbf{j})$
$\hat{\mathbf{v}}=\dfrac{1}{\sqrt{5}}(\mathbf{j}+2\mathbf{k})$

d) $\mathbf{u}\bullet\mathbf{v}=0-1+0=-1$

e) The angle between $\mathbf{u}$ and $\mathbf{v}$ is $\cos^{-1}\dfrac{-1}{\sqrt{10}}\approx 108.4°$.

f) The scalar projection of $\mathbf{u}$ in the direction of $\mathbf{v}$ is $\dfrac{\mathbf{u}\bullet\mathbf{v}}{|\mathbf{v}|}=\dfrac{-1}{\sqrt{5}}$.

g) The vector projection of $\mathbf{v}$ along $\mathbf{u}$ is $\dfrac{(\mathbf{v}\bullet\mathbf{u})\mathbf{u}}{|\mathbf{u}|^2}=-\dfrac{1}{2}(\mathbf{i}-\mathbf{j})$.

**4.** $2\mathbf{x}+3\mathbf{i}-5\mathbf{j}=5\mathbf{x}+\mathbf{j}-9\mathbf{k}$
$2\mathbf{x}-5\mathbf{x}=-3\mathbf{i}+5\mathbf{j}+\mathbf{j}-9\mathbf{k}$
$-3\mathbf{x}=-3\mathbf{i}+6\mathbf{j}-9\mathbf{k}$
$\mathbf{x}=\mathbf{i}-2\mathbf{j}+3\mathbf{k}.$

**6.**

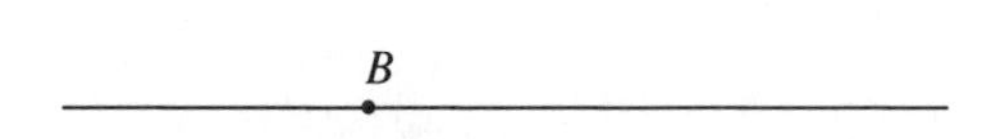


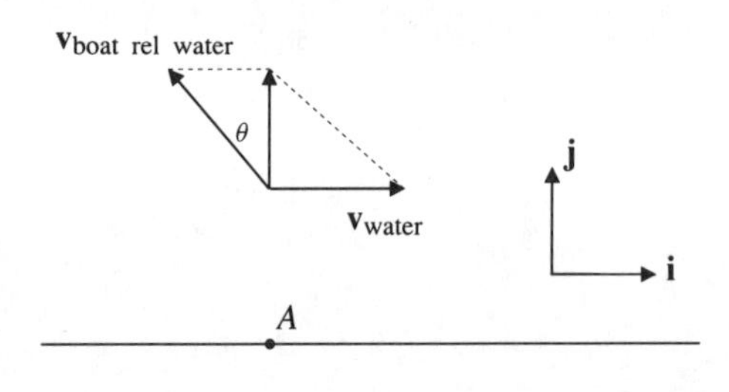


Fig. 3.2.6

Let the $x$-axis point east and the $y$-axis north. The velocity of the water is

$$\mathbf{v}_{\text{water}}=3\mathbf{i}.$$

If you row through the water with speed 5 in the direction making angle $\theta$ west of north, then your velocity relative to the water will be

$$\mathbf{v}_{\text{boat rel water}}=-5\sin\theta\mathbf{i}+5\cos\theta\mathbf{j}.$$

Therefore, your velocity relative to the land will be

$$\begin{aligned}\mathbf{v}_{\text{boat rel land}}&=\mathbf{v}_{\text{boat rel water}}+\mathbf{v}_{\text{water}}\\&=(3-5\sin\theta)\mathbf{i}+5\cos\theta\mathbf{j}.\end{aligned}$$

To make progress in the direction $\mathbf{j}$, choose $\theta$ so that $3=5\sin\theta$. Thus $\theta=\sin^{-1}(3/5)\approx 36.87^\circ$. In this case, your actual speed relative to the land will be
$5\cos\theta=\dfrac{4}{5}\times 5=4$ km/h.
To row from $A$ to $B$, head in the direction $36.87^\circ$ west of north. The 1/2 km crossing will take $(1/2)/4=1/8$ of an hour, or about $7\frac{1}{2}$ minutes.

**8.**

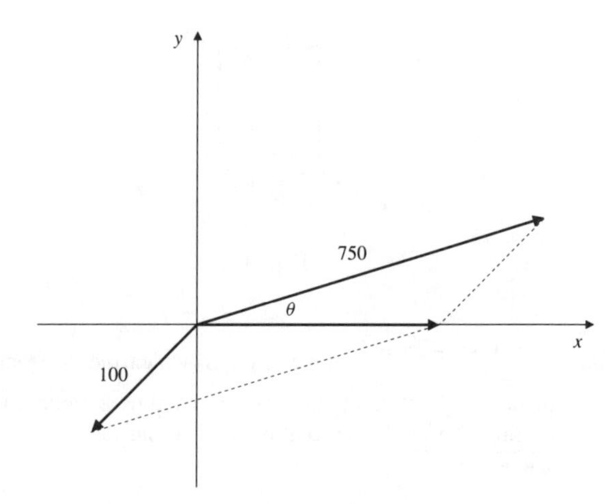


Fig. 3.2.8

Let $\mathbf{i}$ point east and $\mathbf{j}$ point north. If the aircraft heads in a direction $\theta$ north of east, then its velocity relative to the air is

$$750\cos\theta\mathbf{i}+750\sin\theta\mathbf{j}.$$

The velocity of the air relative to the ground is

$$-\frac{100}{\sqrt{2}}\mathbf{i}+-\frac{100}{\sqrt{2}}\mathbf{j}.$$

Thus the velocity of the aircraft relative to the ground is

$$\left(750\cos\theta-\frac{100}{\sqrt{2}}\right)\mathbf{i}+\left(750\sin\theta-\frac{100}{\sqrt{2}}\right)\mathbf{j}.$$

If this velocity is true easterly, then

$$750\sin\theta=\frac{100}{\sqrt{2}},$$

so $\theta\approx 5.41^\circ$. The speed relative to the ground is

$$750\cos\theta-\frac{100}{\sqrt{2}}\approx 675.9 \text{ km/h}.$$

The time for the 1500 km trip is $\dfrac{1500}{675.9}\approx 2.22$ hours.

**10.** The cube with edges $\mathbf{i}$, $\mathbf{j}$, and $\mathbf{k}$ has diagonal $\mathbf{i}+\mathbf{j}+\mathbf{k}$. The angle between $\mathbf{i}$ and the diagonal is

$$\cos^{-1}\frac{\mathbf{i}\bullet(\mathbf{i}+\mathbf{j}+\mathbf{k})}{\sqrt{3}}=\cos^{-1}\frac{1}{\sqrt{3}}\approx 54.7^\circ.$$

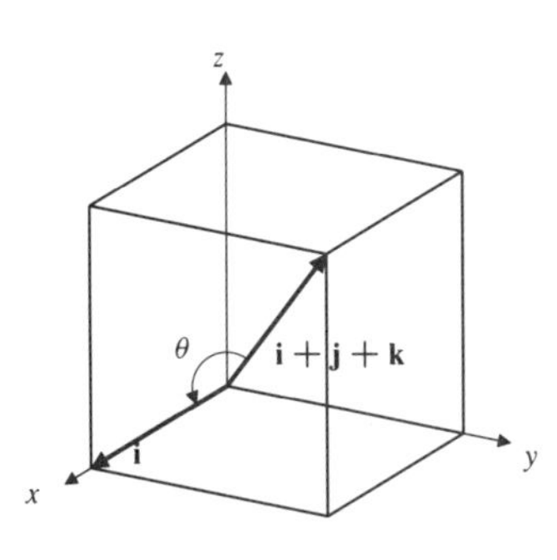


Fig. 3.2.10

**12.** If $\mathbf{u} = u_1\mathbf{i} + u_2\mathbf{j} + u_3\mathbf{k}$, then $\cos\alpha \dfrac{\mathbf{u} \bullet \mathbf{i}}{|\mathbf{u}|} = \dfrac{u_1}{|\mathbf{u}|}$.
Similarly, $\cos\beta = \dfrac{u_2}{|\mathbf{u}|}$ and $\cos\gamma = \dfrac{u_3}{|\mathbf{u}|}$.
Thus, the unit vector in the direction of $\mathbf{u}$ is

$$\hat{\mathbf{u}} = \frac{\mathbf{u}}{|\mathbf{u}|} = \cos\alpha\mathbf{i} + \cos\beta\mathbf{j} + \cos\gamma\mathbf{k},$$

and so $\cos^2\alpha + \cos^2\beta + \cos^2\gamma = |\hat{\mathbf{u}}|^2 = 1$.

**14.** If $A = (1, 0, 0)$, $B = (0, 2, 0)$, and $C = (0, 0, 3)$, then

$$\angle ABC = \cos^{-1}\frac{\overrightarrow{BA} \bullet \overrightarrow{BC}}{|BA||BC|} = \cos^{-1}\frac{4}{\sqrt{5}\sqrt{13}} \approx 60.26^\circ$$
$$\angle BCA = \cos^{-1}\frac{\overrightarrow{CB} \bullet \overrightarrow{CA}}{|CB||CA|} = \cos^{-1}\frac{9}{\sqrt{10}\sqrt{13}} \approx 37.87^\circ$$
$$\angle CAB = \cos^{-1}\frac{\overrightarrow{AC} \bullet \overrightarrow{AB}}{|AC||AB|} = \cos^{-1}\frac{1}{\sqrt{10}\sqrt{5}} \approx 81.87^\circ.$$

**16.** If $\mathbf{a} \neq \mathbf{0}$, then $\mathbf{a} \bullet \mathbf{r} = 0$ implies that the position vector $\mathbf{r}$ is perpenducular to $\mathbf{a}$. Thus the equation is satisfied by all points on the plane through the origin that is normal (perpendicular) to $\mathbf{a}$.

**18.** Vector $\mathbf{x} = x\mathbf{i} + y\mathbf{j} + z\mathbf{k}$ is perpendicular to both $\mathbf{u}$ and $\mathbf{v}$ if

$$\mathbf{u} \bullet \mathbf{x} = 0 \quad \Leftrightarrow \quad 2x + y - 2z = 0$$
$$\mathbf{v} \bullet \mathbf{x} = 0 \quad \Leftrightarrow \quad x + 2y - 2z = 0.$$

Subtracting these equations, we get $x - y = 0$, so $x = y$. The first equation now gives $3x = 2z$. Now $\mathbf{x}$ is a unit vector if $x^2 + y^2 + z^2 = 1$, that is, if $x^2 + x^2 + \frac{9}{4}x^2 = 1$, or $x = \pm 2/\sqrt{17}$. The two unit vectors are

$$\mathbf{x} = \pm\left(\frac{2}{\sqrt{17}}\mathbf{i} + \frac{2}{\sqrt{17}}\mathbf{j} + \frac{3}{\sqrt{17}}\mathbf{k}\right).$$

**20.** Since $\mathbf{u}$, $\mathbf{v}$, and $\mathbf{w}$ all have the same length (3), a vector $\mathbf{x} = x\mathbf{i} + y\mathbf{j} + z\mathbf{k}$ will make equal angles with all three if it has equal dot products with all three, that is, if

$$2x + y - 2z = x + 2y - 2z \quad \Leftrightarrow \quad x = y = 0$$
$$2x + y - 2z = 2x - 2y + z \quad \Leftrightarrow \quad 3y - 3z = 0.$$

Thus $x = y = z$. Two unit vectors satisfying this condition are

$$\mathbf{x} = \pm\left(\frac{1}{\sqrt{3}}\mathbf{i} + \frac{1}{\sqrt{3}}\mathbf{j} + \frac{1}{\sqrt{3}}\mathbf{k}\right).$$

**22.** If $\mathbf{u}$ and $\mathbf{v}$ are not parallel, then neither is the zero vector, and the origin and the two points with position vectors $\mathbf{u}$ and $\mathbf{v}$ lie on a unique plane. The equation $\mathbf{r} = \lambda\mathbf{u} + \mu\mathbf{v}$ ($\lambda$, $\mu$ real) gives the position vector of an arbitrary point on that plane.

**24.** a) $\mathbf{u}$, $\mathbf{v}$, and $\mathbf{u} + \mathbf{v}$ are the sides of a triangle. The triangle inequality says that the length of one side cannot exceed the sum of the lengths of the other two sides.

b) If $\mathbf{u}$ and $\mathbf{v}$ are parallel and point in the *same direction*, (or if at least one of them is the zero vector), then $|\mathbf{u} + \mathbf{v}| = |\mathbf{u}| + |\mathbf{v}|$.

**26.** Suppose $|\mathbf{u}| = |\mathbf{v}| = |\mathbf{w}| = 1$, and $\mathbf{u}\bullet\mathbf{v} = \mathbf{u}\bullet\mathbf{w} = \mathbf{v}\bullet\mathbf{w} = 0$, and let $\mathbf{r} = a\mathbf{u} + b\mathbf{v} + w\mathbf{w}$. Then

$$\mathbf{r} \bullet \mathbf{u} = a\mathbf{u} \bullet \mathbf{u} + b\mathbf{v} \bullet \mathbf{u} + c\mathbf{w} \bullet \mathbf{u} = a|\mathbf{u}|^2 + 0 + 0 = a.$$

Similarly, $\mathbf{r} \bullet \mathbf{v} = b$ and $\mathbf{r} \bullet \mathbf{w} = c$.

**28.** Let $\hat{\mathbf{n}}$ be a unit vector that is perpendicular to $\mathbf{u}$ and lies in the plane containing the origin and the points $U$, $V$, and $P$. Then $\hat{\mathbf{u}} = \mathbf{u}/|\mathbf{u}|$ and $\hat{\mathbf{n}}$ constitute a standard basis in that plane, so each of the vectors $\mathbf{v}$ and $\mathbf{r}$ can be expressed in terms of them:

$$\mathbf{v} = s\hat{\mathbf{u}} + t\hat{\mathbf{n}}$$
$$\mathbf{r} = x\hat{\mathbf{u}} + y\hat{\mathbf{n}}.$$

Since $\mathbf{v}$ is not parallel to $\mathbf{u}$, we have $t \neq 0$. Thus $\hat{\mathbf{n}} = (1/t)(\mathbf{v} - s\hat{\mathbf{u}})$ and

$$\mathbf{r} = x\hat{\mathbf{u}} + \frac{y}{t}(\mathbf{v} - s\hat{\mathbf{u}}) = \lambda\mathbf{u} + \mu\mathbf{v},$$

where $\lambda = (tx - ys)/(t|\mathbf{u}|)$ and $\mu = y/t$.

## Section 3.3 The Cross Product in 3-Space (page 127)

**2.** $(\mathbf{j} + 2\mathbf{k}) \times (-\mathbf{i} - \mathbf{j} + \mathbf{k}) = 3\mathbf{i} - 2\mathbf{j} + \mathbf{k}$

**4.** A vector perpendicular to the plane containing the three given points is

$$(-a\mathbf{i} + b\mathbf{j}) \times (-a\mathbf{i} + c\mathbf{k}) = bc\mathbf{i} + ac\mathbf{j} + ab\mathbf{k}.$$

A unit vector in this direction is

$$\frac{bc\mathbf{i} + ac\mathbf{j} + ab\mathbf{k}}{\sqrt{b^2c^2 + a^2c^2 + a^2b^2}}.$$

The triangle has area $\frac{1}{2}\sqrt{b^2c^2 + a^2c^2 + a^2b^2}$.

**6.** A vector perpendicular to $\mathbf{u} = 2\mathbf{i} - \mathbf{j} - 2\mathbf{k}$ and to $\mathbf{v} = 2\mathbf{i} - 3\mathbf{j} + \mathbf{k}$ is the cross product

$$\mathbf{u} \times \mathbf{v} = \begin{vmatrix} \mathbf{i} & \mathbf{j} & \mathbf{k} \\ 2 & -1 & -2 \\ 2 & -3 & 1 \end{vmatrix} = -7\mathbf{i} - 6\mathbf{j} - 4\mathbf{k},$$

which has length $\sqrt{101}$. A unit vector with positive $\mathbf{k}$ component that is perpenducular to $\mathbf{u}$ and $\mathbf{v}$ is

$$\frac{-1}{\sqrt{101}}\mathbf{u} \times \mathbf{v} = \frac{1}{\sqrt{101}}(7\mathbf{i} + 6\mathbf{j} + 4\mathbf{k}).$$

**8.**
$$\begin{aligned} \mathbf{u} \times \mathbf{v} &= \begin{vmatrix} \mathbf{i} & \mathbf{j} & \mathbf{k} \\ u_1 & u_2 & u_3 \\ v_1 & v_2 & v_3 \end{vmatrix} \\ &= -\begin{vmatrix} \mathbf{i} & \mathbf{j} & \mathbf{k} \\ v_1 & v_2 & v_3 \\ u_1 & u_2 & u_3 \end{vmatrix} = -\mathbf{v} \times \mathbf{u}. \end{aligned}$$

**10.**
$$\begin{aligned} (t\mathbf{u}) \times \mathbf{v} &= \begin{vmatrix} \mathbf{i} & \mathbf{j} & \mathbf{k} \\ tu_1 & tu_2 & tu_3 \\ v_1 & v_2 & v_3 \end{vmatrix} \\ &= t\begin{vmatrix} \mathbf{i} & \mathbf{j} & \mathbf{k} \\ u_1 & u_2 & u_3 \\ v_1 & v_2 & v_3 \end{vmatrix} = t(\mathbf{u} \times \mathbf{v}), \\ \mathbf{u} \times (t\mathbf{v}) &= -(t\mathbf{v}) \times \mathbf{u} \\ &= -t(\mathbf{v} \times \mathbf{u}) = t(\mathbf{u} \times \mathbf{v}). \end{aligned}$$

**12.** Both $\mathbf{u} = \cos\beta\,\mathbf{i} + \sin\beta\,\mathbf{j}$ and $\mathbf{v} = \cos\alpha\,\mathbf{i} + \sin\alpha\,\mathbf{j}$ are unit vectors. They make angles $\beta$ and $\alpha$, respectively, with the positive $x$-axis, so the angle between them is $|\alpha - \beta| = \alpha - \beta$, since we are told that $0 \le \alpha - \beta \le \pi$. They span a parallelogram (actually a rhombus) having area

$$|\mathbf{u} \times \mathbf{v}| = |\mathbf{u}||\mathbf{v}|\sin(\alpha - \beta) = \sin(\alpha - \beta).$$

But

$$\mathbf{u} \times \mathbf{v} = \begin{vmatrix} \mathbf{i} & \mathbf{j} & \mathbf{k} \\ \cos\beta & \sin\beta & 0 \\ \cos\alpha & \sin\alpha & 0 \end{vmatrix} = (\sin\alpha\cos\beta - \cos\alpha\sin\beta)\mathbf{k}.$$

Because $\mathbf{v}$ is displaced counterclockwise from $\mathbf{u}$, the cross product above must be in the positive $k$ direction. Therefore its length is the $k$ component. Therefore

$$\sin(\alpha - \beta) = \sin\alpha\cos\beta - \cos\alpha\sin\beta.$$

**14.**

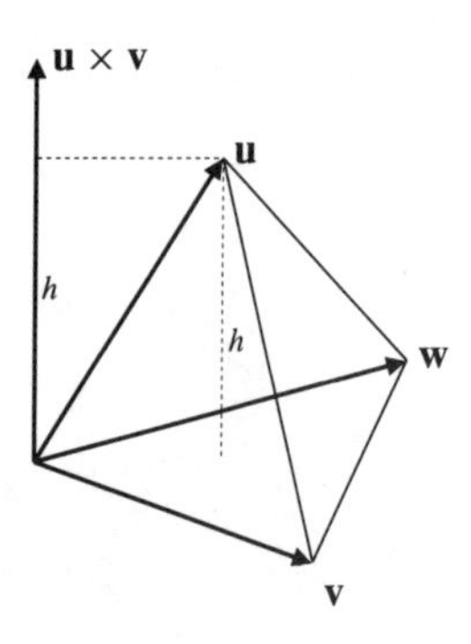


Fig. 3.3.14

The base of the tetrahedron is a triangle spanned by $\mathbf{v}$ and $\mathbf{w}$, which has area

$$A = \frac{1}{2}|\mathbf{v} \times \mathbf{w}|.$$

The altitude $h$ of the tetrahedron (measured perpendicular to the plane of the base) is equal to the length of the projection of $\mathbf{u}$ onto the vector $\mathbf{v} \times \mathbf{w}$ (which is perpendicular to the base). Thus

$$h = \frac{|\mathbf{u} \bullet (\mathbf{v} \times \mathbf{w})|}{|\mathbf{v} \times \mathbf{w}|}.$$

The volume of the tetrahedron is

$$\begin{aligned} V &= \frac{1}{3}Ah = \frac{1}{6}|\mathbf{u} \bullet (\mathbf{v} \times \mathbf{w})| \\ &= \frac{1}{6}\left|\begin{vmatrix} u_1 & u_2 & u_3 \\ v_1 & v_2 & v_3 \\ w_1 & w_2 & w_3 \end{vmatrix}\right|. \end{aligned}$$

**16.**

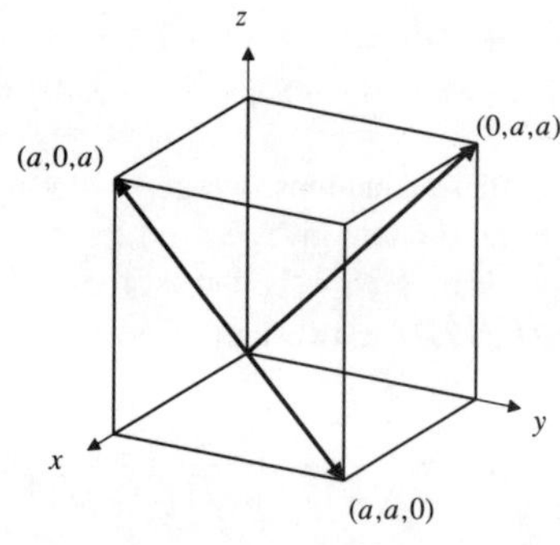


Fig. 3.3.16

Let the cube be as shown in the figure. The required parallelepiped is spanned by $a\mathbf{i}+a\mathbf{j}$, $a\mathbf{j}+a\mathbf{k}$, and $a\mathbf{i}+a\mathbf{k}$. Its volume is

$$V = \left| \begin{vmatrix} a & a & 0 \\ 0 & a & a \\ a & 0 & a \end{vmatrix} \right| = 2a^3 \text{ cu. units.}$$

**18.**
$$\begin{aligned}\mathbf{u} \bullet (\mathbf{v} \times \mathbf{w}) &= \begin{vmatrix} u_1 & u_2 & u_3 \\ v_1 & v_2 & v_3 \\ w_1 & w_2 & w_3 \end{vmatrix} \\ &= -\begin{vmatrix} v_1 & v_2 & v_3 \\ u_1 & u_2 & u_3 \\ w_1 & w_2 & w_3 \end{vmatrix} \\ &= \begin{vmatrix} v_1 & v_2 & v_3 \\ w_1 & w_2 & w_3 \\ u_1 & u_2 & u_3 \end{vmatrix} \\ &= \mathbf{v} \bullet (\mathbf{w} \times \mathbf{u}) \\ &= \mathbf{w} \bullet (\mathbf{u} \times \mathbf{v}) \qquad \text{(by symmetry).}\end{aligned}$$

**20.** If $\mathbf{v} \times \mathbf{w} \neq \mathbf{0}$, then $(\mathbf{v} \times \mathbf{w}) \bullet (\mathbf{v} \times \mathbf{w}) \neq 0$. By the previous exercise, there exist constants $\lambda$, $\mu$ and $\nu$ such that

$$\mathbf{u} = \lambda\mathbf{v} + \mu\mathbf{w} + \nu(\mathbf{v} \times \mathbf{w}).$$

But $\mathbf{v} \times \mathbf{w}$ is perpendicular to both $\mathbf{v}$ and $\mathbf{w}$, so

$$\mathbf{u} \bullet (\mathbf{v} \times \mathbf{w}) = 0 + 0 + \nu(\mathbf{v} \times \mathbf{w}) \bullet (\mathbf{v} \times \mathbf{w}).$$

If $\mathbf{u} \bullet (\mathbf{v} \times \mathbf{w}) = 0$, then $\nu = 0$, and

$$\mathbf{u} = \lambda\mathbf{v} + \mu\mathbf{w}.$$

**22.** $\mathbf{u} \bullet \mathbf{v} \times \mathbf{w}$ makes sense in that it must mean $\mathbf{u} \bullet (\mathbf{v} \times \mathbf{w})$. ($(\mathbf{u} \bullet \mathbf{v}) \times \mathbf{w}$ makes no sense since it is the cross product of a scalar and a vector.)

$\mathbf{u} \times \mathbf{v} \times \mathbf{w}$ makes no sense. It is ambiguous, since $(\mathbf{u} \times \mathbf{v}) \times \mathbf{w}$ and $\mathbf{u} \times (\mathbf{v} \times \mathbf{w})$ are not in general equal.

**24.** If $\mathbf{u}$, $\mathbf{v}$, and $\mathbf{w}$ are mutually perpendicular, then $\mathbf{v} \times \mathbf{w}$ is parallel to $\mathbf{u}$, so $\mathbf{u} \times (\mathbf{v} \times \mathbf{w}) = \mathbf{0}$. In this case, $\mathbf{u} \bullet (\mathbf{v} \times \mathbf{w}) = \pm|\mathbf{u}||\mathbf{v}||\mathbf{w}|$; the sign depends on whether $\mathbf{u}$ and $\mathbf{v} \times \mathbf{w}$ are in the same or opposite directions.

**26.** If $\mathbf{a} = -\mathbf{i} + 2\mathbf{j} + 3\mathbf{k}$ and $\mathbf{x} = x\mathbf{i} + y\mathbf{j} + z\mathbf{k}$, then

$$\begin{aligned}\mathbf{a} \times \mathbf{x} &= \begin{vmatrix} \mathbf{i} & \mathbf{j} & \mathbf{k} \\ -1 & 2 & 3 \\ x & y & z \end{vmatrix} \\ &= (2z - 3y)\mathbf{i} + (3x + z)\mathbf{y} - (y + 2x)\mathbf{k} \\ &= \mathbf{i} + 5\mathbf{j} - 3\mathbf{k},\end{aligned}$$

provided $2z - 3y = 1$, $3x + z = 5$, and $-y - 2x = -3$. This system is satisfied by $x = t$, $y = 3 - 2t$, $z = 5 - 3t$, for any real number $t$. Thus

$$x = t\mathbf{i} + (3 - 2t)\mathbf{j} + (5 - 3t)\mathbf{k}$$

gives a solution of $\mathbf{a} \times \mathbf{x} = \mathbf{i} + 5\mathbf{j} - 3\mathbf{k}$ for any $t$. These solutions constitute a line parallel to $\mathbf{a}$.

**28.** The equation $\mathbf{a} \times \mathbf{x} = \mathbf{b}$ can be solved for $\mathbf{x}$ if and only if $\mathbf{a} \bullet \mathbf{b} = 0$. The "only if" part is demonstrated in the previous solution. For the "if" part, observe that if $\mathbf{a} \bullet \mathbf{b} = 0$ and $\mathbf{x}_0 = (\mathbf{b} \times \mathbf{a})/|\mathbf{a}|^2$, then by Exercise 23,

$$\mathbf{a} \times \mathbf{x}_0 = \frac{1}{|a|^2}\mathbf{a} \times (\mathbf{b} \times \mathbf{a}) = \frac{(\mathbf{a} \bullet \mathbf{a})\mathbf{b} - (\mathbf{a} \bullet \mathbf{b})\mathbf{a}}{|a|^2} = \mathbf{b}.$$

The solution $\mathbf{x}_0$ is not unique; as suggested by the example in Exercise 26, any multiple of $\mathbf{a}$ can be added to it and the result will still be a solution. If $\mathbf{x} = \mathbf{x}_0 + t\mathbf{a}$, then

$$\mathbf{a} \times \mathbf{x} = \mathbf{a} \times \mathbf{x}_0 + t\mathbf{a} \times \mathbf{a} = \mathbf{b} + 0 = \mathbf{b}.$$

## Section 3.4 Planes and Lines (page 135)

**2.** The plane through $(0, 2, -3)$ normal to $4\mathbf{i} - \mathbf{j} - 2\mathbf{k}$ has equation

$$4(x - 0) - (y - 2) - 2(z + 3) = 0,$$

or $4x - y - 2z = 4$.

**4.** The plane passing through $(1, 2, 3)$, parallel to the plane $3x + y - 2z = 15$, has equation $3z + y - 2z = 3 + 2 - 6$, or $3x + y - 2z = -1$.

**6.** The plane passing through $(-2, 0, 0)$, $(0, 3, 0)$, and $(0, 0, 4)$ has equation

$$\frac{x}{-2} + \frac{y}{3} + \frac{z}{4} = 1,$$

or $6x - 4y - 3z = -12$.

**8.** Since $(-2, 0, -1)$ does not lie on $x - 4y + 2z = -5$, the required plane will have an equation of the form

$$2x + 3y - z + \lambda(x - 4y + 2z + 5) = 0$$

for some $\lambda$. Thus

$$-4 + 1 + \lambda(-2 - 2 + 5) = 0,$$

so $\lambda = 3$. The required plane is $5x - 9y + 5z = -15$.

**10.** Three distinct points will not determine a unique plane through them if they all lie on a straight line. If the points have position vectors $\mathbf{r}_1$, $\mathbf{r}_2$, and $\mathbf{r}_3$, then they will all lie on a straight line if

$$(\mathbf{r}_2 - \mathbf{r}_1) \times (\mathbf{r}_3 - \mathbf{r}_1) = \mathbf{0}.$$

**12.** $x + y + z = \lambda$ is the family of all (parallel) planes normal to the vector $\mathbf{i} + \mathbf{j} + \mathbf{k}$.

**14.** The distance from the planes

$$\lambda x + \sqrt{1 - \lambda^2}\, y = 1$$

to the origin is $1/\sqrt{\lambda^2 + 1 - \lambda^2} = 1$. Hence the equation represents the family of all vertical planes at distance 1 from the origin. All such planes are tangent to the cylinder $x^2 + y^2 = 1$.

**16.** The line through $(-1, 0, 1)$ perpendicular to the plane $2x - y + 7z = 12$ is parallel to the normal vector $2\mathbf{i} - \mathbf{j} + 7\mathbf{k}$ to that plane. The equations of the line are, in vector parametric form,

$$\mathbf{r} = (-1 + 2t)\mathbf{i} - t\mathbf{j} + (1 + 7t)\mathbf{k},$$

or in scalar parametric form,

$$x = -1 + 2t, \quad y = -t, \quad z = 1 + 7t,$$

or in standard form

$$\frac{x + 1}{2} = \frac{y}{-1} = \frac{z - 1}{7}.$$

**18.** A line parallel to $x + y = 0$ and to $x - y + 2z = 0$ is parallel to the cross product of the normal vectors to these two planes, that is, to the vector

$$(\mathbf{i} + \mathbf{j}) \times (\mathbf{i} - \mathbf{j} + 2\mathbf{k}) = 2(\mathbf{i} - \mathbf{j} - \mathbf{k}).$$

Since the line passes through $(2, -1, -1)$, its equations are, in vector parametric form

$$\mathbf{r} = (2 + t)\mathbf{i} - (1 + t)\mathbf{j} - (1 + t)\mathbf{k},$$

or in scalar parametric form

$$x = 2 + t, \quad y = -(1 + t), \quad z = -(1 + t),$$

or in standard form

$$x - 2 = -(y + 1) = -(z + 1).$$

**20.** The line $\mathbf{r} = (1 - 2t)\mathbf{i} + (4 + 3t)\mathbf{j} + (9 - 4t)\mathbf{k}$ has standard form

$$\frac{x - 1}{-2} = \frac{y - 4}{3} = \frac{z - 9}{-4}.$$

**22.** The line $\begin{cases} x - 2y + 3z = 0 \\ 2x + 3y - 4z = 4 \end{cases}$ is parallel to the vector

$$(\mathbf{i} - 2\mathbf{j} + 3\mathbf{k}) \times (2\mathbf{i} + 3\mathbf{j} - 4\mathbf{k}) = -\mathbf{i} + 10\mathbf{j} + 7\mathbf{k}.$$

We need a point on this line. Putting $z = 0$, we get

$$x - 2y = 0, \qquad 2x + 3y = 4.$$

The solution of this system is $y = 4/7$, $x = 8/7$. A possible standard form for the given line is

$$\frac{x - \frac{8}{7}}{-1} = \frac{y - \frac{4}{7}}{10} = \frac{z}{7},$$

though, of course, this answer is not unique as the coordinates of any point on the line could have been used.

**24.** The distance from $(0, 0, 0)$ to $x + 2y + 3z = 4$ is

$$\frac{4}{\sqrt{1^2 + 2^2 + 3^2}} = \frac{4}{\sqrt{14}} \text{ units.}$$

**26.** A vector parallel to the line $x + y + z = 0$, $2x - y - 5z = 1$ is

$$\mathbf{a} = (\mathbf{i} + \mathbf{j} + \mathbf{k}) \times (2\mathbf{i} - \mathbf{j} - 5\mathbf{k}) = -4\mathbf{i} + 7\mathbf{j} - 3\mathbf{k}.$$

We need a point on this line: if $z = 0$ then $x + y = 0$ and $2x - y = 1$, so $x = 1/3$ and $y = -1/3$. The position vector of this point is

$$\mathbf{r}_1 = \frac{1}{3}\mathbf{i} - \frac{1}{3}\mathbf{j}.$$

The distance from the origin to the line is

$$s = \frac{|\mathbf{r}_1 \times \mathbf{a}|}{|\mathbf{a}|} = \frac{|\mathbf{i} + \mathbf{j} + \mathbf{k}|}{\sqrt{74}} = \sqrt{\frac{3}{74}} \text{ units.}$$

**28.** The line $x - 2 = \dfrac{y + 3}{2} = \dfrac{z - 1}{4}$ passes through the point $(2, -3, 1)$, and is parallel to $\mathbf{a} = \mathbf{i} + 2\mathbf{j} + 4\mathbf{k}$.
The plane $2y - z = 1$ has normal $\mathbf{n} = 2\mathbf{j} - \mathbf{k}$.
Since $\mathbf{a} \bullet \mathbf{n} = 0$, the line is parallel to the plane.
The distance from the line to the plane is equal to the distance from $(2, -3, 1)$ to the plane $2y - z = 1$, so is

$$D = \frac{|-6 - 1 - 1|}{\sqrt{4 + 1}} = \frac{8}{\sqrt{5}} \text{ units.}$$

**30.** $\dfrac{x - x_0}{\sqrt{1 - \lambda^2}} = \dfrac{y - y_0}{\lambda} = z - z_0$ represents all lines through $(x_0, y_0, z_0)$ parallel to the vectors

$$\mathbf{a} = \sqrt{1 - \lambda^2}\mathbf{i} + \lambda\mathbf{j} + \mathbf{k}.$$

All such lines are generators of the circular cone

$$(z - z_0)^2 = (x - x_0)^2 + (y - y_0)^2,$$

so the given equations specify all straight lines lying on that cone.

## Section 3.5 Quadric Surfaces (page 139)

**2.** $x^2 + y^2 + 4z^2 = 4$ represents an oblate spheroid, that is, an ellipsoid with its two longer semi-axes equal. In this case the longer semi-axes have length 2, and the shorter one (in the $z$ direction) has length 1. Cross-sections in planes perpendicular to the $z$-axis between $z = -1$ and $z = 1$ are circles.

**4.** $x^2 + 4y^2 + 9z^2 + 4x - 8y = 8$

$(x+2)^2 + 4(y-1)^2 + 9z^2 = 8 + 8 = 16$

$\dfrac{(x+2)^2}{4^2} + \dfrac{(y-1)^2}{2^2} + \dfrac{z^2}{(4/3)^2} = 1$

This is an ellipsoid with centre $(-2, 1, 0)$ and semi-axes 4, 2, and 4/3.

**6.** $z = x^2 - 2y^2$ represents a hyperbolic paraboloid.

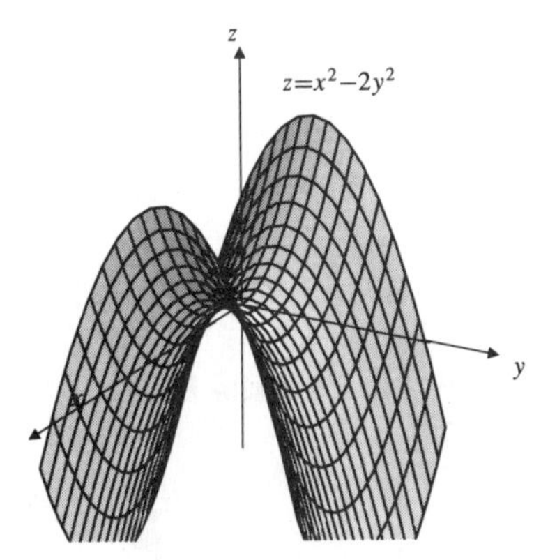


Fig. 3.5.6

**8.** $-x^2 + y^2 + z^2 = 4$ represents a hyperboloid of one sheet, with circular cross-sections in all planes perpendicular to the $x$-axis.

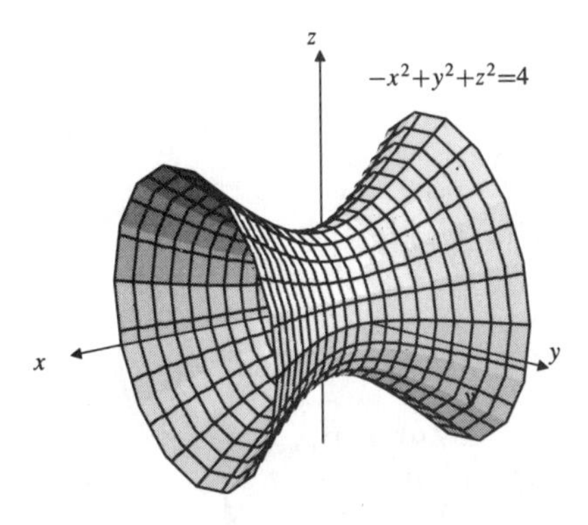


Fig. 3.5.8

**10.** $x^2 + 4z^2 = 4$ represents an elliptic cylinder with axis along the $y$-axis.

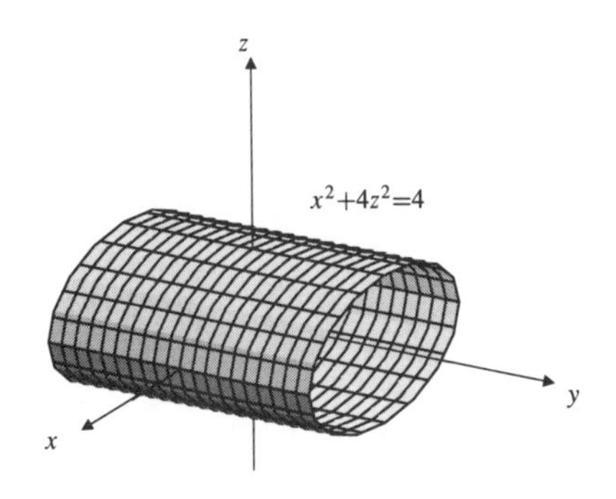


Fig. 3.5.10

**12.** $y = z^2$ represents a parabolic cylinder with vertex line along the $x$-axis.

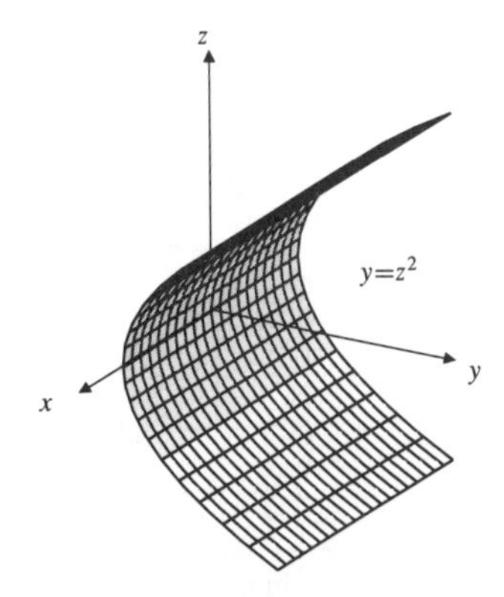


Fig. 3.5.12

**14.** $x^2 = y^2 + 2z^2$ represents an elliptic cone with vertex at the origin and axis along the $x$-axis.

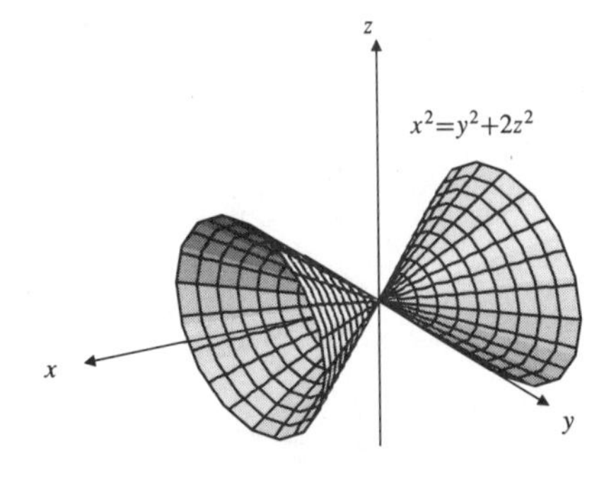


Fig. 3.5.14

**16.** $(z-1)^2 = (x-2)^2 + (y-3)^2 + 4$ represents a hyperboloid of two sheets with centre at $(2, 3, 1)$, axis along the line $x = 2$, $y = 3$, and vertices at $(2, 3, -1)$ and $(2, 3, 3)$.

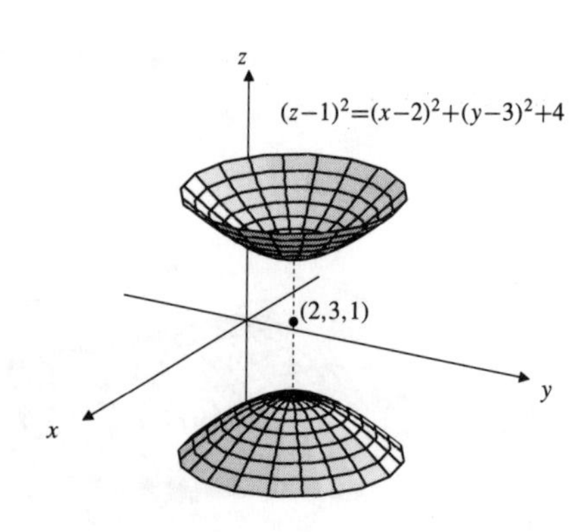


Fig. 3.5.16

**18.** $\begin{cases} x^2+y^2=1 \\ z=x+y \end{cases}$ is the ellipse of intersection of the plane $z = x+y$ and the circular cylinder $x^2+y^2=1$. The centre of the ellipse is at the origin, and the ends of the major axis are $\pm(1/\sqrt{2}, 1/\sqrt{2}, \sqrt{2})$.

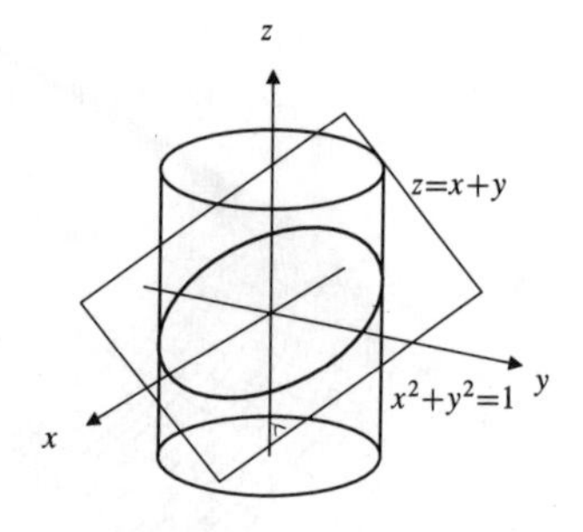


Fig. 3.5.18

**20.** $\begin{cases} x^2+2y^2+3z^2=6 \\ y=1 \end{cases}$ is an ellipse in the plane $y=1$. Its projection onto the $xz$-plane is the ellipse $x^2+3z^2=4$. One quarter of the ellipse is shown in the figure.

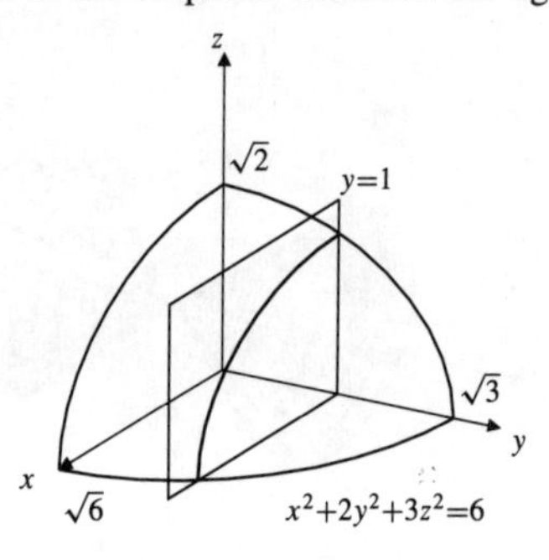


Fig. 3.5.20

**22.** $z = xy$

Family 1: $\begin{cases} z = \lambda x \\ \lambda = y. \end{cases}$

Family 2: $\begin{cases} z = \mu y \\ \mu = x. \end{cases}$

**24.** The plane $z = cx+k$ intersects the elliptic cone $z^2 = 2x^2+y^2$ on the cylinder

$$c^2x^2+2ckx+k^2 = 2x^2+y^2$$
$$(2-c^2)x^2-2ckx+y^2 = k^2$$
$$(2-c^2)\left(x-\frac{ck}{2-c^2}\right)^2+y^2 = k^2+\frac{c^2k^2}{2-c^2} = \frac{2k^2}{2-c^2}$$
$$\frac{(x-x_0)^2}{a^2}+\frac{y^2}{b^2} = 1,$$

where $x_0 = \dfrac{ck}{2-c^2}$, $a^2 = \dfrac{2k^2}{(2-c^2)^2}$, and $b^2 = \dfrac{2k^2}{2-c^2}$.

As in the previous exercise, $z = cx+k$ intersects the cylinder (and hence the cone) in an ellipse with principal axes joining the points

$$(x_0-a, 0, c(x_0-a)+k) \quad \text{to} \quad (x_0+a, 0, c(x_0+a)+k),$$
$$\text{and } (x_0, -b, cx_0+k) \quad \text{to} \quad (x_0, b, cx_0+k).$$

The centre of this ellipse is $(x_0, 0, cx_0+k)$. The ellipse is a circle if its two semi-axes have equal lengths, that is, if

$$a^2+c^2a^2 = b^2,$$

that is,

$$(1+c^2)\frac{2k^2}{(2-c^2)^2} = \frac{2k^2}{2-c^2},$$

or $1+c^2 = 2-c^2$. Thus $c = \pm 1/\sqrt{2}$. A vector normal to the plane $z = \pm(x/\sqrt{2})+k$ is $\mathbf{a} = \mathbf{i} \pm \sqrt{2}\mathbf{k}$.

## Section 3.6 A Little Matrix Algebra (page 147)

**2.** $$\begin{pmatrix} 1 & 1 & 1 \\ 0 & 1 & 1 \\ 0 & 0 & 1 \end{pmatrix}\begin{pmatrix} 1 & 1 & 1 \\ 0 & 1 & 1 \\ 0 & 0 & 1 \end{pmatrix} = \begin{pmatrix} 1 & 2 & 3 \\ 0 & 1 & 2 \\ 0 & 0 & 1 \end{pmatrix}$$

**4.** $$\begin{pmatrix} w & x \\ y & z \end{pmatrix}\begin{pmatrix} a & b \\ c & d \end{pmatrix} = \begin{pmatrix} aw+cx & bw+dx \\ ay+cz & by+dz \end{pmatrix}$$

**6.** $$\mathbf{x} = \begin{pmatrix} x \\ y \\ z \end{pmatrix}, \qquad \mathcal{A} = \begin{pmatrix} a & p & q \\ p & b & r \\ q & r & c \end{pmatrix}$$

$$\mathbf{x}\mathbf{x}^T = \begin{pmatrix} x \\ y \\ z \end{pmatrix}(x, y, z) = \begin{pmatrix} x^2 & xy & xz \\ xy & y^2 & yz \\ xz & yz & z^2 \end{pmatrix}$$

$$\mathbf{x}^T\mathbf{x} = (x, y, z)\begin{pmatrix} x \\ y \\ z \end{pmatrix} = (x^2+y^2+z^2)$$

$$\mathbf{x}^T\mathcal{A}\mathbf{x} = (x, y, z)\begin{pmatrix} a & p & q \\ p & b & r \\ q & r & c \end{pmatrix}\begin{pmatrix} x \\ y \\ z \end{pmatrix}$$
$$= (x, y, z)\begin{pmatrix} ax+py+qz \\ px+by+rz \\ qx+ry+cz \end{pmatrix}$$
$$= ax^2+by^2+cz^2+2pxy+2qxz+2ryz$$

**8.** $\begin{vmatrix} 1 & 1 & 1 & 1 \\ 1 & 2 & 3 & 4 \\ -2 & 0 & 2 & 4 \\ 3 & -3 & 2 & -2 \end{vmatrix}$

$$= -2\begin{vmatrix} 1 & 1 & 1 \\ 2 & 3 & 4 \\ -3 & 2 & -2 \end{vmatrix} + 2\begin{vmatrix} 1 & 1 & 1 \\ 1 & 2 & 4 \\ 3 & -3 & -2 \end{vmatrix} - 4\begin{vmatrix} 1 & 1 & 1 \\ 1 & 2 & 3 \\ 3 & -3 & 2 \end{vmatrix}$$

$$= -2\begin{vmatrix} 1 & 1 & 1 \\ 0 & 1 & 2 \\ 0 & 5 & 1 \end{vmatrix} + 2\begin{vmatrix} 1 & 1 & 1 \\ 0 & 1 & 3 \\ 0 & -6 & -5 \end{vmatrix} - 4\begin{vmatrix} 1 & 1 & 1 \\ 0 & 1 & 2 \\ 0 & -6 & -1 \end{vmatrix}$$

$$= -2\begin{vmatrix} 1 & 2 \\ 5 & 1 \end{vmatrix} + 2\begin{vmatrix} 1 & 3 \\ -6 & -5 \end{vmatrix} - 4\begin{vmatrix} 1 & 2 \\ -6 & -1 \end{vmatrix}$$

$$= -2(-9) + 2(13) - 4(11) = 0$$

**10.** $\begin{vmatrix} 1 & 1 \\ x & y \end{vmatrix} = y - x$. If

$$f(x, y, z) = \begin{vmatrix} 1 & 1 & 1 \\ x & y & z \\ x^2 & y^2 & z^2 \end{vmatrix},$$

then $f$ is a polynomial of degree 2 in $z$.
Since $f(x, y, x) = 0$ and $f(x, y, y) = 0$, we must have $f(x, y, z) = A(z - x)(z - y)$ for some $A$ independent of $z$. But

$$Axy = f(x, y, 0) = \begin{vmatrix} 1 & 1 & 1 \\ x & y & 0 \\ x^2 & y^2 & 0 \end{vmatrix} = xy(y - x),$$

so $A = y - x$ and

$$f(x, y, z) = (y - x)(z - x)(z - y).$$

Generalization:

$$\begin{vmatrix} 1 & 1 & 1 & \cdots & 1 \\ x_1 & x_2 & x_3 & \cdots & x_n \\ x_1^2 & x_2^2 & x_3^2 & \cdots & x_n^2 \\ \vdots & \vdots & \vdots & \ddots & \vdots \\ x_1^{n-1} & x_2^{n-1} & x_3^{n-1} & \cdots & x_n^{n-1} \end{vmatrix} = \prod_{1 \le i < j \le n} (x_j - x_i).$$

**12.** If $\mathcal{A} = \begin{pmatrix} a & b \\ c & d \end{pmatrix}$, then $\mathcal{A}^T = \begin{pmatrix} a & c \\ b & d \end{pmatrix}$, and

$$\det(\mathcal{A}) = ad - bc = \det(\mathcal{A}^T).$$

We generalize this by induction.
Suppose $\det(\mathcal{B}^T) = \det(\mathcal{B})$ for any $(n - 1) \times (n - 1)$ matrix, where $n \ge 3$. Let

$$\mathcal{A} = \begin{pmatrix} a_{11} & a_{12} & \cdots & a_{1n} \\ a_{21} & a_{22} & \cdots & a_{2n} \\ \vdots & \vdots & \ddots & \vdots \\ a_{n1} & a_{n2} & \cdots & a_{nn} \end{pmatrix}$$

be an $n \times n$ matrix. If $\det(\mathcal{A})$ is expanded in minors about the first row, and $\det(\mathcal{A}^T)$ is expanded in minors about the first column, the corresponding terms in these expansions are equal by the induction hypothesis. (The $(n-1)\times(n-1)$ matrices whose determinants appear in one expansion are the transposes of those in the other expansion.) Therefore $\det(\mathcal{A}^T) = \det(\mathcal{A})$ for any square matrix $\mathcal{A}$.

**14.** If $\mathcal{A}_\theta = \begin{pmatrix} \cos\theta & \sin\theta \\ -\sin\theta & \cos\theta \end{pmatrix}$, then

$$\mathcal{A}_{-\theta} = \begin{pmatrix} \cos(-\theta) & \sin(-\theta) \\ -\sin(-\theta) & \cos(-\theta) \end{pmatrix} = \begin{pmatrix} \cos\theta & -\sin\theta \\ \sin\theta & \cos\theta \end{pmatrix},$$

and

$$\mathcal{A}_\theta\mathcal{A}_{-\theta} = \begin{pmatrix} 1 & 0 \\ 0 & 1 \end{pmatrix} = I.$$

Thus $\mathcal{A}_{-\theta} = (\mathcal{A}_\theta)^{-1}$.

**16.** Let $\mathcal{A} = \begin{pmatrix} 1 & 0 & -1 \\ -1 & 1 & 0 \\ 2 & 1 & 3 \end{pmatrix}$, $\mathcal{A}^{-1} = \begin{pmatrix} a & b & c \\ d & e & f \\ g & h & i \end{pmatrix}$. Since $\mathcal{A}\mathcal{A}^{-1} = I$ we must have

$$\begin{array}{lll} a - g = 1 & b - h = 0 & c - i = 0 \\ -a + d = 0 & -b + e = 1 & -c + f = 0 \\ 2a + d + 3g = 0 & 2b + e + 3h = 0 & 2c + f + 3i = 1. \end{array}$$

Solving these three systems of equations, we get

$$\mathcal{A}^{-1} = \begin{pmatrix} \frac{1}{2} & -\frac{1}{6} & \frac{1}{6} \\ \frac{1}{2} & \frac{5}{6} & \frac{1}{6} \\ -\frac{1}{2} & -\frac{1}{6} & \frac{1}{6} \end{pmatrix}.$$

**18.** If $\mathcal{A}$ is the matrix of Exercises 16 and 17 then $\det(\mathcal{A}) = 6$. By Cramer's Rule,

$$x = \frac{1}{6}\begin{vmatrix} -2 & 0 & -1 \\ 1 & 1 & 0 \\ 13 & 1 & 3 \end{vmatrix} = \frac{6}{6} = 1$$

$$y = \frac{1}{6}\begin{vmatrix} 1 & -2 & -1 \\ -1 & 1 & 0 \\ 2 & 13 & 3 \end{vmatrix} = \frac{12}{6} = 2$$

$$z = \frac{1}{6}\begin{vmatrix} 1 & 0 & -2 \\ -1 & 1 & 1 \\ 2 & 1 & 13 \end{vmatrix} = \frac{18}{6} = 3.$$

**20.** Let $F(x_1, x_2) = \mathcal{F}\begin{pmatrix} x_1 \\ x_2 \end{pmatrix}$, where $\mathcal{F} = \begin{pmatrix} a & b \\ c & d \end{pmatrix}$.
Let $G(y_1, y_2) = \mathcal{G}\begin{pmatrix} y_1 \\ y_2 \end{pmatrix}$, where $\mathcal{G} = \begin{pmatrix} p & q \\ r & s \end{pmatrix}$.
If $y_1 = ax_1 + bx_2$ and $y_2 = cx_1 + dx_2$, then

$$\begin{aligned} G \circ F(x_1, x_2) &= G(y_1, y_2) \\ &= \begin{pmatrix} p & q \\ r & s \end{pmatrix} \begin{pmatrix} ax_1 + bx_2 \\ cx_1 + dx_2 \end{pmatrix} \\ &= \begin{pmatrix} pax_1 + pbx_2 + qcx_1 + qdx_2 \\ rax_1 + rbx_2 + scx_1 + sdx_2 \end{pmatrix} \\ &= \begin{pmatrix} pa + qc & pb + qd \\ ra + sc & rb + sd \end{pmatrix} \begin{pmatrix} x_1 \\ x_2 \end{pmatrix} \\ &= \begin{pmatrix} p & q \\ r & s \end{pmatrix} \begin{pmatrix} a & b \\ c & d \end{pmatrix} \begin{pmatrix} x_1 \\ x_2 \end{pmatrix} \\ &= \mathcal{G}\mathcal{F}\begin{pmatrix} x_1 \\ x_2 \end{pmatrix}. \end{aligned}$$

Thus, $G \circ F$ is represented by the matrix $\mathcal{G}\mathcal{F}$.

## Review Exercises 3 (page 148)

**2.** $y - z \geq 1$ represents all points on or below the plane parallel to the $x$-axis that passes through the points $(0, 1, 0)$ and $(0, 0, -1)$.

**4.** $x - 2y - 4z = 8$ represents all points on the plane passing through the three points $(8, 0, 0)$, $(0, -4, 0)$, and $(0, 0, -2)$.

**6.** $y = z^2$ represents the parabolic cylinder parallel to the $x$-axis containing the curve $y = z^2$ in the $yz$-plane.

**8.** $z = xy$ is the hyperbolic paraboloid containing the $x$- and $y$-axes that results from rotating the hyperbolic paraboloid $z = (x^2 - y^2)/2$ through 45° about the $z$-axis.

**10.** $x^2 + y^2 - 4z^2 = 4$ represents a hyperboloid of one sheet with circular cross-sections in planes perpendicular to the $z$-axis, and asymptotic to the cone obtained by rotating the line $x = 2z$ about the $z$-axis.

**12.** $x^2 - y^2 - 4z^2 = 4$ represents a hyperboloid of two sheets asymptotic to the cone of the previous exercise.

**14.** $(x - z)^2 + y^2 = z^2$ represents an elliptic cone with oblique axis along the line $z = x$ in the $xz$-plane, having circular cross-sections of radius $|k|$ in horizontal planes $z = k$. The $z$-axis lies on the cone.

**16.** $x+y+2z = 1$, $x+y+z = 0$ together represent the straight line through the points $(-1, 0, 1)$ and $(0, -1, 1)$.

**18.** $x^2 + z^2 \leq 1$, $x - y \geq 0$ together represent all points that lie inside or on the circular cylinder of radius 1 and axis along the $y$-axis and also either on the vertical plane $x - y = 0$ or on the side of that plane containing the positive $x$-axis.

**20.** A plane through $(2, -1, 1)$ and $(1, 0, -1)$ is parallel to $\mathbf{b} = (2 - 1)\mathbf{i} + (-1 - 0)\mathbf{j} + (1 - (-1))\mathbf{k} = \mathbf{i} - \mathbf{j} + 2\mathbf{k}$. If it is also parallel to the vector $\mathbf{a}$ in the previous solution, then it is normal to

$$\mathbf{a} \times \mathbf{b} = \begin{vmatrix} \mathbf{i} & \mathbf{j} & \mathbf{k} \\ 2 & -1 & 3 \\ 1 & -1 & 2 \end{vmatrix} = \mathbf{i} - \mathbf{j} - \mathbf{k}.$$

The plane has equation $(x - 1) - (y - 0) - (z + 1) = 0$, or $x - y - z = 2$.

**22.** The plane through $A = (-1, 1, 0)$, $B = (0, 4, -1)$ and $C = (2, 0, 0)$ has normal

$$\overrightarrow{AC} \times \overrightarrow{AB} = \begin{vmatrix} \mathbf{i} & \mathbf{j} & \mathbf{k} \\ 3 & -1 & 0 \\ 1 & 3 & -1 \end{vmatrix} = \mathbf{i} + 3\mathbf{j} + 10\mathbf{k}.$$

Its equation is $(x - 2) + 3y + 10z = 0$, or $x + 3y + 10z = 2$.

**24.** A plane containing the line of intersection of the planes $x + y + z = 0$ and $2x + y - 3z = 2$ has equation

$$2x + y - 3z - 2 + \lambda(x + y + z - 0) = 0.$$

This plane is perpendicular to $x - 2y - 5z = 17$ if their normals are perpendicular, that is, if

$$1(2 + \lambda) - 2(1 + \lambda) - 5(-3 + \lambda) = 0,$$

or $9x + 7y - z = 4$.

**26.** A vector parallel to the planes $x - y = 3$ and $x + 2y + z = 1$ is $(\mathbf{i} - \mathbf{j}) \times (\mathbf{i} + 2\mathbf{j} + \mathbf{k}) = -\mathbf{i} - \mathbf{j} + 3\mathbf{k}$. A line through $(1, 0, -1)$ parallel to this vector is

$$\frac{x - 1}{-1} = \frac{y}{-1} = \frac{z + 1}{3}.$$

**28.** The vector

$$\begin{aligned} \mathbf{a} &= (1 + t)\mathbf{i} - t\mathbf{j} - (2 + 2t)\mathbf{k} - \big(2s\mathbf{i} + (s - 2)\mathbf{j} - (1 + 3s)\mathbf{k}\big) \\ &= (1 + t - 2s)\mathbf{i} - (t + s - 2)\mathbf{j} - (1 + 2t - 3s)\mathbf{k} \end{aligned}$$

joins points on the two lines and is perpendicular to both lines if $\mathbf{a} \bullet (\mathbf{i} - \mathbf{j} - 2\mathbf{k}) = 0$ and $\mathbf{a} \bullet (2\mathbf{i} + \mathbf{j} - 3\mathbf{k}) = 0$, that is, if

$$\begin{aligned} 1 + t - 2s + t + s - 2 + 2 + 4t - 6s &= 0 \\ 2 + 2t - 4s - t - s + 2 + 3 + 6t - 9s &= 0, \end{aligned}$$

or, on simplification,

$$\begin{aligned} 6t - 7s &= -1 \\ 7t - 14s &= -7. \end{aligned}$$

This system has solution $t = 1$, $s = 1$. We would expect to use $\mathbf{a}$ as a vector perpendicular to both lines, but, as it happens, $\mathbf{a} = \mathbf{0}$ if $t = s = 1$, because the two given lines intersect at $(2, -1, -4)$. A nonzero vector perpendicular to both lines is

$$\begin{vmatrix} \mathbf{i} & \mathbf{j} & \mathbf{k} \\ 1 & -1 & -2 \\ 2 & 1 & -3 \end{vmatrix} = 5\mathbf{i} - \mathbf{j} + 3\mathbf{k}.$$

Thus the required line is parallel to this vector and passes through $(2, -1, -4)$, so its equation is

$$\mathbf{r} = (2 + 5t)\mathbf{i} - (1 + t)\mathbf{j} + (-4 + 3t)\mathbf{k}.$$

**30.** The points with position vectors $\mathbf{r}_1$, $\mathbf{r}_2$, $\mathbf{r}_3$, and $\mathbf{r}_4$ are coplanar if the tetrahedron having these points as vertices has zero volume, that is, if

$$\big[(\mathbf{r}_2 - \mathbf{r}_1) \times (\mathbf{r}_3 - \mathbf{r}_1)\big] \bullet (\mathbf{r}_4 - \mathbf{r}_1) = 0.$$

(Any permutation of the subscripts 1, 2, 3, and 4 in the above equation will do as well.)

**32.** The tetrahedron with vertices $A = (1, 2, 1)$, $B = (4, -1, 1)$, $C = (3, 4, -2)$, and $D = (2, 2, 2)$ has volume

$$\begin{aligned} \frac{1}{6}|(\overrightarrow{AB} \times \overrightarrow{AC}) \bullet \overrightarrow{AD}| &= \frac{1}{6}|(9\mathbf{i} + 9\mathbf{j} + 12\mathbf{k}) \bullet (\mathbf{i} + \mathbf{k})| \\ &= \frac{9 + 12}{6} = \frac{7}{2} \text{ cu. units.} \end{aligned}$$

**34.** Let $\mathcal{A} = \begin{pmatrix} 1 & 1 & 1 \\ 2 & 1 & 0 \\ 1 & 0 & -1 \end{pmatrix}$, $\mathbf{x} = \begin{pmatrix} x_1 \\ x_2 \\ x_3 \end{pmatrix}$, and $\mathbf{b} = \begin{pmatrix} b_1 \\ b_2 \\ b_3 \end{pmatrix}$.

Then

$$\mathcal{A}\mathbf{x} = \mathbf{b} \quad \Leftrightarrow \quad \begin{aligned} x_1 + x_2 + x_3 &= b_1 \\ 2x_1 + x_2 \qquad &= b_2 \\ x_1 \qquad - x_3 &= b_3. \end{aligned}$$

The sum of the first and third equations is $2x_1 + x_2 = b_1 + b_3$, which is incompatible with the second equation unless $b_2 = b_1 + b_3$, that is, unless

$$\mathbf{b} \bullet (\mathbf{i} - \mathbf{j} + \mathbf{k}) = 0.$$

If $\mathbf{b}$ satisfies this condition then there will be a line of solutions; if $x_1 = t$, then $x_2 = b_2 - 2t$, and $x_3 = t - b_3$, so

$$\mathbf{x} = \begin{pmatrix} t \\ b_2 - 2t \\ t - b_3 \end{pmatrix}$$

is a solution for any $t$.

## Challenging Problems 3 (page 149)

**2.** By the formula for the vector triple product given in Exercise 23 of Section 3.3,

$$\begin{aligned} (\mathbf{u} \times \mathbf{v}) \times (\mathbf{w} \times \mathbf{x}) &= [(\mathbf{u} \times \mathbf{v}) \bullet \mathbf{x}]\mathbf{w} - [(\mathbf{u} \times \mathbf{v}) \bullet \mathbf{w}]\mathbf{x} \\ (\mathbf{u} \times \mathbf{v}) \times (\mathbf{w} \times \mathbf{x}) &= -(\mathbf{w} \times \mathbf{x}) \times (\mathbf{u} \times \mathbf{v}) \\ &= -[(\mathbf{w} \times \mathbf{x}) \bullet \mathbf{v}]\mathbf{u} + [(\mathbf{w} \times \mathbf{x}) \bullet \mathbf{u}]\mathbf{v}. \end{aligned}$$

In particular, if $\mathbf{w} = \mathbf{u}$, then, since $(\mathbf{u} \times \mathbf{v}) \bullet \mathbf{u} = 0$, we have

$$(\mathbf{u} \times \mathbf{v}) \times (\mathbf{u} \times \mathbf{x}) = [(\mathbf{u} \times \mathbf{v}) \bullet \mathbf{x}]\mathbf{u},$$

or, replacing $x$ with $w$,

$$(\mathbf{u} \times \mathbf{v}) \times (\mathbf{u} \times \mathbf{w}) = [(\mathbf{u} \times \mathbf{v}) \bullet \mathbf{w}]\mathbf{u}.$$

**4.** a) Let $Q_1$ and $Q_2$ be the points on lines $L_1$ and $L_2$, respectively, that are closest together. As observed in Example 9 of Section 3.4, $\overrightarrow{Q_1Q_2}$ is perpendicular to both lines.
Therefore, the plane $P_1$ through $Q_1$ having normal $\overrightarrow{Q_1Q_2}$ contains the line $L_1$. Similarly, the plane $P_2$ through $Q_2$ having normal $\overrightarrow{Q_1Q_2}$ contains the line $L_2$. These planes are parallel since they have the same normal. They are different planes because $Q_1 \neq Q_2$ (because the lines are skew).

b) Line $L_1$ through $(1, 1, 0)$ and $(2, 0, 1)$ is parallel to $\mathbf{i} - \mathbf{j} + \mathbf{k}$, and has parametric equation

$$\mathbf{r}_1 = (1 + t)\mathbf{i} + (1 - t)\mathbf{j} + t\mathbf{k}.$$

Line $L_2$ through $(0, 1, 1)$ and $(1, 2, 2)$ is parallel to $\mathbf{i} + \mathbf{j} + \mathbf{k}$, and has parametric equation

$$\mathbf{r}_2 = s\mathbf{i} + (1 + s)\mathbf{j} + (1 + s)\mathbf{k}.$$

Now $\mathbf{r}_2 - \mathbf{r}_1 = (s - t - 1)\mathbf{i} + (s + t)\mathbf{j} + (1 + s - t)\mathbf{k}$.

To find the points $Q_1$ on $L_1$ and $Q_2$ on $L_2$ for which $\overrightarrow{Q_1Q_2}$ is perpendicular to both lines, we solve

$$\begin{aligned} (s - t - 1) - (s + t) + (1 + s - t) &= 0 \\ (s - t - 1) + (s + t) + (1 + s - t) &= 0. \end{aligned}$$

Subtracting these equations gives $s + t = 0$, so $t = -s$. Then substituting into either equation gives $2s - 1 + 1 + 2s = 0$, so $s = -t = 0$. Thus $Q_1 = (1, 1, 0)$ and $Q_2 = (0, 1, 1)$, and $\overrightarrow{Q_1Q_2} = -\mathbf{i} + \mathbf{k}$. The required planes are $x - z = 1$ (containing $L_1$) and $x - z = -1$ (containing $L_2$).

# CHAPTER 4. PARTIAL DIFFERENTIATION

## Section 4.1 Functions of Several Variables (page 155)

**2.** $f(x, y) = \sqrt{xy}$.
Domain is the set of points $(x, y)$ for which $xy \geq 0$, that is, points on the coordinate axes and in the first and third quadrants.

**4.** $f(x, y) = \dfrac{xy}{x^2 - y^2}$.
The domain consists of all points not on the lines $x = \pm y$.

**6.** $f(x, y) = 1/\sqrt{x^2 - y^2}$.
The domain consists of all points in the part of the plane where $|x| > |y|$.

**8.** $f(x, y) = \sin^{-1}(x + y)$.
The domain consists of all points in the strip $-1 \leq x + y \leq 1$.

**10.** $f(x, y, z) = \dfrac{e^{xyz}}{\sqrt{xyz}}$.
The domain consists of all points $(x, y, z)$ where $xyz > 0$, that is, all points in the four octants $x > 0,\ y > 0,\ z > 0$; $x > 0,\ y < 0,\ z < 0$; $x < 0,\ y > 0,\ z < 0$; and $x < 0$, $y < 0,\ z > 0$.

**12.** $f(x, y) = \sin x,\ 0 \leq x \leq 2\pi,\ 0 \leq y \leq 1$

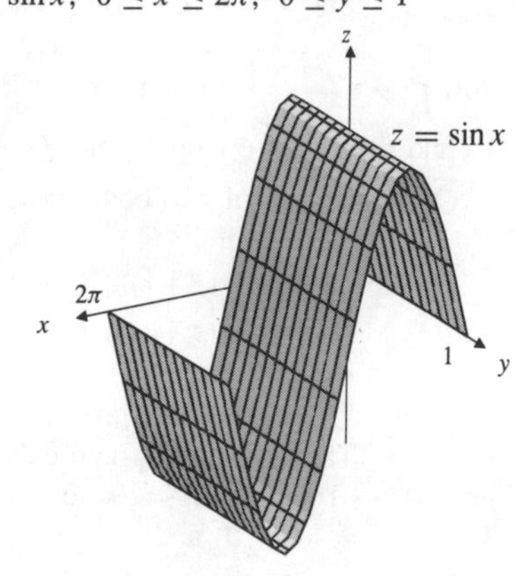


Fig. 4.1.12

**14.** $f(x, y) = 4 - x^2 - y^2,\ (x^2 + y^2 \leq 4,\ x \geq 0,\ y \geq 0)$

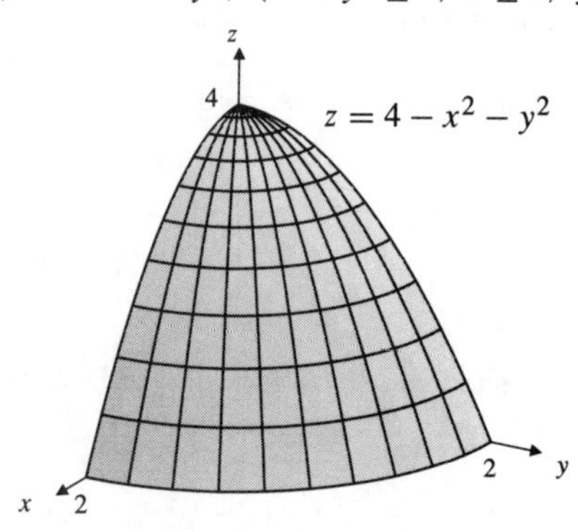


Fig. 4.1.14

**16.** $f(x, y) = 4 - x^2$

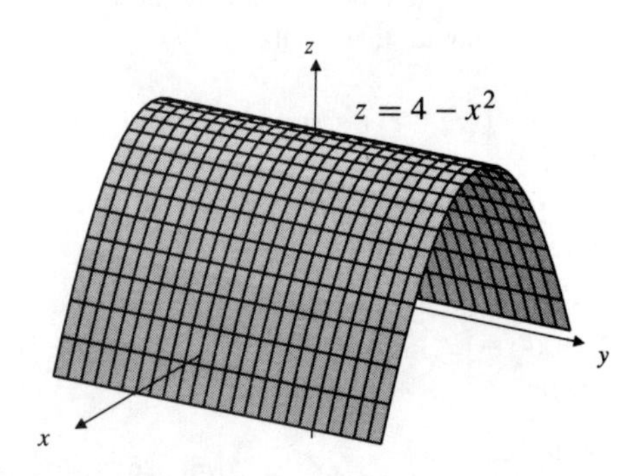


Fig. 4.1.16

**18.** $f(x, y) = 6 - x - 2y$

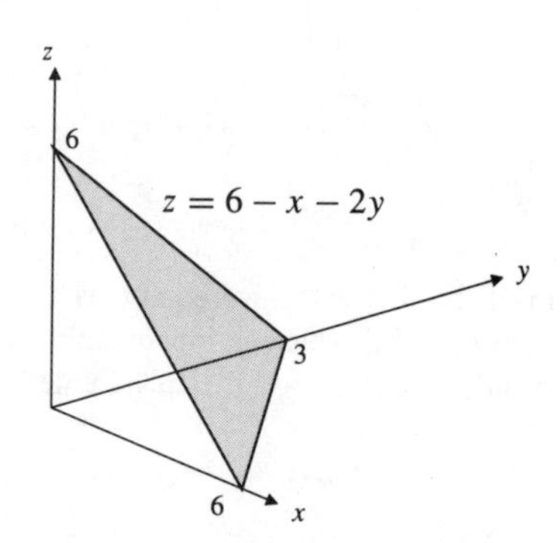


Fig. 4.1.18

**20.** $f(x, y) = x^2 + 2y^2 = C$, a family of similar ellipses centred at the origin.

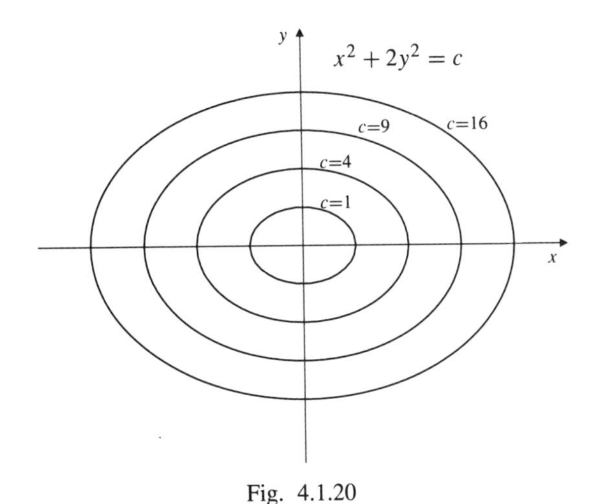


Fig. 4.1.20

**22.** $f(x, y) = \dfrac{x^2}{y} = C$, a family of parabolas, $y = x^2/C$, with vertices at the origin and vertical axes.

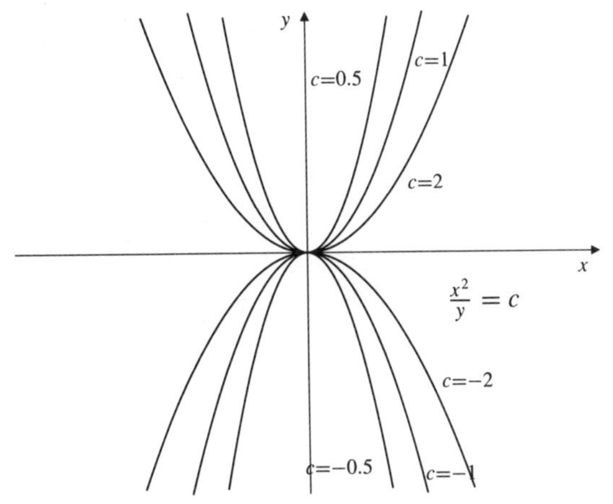


Fig. 4.1.22

**24.**

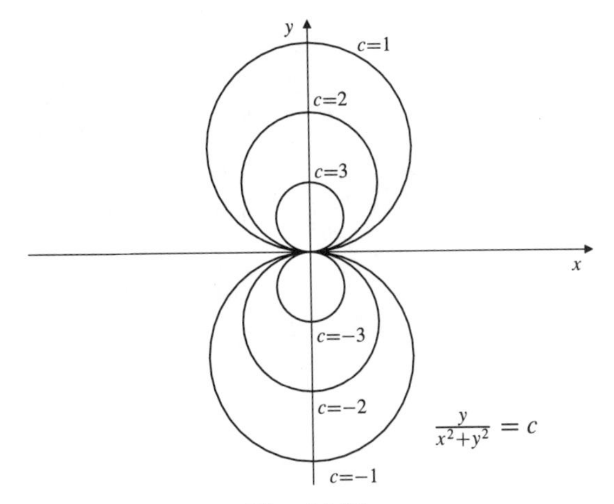


Fig. 4.1.24

$$f(x, y) = \frac{y}{x^2 + y^2} = C.$$

This is the family $x^2 + \left(y - \frac{1}{2C}\right)^2 = \frac{1}{4C^2}$ of circles passing through the origin and having centres on the $y$-axis. The origin itself is, however, not on any of the level curves.

**26.**

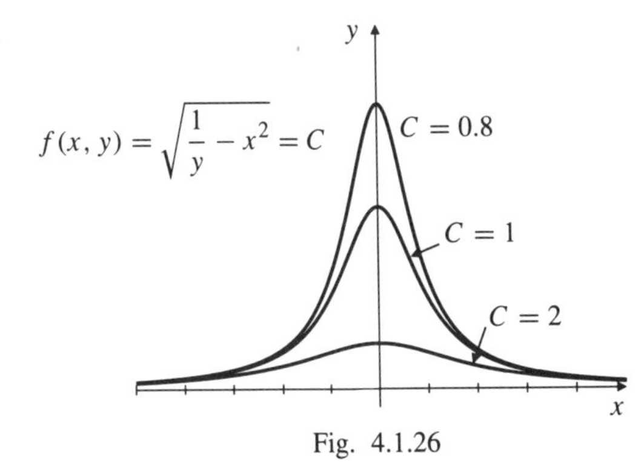


Fig. 4.1.26

$$f(x, y) = \sqrt{\frac{1}{y} - x^2} = C \;\Rightarrow\; y = \frac{1}{x^2 + C^2}.$$

**28.** $C$ is a "pass" between two peaks to the east and west. The land is level at $C$ and rises as you move to the east or west, but falls as you move to the north or south.

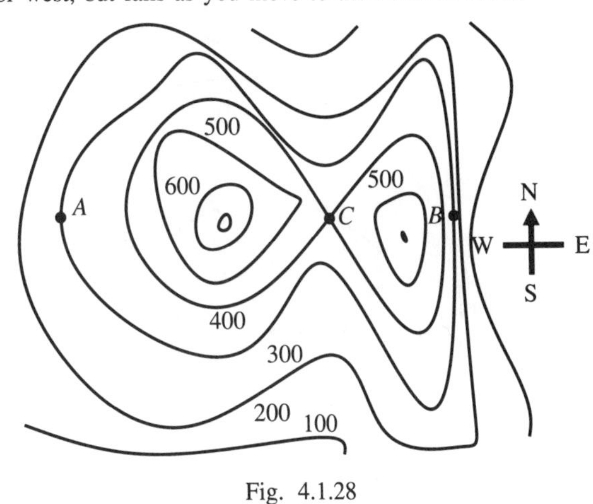


Fig. 4.1.28

**30.** The graph of the function whose level curves are as shown in part (b) of Figure 4.1.30 is a cylinder parallel to the $x$-axis, rising from height zero first steeply and then more and more slowly as $y$ increases. It is consistent with, say, a function of the form $f(x, y) = \sqrt{y + 5}$.

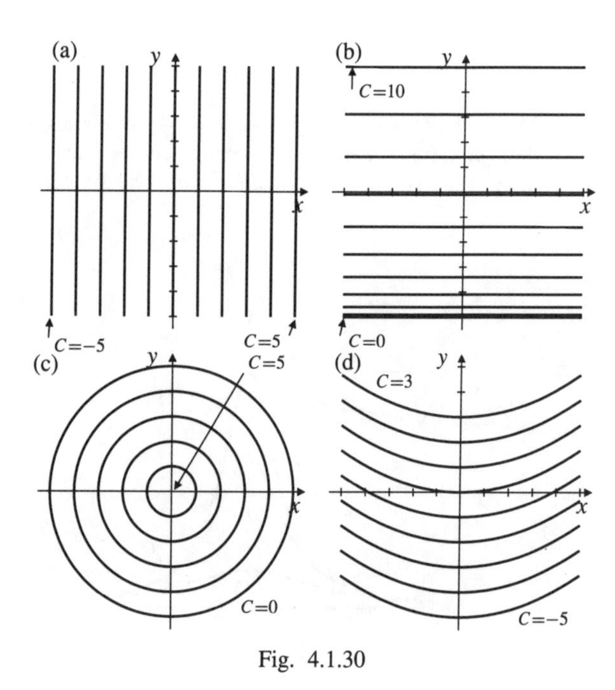


Fig. 4.1.30

**32.** The graph of the function whose level curves are as shown in part (d) of Figure 4.1.30 is a cylinder (possibly parabolic) with axis in the $yz$-plane, sloping upwards in the direction of increasing $y$. It is consistent with, say, a function of the form $f(x, y) = y - x^2$.

**34.** $4z^2 = (x - z)^2 + (y - z)^2$.
If $z = c > 0$, we have $(x - c)^2 + (y - c)^2 = 4c^2$, which is a circle in the plane $z = c$, with centre $(c, c, c)$ and radius $2c$.

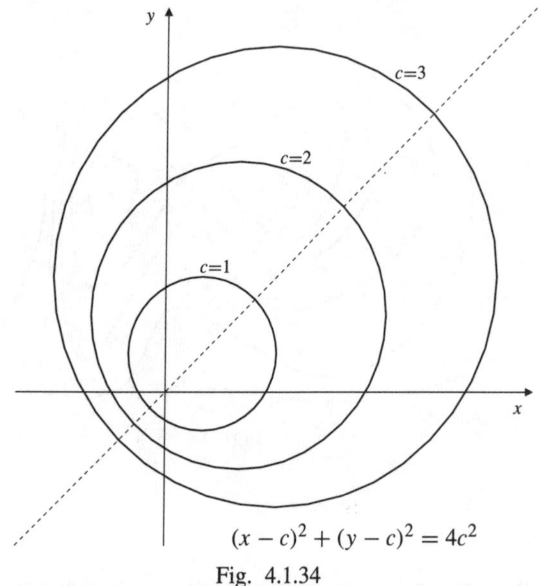


Fig. 4.1.34

The graph of the function $z = z(x, y) \geq 0$ defined by the given equation is (the upper half of) an elliptic cone with axis along the line $x = y = z$, and circular cross-sections in horizontal planes.

**36.** If the level surface $f(x, y, z) = C$ is the plane

$$\frac{x}{C^3} + \frac{y}{2C^3} + \frac{z}{3C^3} = 1,$$

that is, $x + \frac{y}{2} + \frac{z}{3} = C^3$, then

$$f(x, y, z) = \left(x + \frac{y}{2} + \frac{z}{3}\right)^{1/3}.$$

**38.** $f(x, y, z) = x + 2y + 3z$.
The level surfaces are parallel planes having common normal vector $\mathbf{i} + 2\mathbf{j} + 3\mathbf{k}$.

**40.** $f(x, y, z) = \dfrac{x^2 + y^2}{z^2}$.
The equation $f(x, y, z) = c$ can be rewritten $x^2 + y^2 = C^2 z^2$. The level surfaces are circular cones with vertices at the origin and axes along the $z$-axis.

**42.** $f(x, y, z, t) = x^2 + y^2 + z^2 + t^2$.
The "level hypersurface" $f(x, y, z, t) = c > 0$ is the "4-sphere" of radius $\sqrt{c}$ centred at the origin in $\mathbb{R}^4$. That is, it consists of all points in $\mathbb{R}^4$ at distance $\sqrt{c}$ from the origin.

**44.**

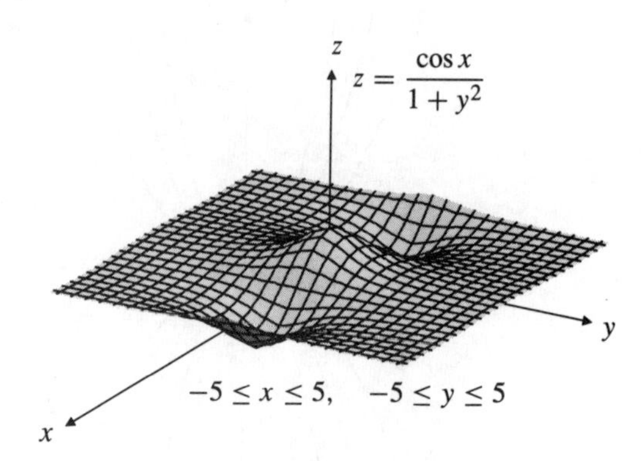


Fig. 4.1.44

**46.**

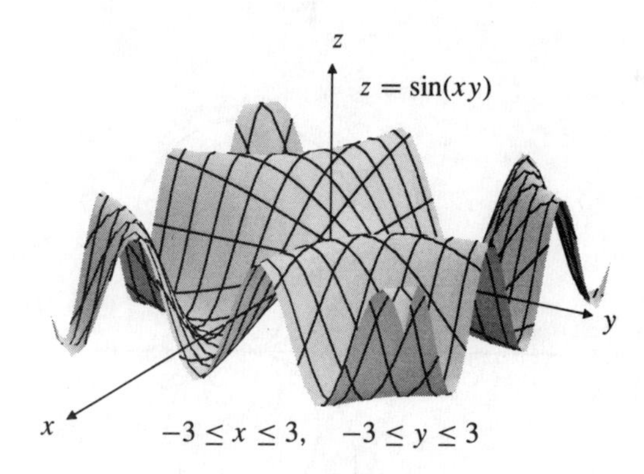


Fig. 4.1.46

**48.**

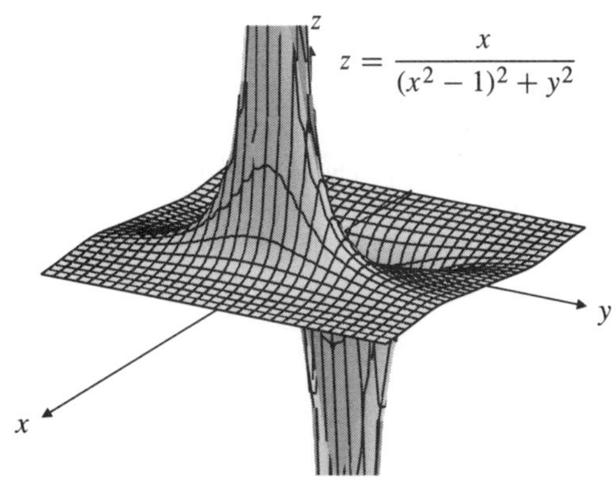


Fig. 4.1.48

**50.** The graph is asymptotic to the coordinate planes.

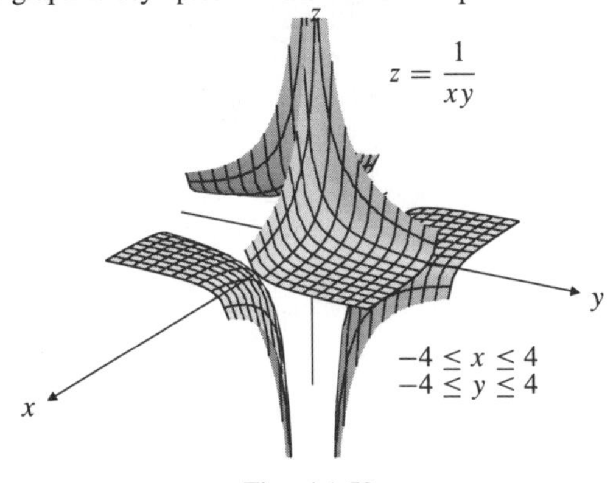


Fig. 4.1.50

## Section 4.2 Limits and Continuity (page 161)

**2.** $S = \{(x, y) : x \ge 0,\ y < 0\}$
The boundary of $S$ consists of points $(x, 0)$ where $x \ge 0$, and points $(0, y)$ where $y \le 0$.
The interior of $S$ consists of all points of $S$ that are not on the $y$-axis, that is, all points $(x, y)$ satisfying $x > 0$ and $y < 0$.
$S$ is neither open nor closed; it contains some, but not all, of its boundary points.
$S$ is not bounded; $(x, -1)$ belongs to $S$ for $0 < x < \infty$.

**4.** $S = \{(x, y) : |x| + |y| \le 1\}$
The boundary of $S$ consists of all points on the edges of the square with vertices $(\pm 1, 0)$ and $(0, \pm 1)$.
The interior of $S$ consists of all points inside that square.
$S$ is closed since it contains all its boundary points. It is bounded since all points in it are at distance not greater than 1 from the origin.

**6.** $S = \{(x, y, z) : x \ge 0,\ y > 1,\ z < 2\}$
Boundary: the quarter planes $x = 0$, $(y \ge 1,\ z \le 2)$, $y = 1$, $(x \ge 0,\ z \le 2)$, and $z = 2$, $(x \ge 0,\ y \ge 1)$.
Interior: the set of points $(x, y, z)$ such that $x > 0$, $y > 1$, $z < 2$.
$S$ is neither open nor closed.

**8.** $S = \{(x, y, z) : x^2 + y^2 < 1,\ y + z > 2\}$
Boundary: the part of the cylinder $x^2 + y^2 = 1$ that lies on or above the plane $y + z = 2$ together with the part of that plane that lies inside the cylinder.
Interior: all points that are inside the cylinder $x^2 + y^2 = 1$ and above the plane $y + z = 2$. $S$ is open.

**10.** $$\lim_{(x,y)\to(0,0)} \sqrt{x^2 + y^2} = 0$$

**12.** Let $f(x, y) = \dfrac{x}{x^2 + y^2}$.
Then $|f(x, 0)| = |1/x| \to \infty$ as $x \to 0$.
But $|f(0, y)| = 0 \to 0$ as $y \to 0$.
Thus $\lim_{(x,y)\to(0,0)} f(x, y)$ does not exist.

**14.** $$\lim_{(x,y)\to(0,1)} \frac{x^2(y-1)^2}{x^2 + (y-1)^2} = 0, \text{ because}$$
$$0 \le \left| \frac{x^2(y-1)^2}{x^2 + (y-1)^2} \right| \le x^2$$
and $x^2 \to 0$ as $(x, y) \to (0, 1)$.

**16.** $$\lim_{(x,y)\to(0,0)} \frac{\sin(x-y)}{\cos(x+y)} = \frac{\sin 0}{\cos 0} = 0.$$

**18.** $\displaystyle\lim_{(x,y)\to(1,2)} \frac{2x^2 - xy}{4x^2 - y^2}$ does not exist, because the fraction is not defined at any points on the line $y = 2x$, which passes through $(1, 2)$, and therefore is not defined at all points sufficiently close to $(1, 2)$.

**20.** If $x = 0$ and $y \ne 0$, then $\dfrac{x^2y^2}{2x^4 + y^4} = 0$.
If $x = y \ne 0$, then $\dfrac{x^2y^2}{2x^4 + y^4} = \dfrac{x^4}{2x^4 + x^4} = \dfrac{1}{3}$.
Therefore $\displaystyle\lim_{(x,y)\to(0,0)} \frac{x^2y^2}{2x^4 + y^4}$ does not exist.

**22.** For $x \ne y$, we have
$$f(x, y) = \frac{x^3 - y^3}{x - y} = x^2 + xy + y^2.$$
The latter expression has the value $3x^2$ at points of the line $x = y$. Therefore, if we extend the definition of $f(x, y)$ so that $f(x, x) = 3x^2$, then the resulting function will be equal to $x^2 + xy + y^2$ everywhere, and so continuous everywhere.

**24.** Let $f$ be the function of Example 3 of Section 4.2:
$$f(x, y) = \begin{cases} \dfrac{2xy}{x^2 + y^2} & \text{if } (x, y) \ne (0, 0) \\ 0 & \text{if } (x, y) = (0, 0). \end{cases}$$

Let $a = b = 0$. If $g(x) = f(x, 0)$ and $h(y) = f(0, y)$, then $g(x) = 0$ for all $x$, and $h(y) = 0$ for all $y$, so $g$ and $h$ are continuous at 0. But, as shown in Example 3 of Section 4.2, $f$ is not continuous at $(0, 0)$.

If $f(x, y)$ is continuous at $(a, b)$, then $g(x) = f(x, b)$ is continuous at $x = a$ because

$$\lim_{x\to a} g(x) = \lim_{\substack{x\to a\\ y=b}} f(x, y) = f(a, b).$$

Similarly, $h(y) = f(a, y)$ is continuous at $y = b$.

**26.** Since $|x| \le \sqrt{x^2 + y^2}$ and $|y| \le \sqrt{x^2 + y^2}$, we have

$$\left|\frac{x^m y^n}{(x^2 + y^2)^p}\right| \le \frac{(x^2 + y^2)^{(m+n)/2}}{(x^2 + y^2)^p} = (x^2 + y^2)^{-p+(m+n)/2}.$$

The expression on the right $\to 0$ as $(x, y) \to (0, 0)$, provided $m + n > 2p$. In this case

$$\lim_{(x,y)\to(0,0)} \frac{x^m y^n}{(x^2 + y^2)^p} = 0.$$

**28.** $f(x, y) = \dfrac{\sin x \sin^3 y}{1 - \cos(x^2 + y^2)}$ cannot be defined at $(0, 0)$ so as to become continuous there, because $f(x, y)$ has no limit as $(x, y) \to (0, 0)$. To see this, observe that $f(x, 0) = 0$, so the limit must be 0 if it exists at all. However,

$$f(x, x) = \frac{\sin^4 x}{1 - \cos(2x^2)} = \frac{\sin^4 x}{2\sin^2(x^2)}$$

which approaches 1/2 as $x \to 0$ by l'Hôpital's Rule or by using Maclaurin series.

**30.** The graphing software is unable to deal effectively with the discontinuity at $(x, y) = (0, 0)$ so it leaves some gaps and rough edges near the $z$-axis. The surface lies between a ridge along $y = x^2$, $z = 1$, and a ridge along $y = -x^2$, $z = -1$. It appears to be creased along the $z$-axis. The level curves are parabolas $y = kx^2$ through the origin. One of the families of rulings on the surface is the family of contours corresponding to level curves.

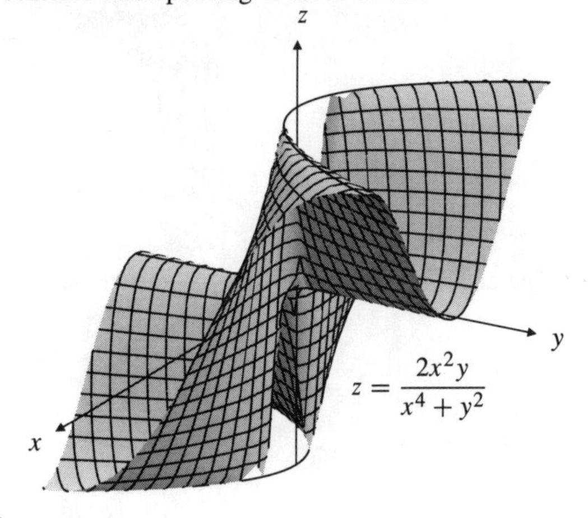


Fig. 4.2.30

## Section 4.3 Partial Derivatives (page 168)

**2.** $f(x, y) = xy + x^2$,
$f_1(x, y) = y + 2x$, $f_2(x, y) = x$,
$f_1(2, 0) = 4$, $f_2(2, 0) = 2$.

**4.** $g(x, y, z) = \dfrac{xz}{y + z}$,
$g_1(x, y, z) = \dfrac{z}{y + z}$, $g_1(1, 1, 1) = \dfrac{1}{2}$,
$g_2(x, y, z) = \dfrac{-xz}{(y + z)^2}$, $g_2(1, 1, 1) = -\dfrac{1}{4}$,
$g_3(x, y, z) = \dfrac{xy}{(y + z)^2}$, $g_3(1, 1, 1) = \dfrac{1}{4}$.

**6.** $w = \ln(1 + e^{xyz})$, $\dfrac{\partial w}{\partial x} = \dfrac{yze^{xyz}}{1 + e^{xyz}}$,
$\dfrac{\partial w}{\partial y} = \dfrac{xze^{xyz}}{1 + e^{xyz}}$, $\dfrac{\partial w}{\partial z} = \dfrac{xye^{xyz}}{1 + e^{xyz}}$,
At $(2, 0, -1)$: $\dfrac{\partial w}{\partial x} = 0$, $\dfrac{\partial w}{\partial y} = -1$, $\dfrac{\partial w}{\partial z} = 0$.

**8.** $f(x, y) = \dfrac{1}{\sqrt{x^2 + y^2}}$,
$f_1(x, y) = -\dfrac{1}{2}(x^2 + y^2)^{-3/2}(2x) = -\dfrac{x}{(x^2 + y^2)^{3/2}}$,
By symmetry, $f_2(x, y) = -\dfrac{y}{(x^2 + y^2)^{3/2}}$,
$f_1(-3, 4) = \dfrac{3}{125}$, $f_2(-3, 4) = -\dfrac{4}{125}$.

**10.** If $g(x_1, x_2, x_3, x_4) = \dfrac{x_1 - x_2^2}{x_3 + x_4^2}$, then

$$\begin{aligned} g_1(x_1, x_2, x_3, x_4) &= \frac{1}{x_3 + x_4^2} & g_1(3, 1, -1, -2) &= \frac{1}{3} \\ g_2(x_1, x_2, x_3, x_4) &= \frac{-2x_2}{x_3 + x_4^2} & g_2(3, 1, -1, -2) &= -\frac{2}{3} \\ g_3(x_1, x_2, x_3, x_4) &= \frac{x_2^2 - x_1}{(x_3 + x_4^2)^2} & g_3(3, 1, -1, -2) &= -\frac{2}{9} \\ g_4(x_1, x_2, x_3, x_4) &= \frac{(x_2^2 - x_1)2x_4}{(x_3 + x_4^2)^2} & g_4(3, 1, -1, -2) &= \frac{8}{9}. \end{aligned}$$

**12.** $f(x, y) = \begin{cases} \dfrac{x^2 - 2y^2}{x - y} & \text{if } x \neq y \\ 0 & \text{if } x = y \end{cases}$

$$f_1(0, 0) = \lim_{h \to 0} \frac{f(h, 0) - f(0, 0)}{h} = \lim_{h \to 0} \frac{h - 0}{h} = 1,$$
$$f_2(0, 0) = \lim_{k \to 0} \frac{f(0, k) - f(0, 0)}{k} = \lim_{k \to 0} \frac{2k}{k} = 2.$$

**14.** $f(x, y) = \dfrac{x - y}{x + y}$, $f(1, 1) = 0$,

$$f_1(x, y) = \frac{(x + y) - (x - y)}{(x + y)^2}, \quad f_1(1, 1) = \frac{1}{2}$$
$$f_2(x, y) = \frac{(x + y)(-1) - (x - y)}{(x + y)^2}, \quad f_2(1, 1) = -\frac{1}{2}.$$

Tangent plane to $z = f(x, y)$ at (1,1) has equation $z = \dfrac{x - 1}{2} - \dfrac{y - 1}{2}$, or $2z = x - y$.
Normal line: $2(x - 1) = -2(y - 1) = -z$.

**16.** $f(x, y) = e^{xy}$, $f_1(x, y) = ye^{xy}$, $f_2(x, y) = xe^{xy}$,
$f(2, 0) = 1$, $f_1(2, 0) = 0$, $f_2(2, 0) = 2$.
Tangent plane to $z = e^{xy}$ at (2,0) has equation $z = 1 + 2y$.
Normal line: $x = 2$, $y = 2 - 2z$.

**18.** $f(x, y) = ye^{-x^2}$, $f_1 = -2xye^{-x^2}$, $f_2 = e^{-x^2}$,
$f(0, 1) = 1$, $f_1(0, 1) = 0$, $f_2(0, 1) = 1$.
Tangent plane to $z = f(x, y)$ at $(0, 1)$ has equation $z = 1 + 1(y - 1)$, or $z = y$.
Normal line: $x = 0$, $y + z = 2$.

**20.** $f(x, y) = \dfrac{2xy}{x^2 + y^2}$, $\quad f(0, 2) = 0$

$$f_1(x, y) = \frac{(x^2 + y^2)2y - 2xy(2x)}{(x^2 + y^2)^2} = \frac{2y(y^2 - x^2)}{(x^2 + y^2)^2}$$
$$f_2(x, y) = \frac{2x(x^2 - y^2)}{(x^2 + y^2)^2} \quad \text{(by symmetry)}$$

$f_1(0, 2) = 1$, $\quad f_2(0, 2) = 0$.
Tangent plane at $(0, 2)$: $z = x$.
Normal line: $z + x = 0$, $y = 2$.

**22.** $f(x, y) = \sqrt{1 + x^3y^2}$ $\quad f(2, 1) = 3$

$$f_1(x, y) = \frac{3x^2y^2}{2\sqrt{1 + x^3y^2}} \quad f_1(2, 1) = 2$$
$$f_2(x, y) = \frac{2x^3y}{2\sqrt{1 + x^3y^2}} \quad f_2(2, 1) = \frac{8}{3}$$

Tangent plane: $z = 3 + 2(x - 2) + \frac{8}{3}(y - 1)$, or $6x + 8y - 3z = 11$.
Normal line: $\dfrac{x - 2}{2} = \dfrac{y - 1}{8/3} = \dfrac{z - 3}{-1}$.

**24.** $z = xye^{-(x^2+y^2)/2}$

$$\frac{\partial z}{\partial x} = ye^{-(x^2+y^2)/2} - x^2ye^{-(x^2+y^2)/2} = y(1 - x^2)e^{-(x^2+y^2)/2}$$
$$\frac{\partial z}{\partial y} = x(1 - y^2)e^{-(x^2+y^2)/2} \quad \text{(by symmetry)}$$

The tangent planes are horizontal at points where both of these first partials are zero, that is, points satisfying

$$y(1 - x^2) = 0 \quad \text{and} \quad x(1 - y^2) = 0.$$

These points are $(0, 0)$, $(1, 1)$, $(-1, -1)$, $(1, -1)$ and $(-1, 1)$.
At (0,0) the tangent plane is $z = 0$.
At $(1, 1)$ and $(-1, -1)$ the tangent plane is $z = 1/e$.
At $(1, -1)$ and $(-1, 1)$ the tangent plane is $z = -1/e$.

**26.** $z = \dfrac{x + y}{x - y}$,

$$\frac{\partial z}{\partial x} = \frac{(x - y)(1) - (x + y)(1)}{(x - y)^2} = \frac{-2y}{(x - y)^2},$$
$$\frac{\partial z}{\partial y} = \frac{(x - y)(1) - (x + y)(-1)}{(x - y)^2} = \frac{2x}{(x - y)^2}.$$

Therefore

$$x\frac{\partial z}{\partial x} + y\frac{\partial z}{\partial y} = -\frac{2xy}{(x - y)^2} + \frac{2xy}{(x - y)^2} = 0.$$

**28.** $w = x^2 + yz$, $\dfrac{\partial w}{\partial x} = 2x$, $\dfrac{\partial w}{\partial y} == z$, $\dfrac{\partial w}{\partial z} = y$.

Therefore

$$\begin{aligned} x\frac{\partial w}{\partial x} + y\frac{\partial w}{\partial y} + z\frac{\partial w}{\partial z} & \\ &= 2x^2 + yz + yz \\ &= 2(x^2 + yz) = 2w. \end{aligned}$$

**30.** $z = f(x^2 + y^2)$,

$$\frac{\partial z}{\partial x} = f'(x^2 + y^2)(2x), \quad \frac{\partial z}{\partial y} = f'(x^2 + y^2)(2y).$$

Thus $y\dfrac{\partial z}{\partial x} - x\dfrac{\partial z}{\partial y} = 2xyf'(x^2 + y^2) - 2xyf'(x^2 + y^2) = 0$.

**32.**

$$f_1(x, y, z) = \lim_{h \to 0} \frac{f(x + h, y, z) - f(x, y, z)}{h}$$
$$f_2(x, y, z) = \lim_{k \to 0} \frac{f(x, y + k, z) - f(x, y, z)}{k}$$
$$f_3(x, y, z) = \lim_{\ell \to 0} \frac{f(x, y, z + \ell) - f(x, y, z)}{\ell}$$

**34.** If $Q = (X, Y, Z)$ is the point on the surface $z = x^2 + y^2$ that is closest to $P = (1, 1, 0)$, then

$$\overrightarrow{PQ} = (X-1)\mathbf{i} + (Y-1)\mathbf{j} + Z\mathbf{k}$$

must be normal to the surface at $Q$, and hence must be parallel to $\mathbf{n} = 2X\mathbf{i} + 2Y\mathbf{j} - \mathbf{k}$. Hence $\overrightarrow{PQ} = t\mathbf{n}$ for some real number $t$, so

$$X - 1 = 2tX, \qquad Y - 1 = 2tY, \qquad Z = -t.$$

Thus $X = Y = \dfrac{1}{1-2t}$, and, since $Z = X^2 + Y^2$, we must have

$$-t = \frac{2}{(1-2t)^2}.$$

Evidently this equation is satisfied by $t = -\frac{1}{2}$. Since the left and right sides of the equation have graphs similar to those in Figure 4.20(b) (in the text), the equation has only this one real solution. Hence $X = Y = \frac{1}{2}$, and so $Z = \frac{1}{2}$. The distance from $(1, 1, 0)$ to $z = x^2$ is the distance from $(1, 1, 0)$ to $\left(\frac{1}{2}, \frac{1}{2}, \frac{1}{2}\right)$, which is $\sqrt{3}/2$ units.

**36.** $f(x, y) = \dfrac{2xy}{x^2+y^2}$ if $(x, y) \neq (0, 0)$, $\qquad f(0, 0) = 0$

$$f_1(0, 0) = \lim_{h\to 0} \frac{f(h, 0) - f(0, 0)}{h} = \lim_{h\to 0} \frac{0-0}{h} = 0$$

$$f_2(0, 0) = \lim_{k\to 0} \frac{f(0, k) - f(0, 0)}{h} = \lim_{k\to 0} \frac{0-0}{k} = 0$$

Thus $f_1(0, 0)$ and $f_2(0, 0)$ both exist even though $f$ is not continuous at $(0, 0)$ (as shown in Example 3 of Section 4.2).

**38.** If $(x, y) \neq (0, 0)$, then

$$f_1(x, y) = 3x^2 \sin\frac{1}{x^2+y^2} - \frac{(x^3+y)2x}{(x^2+y^2)^2}\cos\frac{1}{x^2+y^2}.$$

The first term on the right $\to 0$ as $(x, y) \to (0, 0)$, but the second term has no limit at $(0, 0)$. (It is 0 along $x = 0$, but along $x = y$ it is

$$-\frac{2x^4+2x^2}{4x^4}\cos\frac{1}{2x^2} = -\frac{1}{2}\left(1 + \frac{1}{x^2}\right)\cos\frac{1}{2x^2},$$

which has no limit as $x \to 0$.) Thus $f_1(x, y)$ has no limit at $(0, 0)$ and is not continuous there.

**40.** $f(x, y, z) = \begin{cases} \dfrac{xy^2z}{x^4+y^4+z^4} & \text{if } (x, y, z) \neq (0, 0, 0) \\ 0 & \text{if } (x, y, z) = (0, 0, 0). \end{cases}$

By symmetry we have

$$f_3(0, 0, 0) = f_1(0, 0, 0) = \lim_{h\to 0} \frac{0}{h^5} = 0.$$

Also,

$$f_2(0, 0, 0) = \lim_{k\to 0} \frac{0}{k^5} = 0.$$

$f$ is not continuous at $(0, 0, 0)$; it has different limits as $(x, y, z) \to (0, 0, 0)$ along $x = 0$ and along $x = y = z$. None of $f_1$, $f_2$, and $f_3$ is continuous at $(0, 0, 0)$ either. For example,

$$f_1(x, y, z) = \frac{(y^4 + z^4 - 3x^4)y^2z}{(x^4+y^4+z^4)^2},$$

which has no limit as $(x, y, z) \to (0, 0, 0)$ along the line $x = y = z$.

## Section 4.4 Higher-Order Derivatives (page 173)

**2.** $f(x, y) = x^2 + y^2$, $\quad f_1(x, y) = 2x$, $\quad f_2(x, y) = 2y$,
$f_{11}(x, y) = f_{22}(x, y) = 2$, $\quad f_{12}(x, y) = f_{21}(x, y) = 0$.

**4.** $z = \sqrt{3x^2 + y^2}$,

$$\frac{\partial z}{\partial x} = \frac{3x}{\sqrt{3x^2+y^2}}, \qquad \frac{\partial z}{\partial y} = \frac{y}{\sqrt{3x^2+y^2}},$$

$$\frac{\partial^2 z}{\partial x^2} = \frac{\sqrt{3x^2+y^2}(3) - 3x\dfrac{3x}{\sqrt{3x^2+y^2}}}{3x^2+y^2} = \frac{3y^2}{(3x^2+y^2)^{3/2}},$$

$$\frac{\partial^2 z}{\partial y^2} = \frac{\sqrt{3x^2+y^2} - y\dfrac{y}{\sqrt{3x^2+y^2}}}{3x^2+y^2} = \frac{3x^2}{(3x^2+y^2)^{3/2}},$$

$$\frac{\partial^2 z}{\partial x\partial y} = \frac{\partial^2 z}{\partial y\partial x} = -\frac{3xy}{(3x^2+y^2)^{3/2}}.$$

**6.** $f(x, y) = \ln(1 + \sin(xy))$

$$f_1(x, y) = \frac{y\cos(xy)}{1+\sin(xy)}, \qquad f_2(x, y) = \frac{x\cos(xy)}{1+\sin(xy)}$$

$f_{11}(x, y)$

$$= \frac{(1+\sin(xy))(-y^2\sin(xy)) - (y\cos(xy))(y\cos(xy))}{(1+\sin(xy))^2}$$

$$= -\frac{y^2}{1+\sin(xy)}$$

$$f_{22}(x, y) = -\frac{x^2}{1+\sin(xy)} \quad \text{(by symmetry)}$$

$f_{12}(x, y) =$

$$\frac{(1+\sin(xy))(\cos(xy) - xy\sin(xy)) - (y\cos(xy))(x\cos(xy))}{(1+\sin(xy))^2}$$

$$= \frac{\cos(xy) - xy}{1+\sin(xy)} = f_{21}(x, y).$$

**8.** $f(x, y) = A(x^2 - y^2) + Bxy$, $\quad f_1 = 2Ax + By$,
$f_2 = -2Ay + Bx$,
$f_{11} = 2A$, $\quad f_{22} = -2A$,
Thus $f_{11} + f_{22} = 0$, and $f$ is harmonic.

**10.** $$f(x,y)=\frac{x}{x^2+y^2}$$
$$f_1(x,y)=\frac{x^2+y^2-2x^2}{(x^2+y^2)^2}=\frac{y^2-x^2}{(x^2+y^2)^2}$$
$$f_2(x,y)=-\frac{2xy}{(x^2+y^2)^2}$$
$$f_{11}(x,y)=\frac{(x^2+y^2)^2(-2x)-(y^2-x^2)2(x^2+y^2)(2x)}{(x^2+y^2)^4}=\frac{2x^3-6xy^2}{(x^2+y^2)^3}$$
$$f_{22}(x,y)=-\frac{(x^2+y^2)^2(2x)-2xy2(x^2+y^2)(2y)}{(x^2+y^2)^4}=\frac{-2x^3+6xy^2}{(x^2+y^2)^3}.$$
Evidently $f_{11}(x,y)+f_{22}(x,y)=0$ for $(x,y)\neq(0,0)$. Hence $f$ is harmonic except at the origin.

**12.** $f(x,y)=\tan^{-1}\left(\frac{y}{x}\right)$, $(x\neq 0)$.
$$f_1(x,y)=\frac{1}{1+\frac{y^2}{x^2}}\left(-\frac{y}{x^2}\right)=-\frac{y}{x^2+y^2},$$
$$f_2(x,y)=\frac{1}{1+\frac{y^2}{x^2}}\left(\frac{1}{x}\right)=\frac{x}{x^2+y^2},$$
$$f_{11}=\frac{2xy}{(x^2+y^2)^2},\qquad f_{22}=-\frac{2xy}{(x^2+y^2)^2}.$$
Thus $f_{11}+f_{22}=0$ and $f$ is harmonic.

**14.** Let $g(x,y,z)=zf(x,y)$. Then
$$\begin{aligned}g_1(x,y,z)&=zf_1(x,y), & g_{11}(x,y,z)&=zf_{11}(x,y)\\ g_2(x,y,z)&=zf_2(x,y), & g_{22}(x,y,z)&=zf_{22}(x,y)\\ g_3(x,y,z)&=f(x,y), & g_{33}(x,y,z)&=0.\end{aligned}$$
Thus $g_{11}+g_{22}+g_{33}=z(f_{11}+f_{22})=0$ and $g$ is harmonic because $f$ is harmonic. This proves (a). The proofs of (b) and (c) are similar.

If $h(x,y,z)=f(ax+by,cz)$, then $h_{11}=a^2f_{11}$, $h_{22}=b^2f_{11}$ and $h_{33}=c^2f_{22}$. If $a^2+b^2=c^2$ and $f$ is harmonic then
$$h_{11}+h_{22}+h_{33}=c^2(f_{11}+f_{22})=0,$$
so $h$ is harmonic.

**16.** Let
$$f(x,y)=\begin{cases}\dfrac{2xy}{x^2+y^2} & \text{if } (x,y)\neq(0,0)\\ 0 & \text{if } (x,y)=(0,0).\end{cases}$$
For $(x,y)\neq(0,0)$, we have
$$f_1(x,y)=\frac{(x^2+y^2)2y-2xy(2x)}{(x^2+y^2)^2}=\frac{2y(y^2-x^2)}{(x^2+y^2)^2}$$
$$f_2(x,y)=\frac{2x(x^2-y^2)}{(x^2+y^2)^2}\qquad\text{(by symmetry)}.$$
Let $F(x,y)=(x^2-y^2)f(x,y)$. Then we calculate
$$\begin{aligned}F_1(x,y)&=2xf(x,y)+(x^2-y^2)f_1(x,y)\\&=2xf(x,y)-\frac{2y(y^2-x^2)^2}{(x^2+y^2)^2}\\F_2(x,y)&=-2yf(x,y)+(x^2-y^2)f_2(x,y)\\&=-2yf(x,y)+\frac{2x(x^2-y^2)^2}{(x^2+y^2)^2}\\F_{12}(x,y)&=\frac{2(x^6+9x^4y^2-9x^2y^4-y^6)}{(x^2+y^2)^3}=F_{21}(x,y).\end{aligned}$$
For the values at $(0,0)$ we revert to the definition of derivative to calculate the partials:
$$F_1(0,0)=\lim_{h\to0}\frac{F(h,0)-F(0,0)}{h}=0=F_2(0,0)$$
$$F_{12}(0,0)=\lim_{k\to0}\frac{F_1(0,k)-F_1(0,0)}{k}=\lim_{k\to0}\frac{-2k(k^4)}{k(k^4)}=-2$$
$$F_{21}(0,0)=\lim_{h\to0}\frac{F_2(h,0)-F_2(0,0)}{h}=\lim_{h\to0}\frac{2h(h^4)}{h(h^4)}=2$$
This does not contradict Theorem 1 since the partials $F_{12}$ and $F_{21}$ are not continuous at $(0,0)$. (Observe, for instance, that $F_{12}(x,x)=0$, while $F_{12}(x,0)=2$ for $x\neq0$.)

**18.** $u(x,y,t)=t^{-1}e^{-(x^2+y^2)/4t}$
$$\begin{aligned}\frac{\partial u}{\partial t}&=-\frac{1}{t^2}e^{-(x^2+y^2)/4t}+\frac{x^2+y^2}{4t^3}e^{-(x^2+y^2)/4t}\\\frac{\partial u}{\partial x}&=-\frac{x}{2t^2}e^{-(x^2+y^2)/4t}\\\frac{\partial^2u}{\partial x^2}&=-\frac{1}{2t^2}e^{-(x^2+y^2)/4t}+\frac{x^2}{4t^3}e^{-(x^2+y^2)/4t}\\\frac{\partial^2u}{\partial y^2}&=-\frac{1}{2t^2}e^{-(x^2+y^2)/4t}+\frac{y^2}{4t^3}e^{-(x^2+y^2)/4t}\end{aligned}$$
Thus $\dfrac{\partial u}{\partial t}=\dfrac{\partial^2u}{\partial x^2}+\dfrac{\partial^2u}{\partial y^2}$.

**20.** $u(x,y)$ is biharmonic $\Leftrightarrow \dfrac{\partial^2u}{\partial x^2}+\dfrac{\partial^2u}{\partial y^2}$ is harmonic
$$\Leftrightarrow\left(\frac{\partial^2}{\partial x^2}+\frac{\partial^2}{\partial y^2}\right)\left(\frac{\partial^2u}{\partial x^2}+\frac{\partial^2u}{\partial y^2}\right)=0$$
$$\Leftrightarrow\frac{\partial^4u}{\partial x^4}+2\frac{\partial^4u}{\partial x^2\partial y^2}+\frac{\partial^4u}{\partial y^4}=0$$
by the equality of mixed partials.

**22.** If $u$ is harmonic, then $\dfrac{\partial^2 u}{\partial x^2} + \dfrac{\partial^2 u}{\partial y^2} = 0$. If $v(x, y) = xu(x, y)$, then

$$\begin{aligned}
\frac{\partial^2 v}{\partial x^2} &= \frac{\partial}{\partial x}\left(u + x\frac{\partial u}{\partial x}\right) = 2\frac{\partial u}{\partial x} + x\frac{\partial^2 u}{\partial x^2} \\
\frac{\partial^2 v}{\partial y^2} &= \frac{\partial}{\partial y}\left(x\frac{\partial u}{\partial y}\right) = x\frac{\partial^2 u}{\partial y^2} \\
\frac{\partial^2 v}{\partial x^2} + \frac{\partial^2 v}{\partial y^2} &= 2\frac{\partial u}{\partial x} + x\left(\frac{\partial^2 u}{\partial x^2} + \frac{\partial^2 u}{\partial y^2}\right) = 2\frac{\partial u}{\partial x}.
\end{aligned}$$

Since $u$ is harmonic, so is $\partial u/dx$:

$$\left(\frac{\partial^2}{\partial x^2} + \frac{\partial^2}{\partial y^2}\right)\frac{\partial u}{\partial x} = \frac{\partial}{\partial x}\left(\frac{\partial^2 u}{\partial x^2} + \frac{\partial^2 u}{\partial y^2}\right) = \frac{\partial}{\partial x}(0) = 0.$$

Thus $\dfrac{\partial^2 v}{\partial x^2} + \dfrac{\partial^2 v}{\partial y^2}$ is harmonic, and so $v$ is biharmonic. The proof that $w(x, y) = yu(x, y)$ is biharmonic is similar.

**24.** By Exercise 11, $\ln(x^2 + y^2)$ is harmonic (except at the origin). Therefore $y\ln(x^2 + y^2)$ is biharmonic by Exercise 22.

**26.** $u(x, y, z)$ is biharmonic $\Leftrightarrow \dfrac{\partial^2 u}{\partial x^2} + \dfrac{\partial^2 u}{\partial y^2} + \dfrac{\partial^2 u}{\partial z^2}$ is harmonic

$$\Leftrightarrow \left(\frac{\partial^2}{\partial x^2} + \frac{\partial^2}{\partial y^2} + \frac{\partial^2}{\partial z^2}\right)\left(\frac{\partial^2 u}{\partial x^2} + \frac{\partial^2 u}{\partial y^2} + \frac{\partial^2 u}{\partial z^2}\right) = 0$$

$$\Leftrightarrow \frac{\partial^4 u}{\partial x^4} + \frac{\partial^4 u}{\partial y^4} + \frac{\partial^4 u}{\partial z^4} + 2\left(\frac{\partial^4 u}{\partial x^2 \partial y^2} + \frac{\partial^4 u}{\partial x^2 \partial z^2} + \frac{\partial^4 u}{\partial y^2 \partial z^2}\right) = 0$$

by the equality of mixed partials.

If $u(x, y, z)$ is harmonic then the functions $xu(x, y, z)$, $yu(x, y, z)$, and $zu(x, y, z)$ are all biharmonic. The proof is almost identical to that given in Exercise 22.

## Section 4.5 The Chain Rule (page 183)

**2.** If $w = f(x, y, z)$ where $x = g(s)$, $y = h(s, t)$ and $z = k(t)$, then

$$\frac{\partial w}{\partial t} = f_2(x, y, z)h_2(s, t) + f_3(x, y, z)k'(t).$$

**4.** If $w = f(x, y, z)$ where $y = g(x, z)$ and $z = h(x)$, then

$$\begin{aligned}
\frac{dw}{dx} = f_1(x, y, z) &+ f_2(x, y, z)\big(g_1(x, z) \\
&+ g_2(x, z)h'(x)\big) + f_3(x, y, z)h'(x).
\end{aligned}$$

**6.** If $w = f(x, y)$ where $x = g(r, s)$, $y = h(r, t)$, $r = k(s, t)$ and $s = m(t)$, then

$$\begin{aligned}
\frac{dw}{dt} = f_1(x, y)&\big[g_1(r, s)\big(k_1(s, t)m'(t) \\
&+ k_2(s, t)\big) + g_2(r, s)m'(t)\big] \\
&+ f_2(x, y)\big[h_1(r, t)\big(k_1(s, t)m'(t) \\
&+ k_2(s, t)\big) + h_2(r, t)\big].
\end{aligned}$$

**8.** If $u = \sqrt{x^2 + y^2}$, where $x = e^{st}$ and $y = 1 + s^2\cos t$, then

Method I.

$$\begin{aligned}
\frac{\partial u}{\partial t} &= \frac{x}{\sqrt{x^2 + y^2}}se^{st} + \frac{y}{\sqrt{x^2 + y^2}}(-s^2\sin t) \\
&= \frac{xse^{st} - ys^2\sin t}{\sqrt{x^2 + y^2}}.
\end{aligned}$$

Method II.

$$\begin{aligned}
u &= \sqrt{e^{2st} + (1 + s^2\cos t)^2} \\
\frac{\partial u}{\partial t} &= \frac{2se^{2st} - 2s^2\sin t(1 + s^2\cos t)}{2\sqrt{e^{2st} + (1 + s^2\cos t)^2}} \\
&= \frac{x^2 s - ys^2\sin t}{\sqrt{x^2 + y^2}}.
\end{aligned}$$

**10.** If $z = txy^2$, where $x = t + \ln(y + t^2)$ and $y = e^t$, then

Method I.

$$\begin{aligned}
\frac{dz}{dt} &= \frac{\partial z}{\partial t} + \frac{\partial z}{\partial x}\left(\frac{\partial x}{\partial t} + \frac{\partial x}{\partial y}\frac{\partial y}{\partial t}\right) \\
&\quad + \frac{\partial z}{\partial y}\frac{\partial y}{\partial t} \\
&= xy^2 + ty^2\left(1 + \frac{y + 2t}{y + t^2}\right) + 2txy^2.
\end{aligned}$$

Method II.

$$\begin{aligned}
z &= t\big(t + \ln(e^t + t^2)\big)e^{2t} \\
\frac{\partial z}{\partial t} &= \big(t + \ln(e^t + t^2)\big)e^{2t} + te^{2t}\left(1 + \frac{e^t + 2t}{e^t + t^2}\right) \\
&\quad + 2te^{2t}\big(t + \ln(e^t + t^2)\big) \\
&= xy^2 + ty^2\left(1 + \frac{y + 2t}{y + t^2}\right) + 2txy^2.
\end{aligned}$$

**12.** $\dfrac{\partial}{\partial x}f(2y, 3x) = 3f_2(2y, 3x)$.

**14.** $\dfrac{\partial}{\partial t}f(st^2, s^2 + t)$
$= 2stf_1(st^2, s^2 + t) + f_2(st^2, s^2 + t)$.

**16.** $\frac{\partial}{\partial y} f\big(yf(x,t), f(y,t)\big)$
$$= f(x,t)f_1\big(yf(x,t), f(y,t)\big) + f_1(y,t)f_2\big(yf(x,t), f(y,t)\big).$$

**18.** If $S = S(x, y, z, t)$, where $x = \sin t$, $y = \cos t$ and $z = t$, then the rate of change of $S$ is

$$\frac{dS}{dt} = \frac{\partial S}{\partial x}\cos t - \frac{\partial S}{\partial y}\sin t + \frac{\partial S}{\partial z} + \frac{\partial S}{\partial t}.$$

**20.** Let $u = ax + by$, $v = bx - ay$. Then

$$\begin{aligned}
\frac{\partial^2}{\partial x^2} f(u,v) &= \frac{\partial}{\partial x}\big(af_1(u,v) + bf_2(u,v)\big)\\
&= a^2 f_{11}(u,v) + baf_{12}(u,v) + abf_{21}(u,v) + b^2 f_{22}(u,v),\\
\frac{\partial^2}{\partial y^2} f(u,v) &= \frac{\partial}{\partial y}\big(bf_1(u,v) - af_2(u,v)\big)\\
&= b^2 f_{11}(u,v) - abf_{12}(u,v) - baf_{21}(u,v) + a^2 f_{22}(u,v).
\end{aligned}$$

Thus

$$\begin{aligned}
&\left(\frac{\partial^2}{\partial x^2} + \frac{\partial^2}{\partial y^2}\right) f(ax+by, bx-ay)\\
&= (a^2+b^2)\big(f_{11}(u,v) + f_{22}(u,v)\big) = 0
\end{aligned}$$

if $f$ is harmonic. In this case $f(ax+by, bx-ay)$ also satisfies Laplace's equation and is harmonic.

**22.** Let $u = \dfrac{x}{x^2+y^2}$, $v = -\dfrac{y}{x^2+y^2}$. Then

$$\begin{aligned}
\frac{\partial u}{\partial x} &= \frac{y^2-x^2}{(x^2+y^2)^2} & \frac{\partial v}{\partial x} &= \frac{2xy}{(x^2+y^2)^2}\\
\frac{\partial u}{\partial y} &= -\frac{2xy}{(x^2+y^2)^2} & \frac{\partial v}{\partial y} &= \frac{y^2-x^2}{(x^2+y^2)^2}.
\end{aligned}$$

We have

$$\begin{aligned}
\frac{\partial}{\partial x} f(u,v) &= f_1(u,v)\frac{\partial u}{\partial x} + f_2(u,v)\frac{\partial v}{\partial x}\\
\frac{\partial}{\partial y} f(u,v) &= f_1(u,v)\frac{\partial u}{\partial y} + f_2(u,v)\frac{\partial v}{\partial y}\\
\frac{\partial^2}{\partial x^2} f(u,v) &= f_{11}\left(\frac{\partial u}{\partial x}\right)^2 + f_{12}\frac{\partial u}{\partial x}\frac{\partial v}{\partial x} + f_1\frac{\partial^2 u}{\partial x^2}\\
&\quad + f_{21}\frac{\partial u}{\partial x}\frac{\partial v}{\partial x} + f_{22}\left(\frac{\partial v}{\partial x}\right)^2 + f_2\frac{\partial^2 v}{\partial x^2}\\
\frac{\partial^2}{\partial y^2} f(u,v) &= f_{11}\left(\frac{\partial u}{\partial y}\right)^2 + f_{12}\frac{\partial u}{\partial y}\frac{\partial v}{\partial y} + f_1\frac{\partial^2 u}{\partial y^2}\\
&\quad + f_{21}\frac{\partial u}{\partial y}\frac{\partial v}{\partial y} + f_{22}\left(\frac{\partial v}{\partial y}\right)^2 + f_2\frac{\partial^2 v}{\partial y^2}.
\end{aligned}$$

Noting that

$$\begin{aligned}
&\left(\frac{\partial u}{\partial x}\right)^2 + \left(\frac{\partial u}{\partial y}\right)^2 = \frac{1}{(x^2+y^2)^2} = \left(\frac{\partial v}{\partial x}\right)^2 + \left(\frac{\partial v}{\partial y}\right)^2\\
&\frac{\partial u}{\partial x}\frac{\partial v}{\partial x} + \frac{\partial u}{\partial y}\frac{\partial v}{\partial y} = 0,
\end{aligned}$$

we have

$$\begin{aligned}
&\frac{\partial^2}{\partial x^2} f(u,v) + \frac{\partial^2}{\partial y^2} f(u,v)\\
&= f_{11}\left[\left(\frac{\partial u}{\partial x}\right)^2 + \left(\frac{\partial u}{\partial y}\right)^2\right]\\
&\quad + f_{22}\left[\left(\frac{\partial v}{\partial x}\right)^2 + \left(\frac{\partial v}{\partial y}\right)^2\right]\\
&\quad + 2f_{12}\left[\frac{\partial u}{\partial x}\frac{\partial v}{\partial x} + \frac{\partial u}{\partial y}\frac{\partial v}{\partial y}\right]\\
&\quad + f_1\left[\frac{\partial^2 u}{\partial x^2} + \frac{\partial^2 u}{\partial y^2}\right] + f_2\left[\frac{\partial^2 v}{\partial x^2} + \frac{\partial^2 v}{\partial y^2}\right]\\
&= f_1\left[\frac{\partial^2 u}{\partial x^2} + \frac{\partial^2 u}{\partial y^2}\right] + f_2\left[\frac{\partial^2 v}{\partial x^2} + \frac{\partial^2 v}{\partial y^2}\right],
\end{aligned}$$

because we are given that $f$ is harmonic, that is, $f_{11}(u,v) + f_{22}(u,v) = 0$.

Finally, $u$ is harmonic by Exercise 10 of Section 4.4, and, by symmetry, so is $v$. Thus

$$\frac{\partial^2}{\partial x^2} f(u,v) + \frac{\partial^2}{\partial y^2} f(u,v) = 0$$

and $f\left(\dfrac{x}{x^2+y^2}, -\dfrac{y}{x^2+y^2}\right)$ is harmonic for $(x,y) \neq (0,0)$.

**24.** If $r^2 = x^2 + y^2 + z^2$, then $2r\dfrac{\partial r}{\partial x} = 2x$, so $\dfrac{\partial r}{\partial x} = \dfrac{x}{r}$. Similarly, $\dfrac{\partial r}{\partial y} = \dfrac{y}{r}$ and $\dfrac{\partial r}{\partial z} = \dfrac{z}{r}$. If $u = \dfrac{1}{r}$, then

$$\begin{aligned}
\frac{\partial u}{\partial x} &= -\frac{1}{r^2}\frac{\partial r}{\partial x} = -\frac{x}{r^3}\\
\frac{\partial^2 u}{\partial x^2} &= -\frac{1}{r^3} + \frac{3x}{r^4}\frac{x}{r} = \frac{3x^2 - r^2}{r^5}.
\end{aligned}$$

Similarly,

$$\frac{\partial^2 u}{\partial y^2} = \frac{3y^2 - r^2}{r^5}, \qquad \frac{\partial^2 u}{\partial z^2} = \frac{3z^2 - r^2}{r^5}.$$

Adding these three expressions, we get

$$\frac{\partial^2 u}{\partial x^2} + \frac{\partial^2 u}{\partial y^2} + \frac{\partial^2 u}{\partial z^2} = 0,$$

so $u$ is harmonic except at $r = 0$.

**26.** 
$$\begin{aligned}
\frac{\partial^3}{\partial x \partial y^2} f(2x+3y, xy) &= \frac{\partial^2}{\partial x \partial y}(3f_1 + xf_2) \\
&= \frac{\partial}{\partial x}(9f_{11} + 3xf_{12} + 3xf_{21} + x^2 f_{22}) \\
&= \frac{\partial}{\partial x}(9f_{11} + 6xf_{12} + x^2 f_{22}) \\
&= 18f_{111} + 9yf_{112} + 6f_{12} + 12xf_{121} + 6xyf_{122} \\
&\quad + 2xf_{22} + 2x^2 f_{221} + x^2 y f_{222} \\
&= 18f_{111} + (12x + 9y)f_{112} + (6xy + 2x^2)f_{122} + x^2 y f_{222} \\
&\quad + 6f_{12} + 2xf_{22},
\end{aligned}$$
where all partials are evaluated at $(2x+3y, xy)$.

**28.** 
$$\begin{aligned}
\frac{\partial^3}{\partial t^2 \partial s} f(s^2 - t, s + t^2) &= \frac{\partial^2}{\partial t^2}(2sf_1 + f_2) \\
&= \frac{\partial}{\partial t}(-2sf_{11} + 4stf_{12} - f_{21} + 2tf_{22}) \\
&= \frac{\partial}{\partial t}(-2sf_{11} + (4st - 1)f_{12} + 2tf_{22}) \\
&= 2sf_{111} - 4stf_{112} + 4sf_{12} - (4st - 1)f_{121} \\
&\quad + 2t(4st - 1)f_{122} + 2f_{22} - 2tf_{221} + 4t^2 f_{222} \\
&= 2sf_{111} + (1 - 8st)f_{112} + 4t(2st - 1)f_{122} + 4t^2 f_{222} \\
&\quad + 4sf_{12} + 2f_{22},
\end{aligned}$$
where all partials are evaluated at $(s^2 - t, s + t^2)$.

**30.** If $x = r\cos\theta$ and $y = r\sin\theta$, then $r^2 = x^2 + y^2$ and $\tan\theta = y/x$. Thus $2r\frac{\partial r}{\partial x} = 2x$, so $\frac{\partial r}{\partial x} = \frac{x}{r} = \cos\theta$, and similarly, $\frac{\partial r}{\partial y} = \frac{y}{r} = \sin\theta$. Also

$$\begin{aligned}
\sec^2\theta \frac{\partial \theta}{\partial x} &= -\frac{y}{x^2} & \sec^2\theta \frac{\partial \theta}{\partial y} &= \frac{1}{x} \\
\frac{\partial \theta}{\partial x} &= -\frac{y}{x^2 + y^2} & \frac{\partial \theta}{\partial x} &= \frac{x}{x^2 + y^2} \\
&= -\frac{\sin\theta}{r} & &= \frac{\cos\theta}{r}.
\end{aligned}$$

Now

$$\begin{aligned}
\frac{\partial u}{\partial x} &= \frac{\partial u}{\partial r}\frac{\partial r}{\partial x} + \frac{\partial u}{\partial \theta}\frac{\partial \theta}{\partial x} = \cos\theta \frac{\partial u}{\partial r} - \frac{\sin\theta}{r}\frac{\partial u}{\partial \theta} \\
\frac{\partial u}{\partial y} &= \frac{\partial u}{\partial r}\frac{\partial r}{\partial y} + \frac{\partial u}{\partial \theta}\frac{\partial \theta}{\partial y} = \sin\theta \frac{\partial u}{\partial r} + \frac{\cos\theta}{r}\frac{\partial u}{\partial \theta} \\
\frac{\partial^2 u}{\partial x^2} &= \left(\frac{\partial}{\partial x}\cos\theta\right)\frac{\partial u}{\partial r} + \cos\theta\left(\cos\theta \frac{\partial^2 u}{\partial r^2} - \frac{\sin\theta}{r}\frac{\partial^2 u}{\partial \theta \partial r}\right) \\
&\quad - \left(\frac{\partial}{\partial x}\frac{\sin\theta}{r}\right)\frac{\partial u}{\partial \theta} - \frac{\sin\theta}{r}\left(\cos\theta \frac{\partial^2 u}{\partial r \partial \theta} - \frac{\sin\theta}{r}\frac{\partial^2 u}{\partial \theta^2}\right) \\
&= \frac{\sin^2\theta}{r}\frac{\partial u}{\partial r} + \frac{2\sin\theta\cos\theta}{r^2}\frac{\partial u}{\partial \theta} + \cos^2\theta \frac{\partial^2 u}{\partial r^2} \\
&\quad - \frac{2\sin\theta\cos\theta}{r}\frac{\partial^2 u}{\partial r \partial \theta} + \frac{\sin^2\theta}{r^2}\frac{\partial^2 u}{\partial \theta^2} \\
\frac{\partial^2 u}{\partial y^2} &= \left(\frac{\partial}{\partial y}\sin\theta\right)\frac{\partial u}{\partial r} + \sin\theta\left(\sin\theta \frac{\partial^2 u}{\partial r^2} + \frac{\cos\theta}{r}\frac{\partial^2 u}{\partial \theta \partial r}\right) \\
&\quad + \left(\frac{\partial}{\partial y}\frac{\cos\theta}{r}\right)\frac{\partial u}{\partial \theta} + \frac{\cos\theta}{r}\left(\sin\theta \frac{\partial^2 u}{\partial r \partial \theta} + \frac{\cos\theta}{r}\frac{\partial^2 u}{\partial \theta^2}\right) \\
&= \frac{\cos^2\theta}{r}\frac{\partial u}{\partial r} - \frac{2\sin\theta\cos\theta}{r^2}\frac{\partial u}{\partial \theta} + \sin^2\theta \frac{\partial^2 u}{\partial r^2} \\
&\quad + \frac{2\sin\theta\cos\theta}{r}\frac{\partial^2 u}{\partial r \partial \theta} + \frac{\cos^2\theta}{r^2}\frac{\partial^2 u}{\partial \theta^2}.
\end{aligned}$$

Therefore

$$\frac{\partial^2 u}{\partial x^2} + \frac{\partial^2 u}{\partial y^2} = \frac{\partial^2 u}{\partial r^2} + \frac{1}{r}\frac{\partial u}{\partial r} + \frac{1}{r^2}\frac{\partial^2 u}{\partial \theta^2},$$

as was to be shown.

**32.** 
$$\begin{aligned}
&f(tx, ty) = t^k f(x, y) \\
&xf_1(tx, ty) + yf_2(tx, ty) = kt^{k-1} f(x, y) \\
&x\big(xf_{11}(tx, ty) + yf_{12}(tx, ty)\big) \\
&\quad + y\big(xf_{21}(tx, ty) + yf_{22}(tx, ty)\big) \\
&\qquad = k(k-1)t^{k-2} f(x, y)
\end{aligned}$$
Put $t = 1$ and get

$$x^2 f_{11}(x, y) + 2xyf_{12}(x, y) + y^2 f_{22}(x, y) = k(k-1)f(x, y).$$

**34.** If $f(x_1, \cdots, x_n)$ is positively homogeneous of degree $k$ and has continuous partial derivatives of $m$th order, then

$$\begin{aligned}
&\sum_{i_1, \ldots, i_m = 1}^{n} x_{i_1} \cdots x_{i_m} f_{i_1 \ldots i_m}(x_1, \cdots, x_n) \\
&\qquad = k(k-1)\cdots(k-m+1)f(x_1, \cdots, x_n).
\end{aligned}$$

The proof is identical to that of Exercise 33, except that you differentiate $m$ times before putting $t = 1$.

**36.** a) Since $F_{12}(x, y) = -F_{21}(y, x)$ for $(x, y) \neq (0, 0)$, we have $F_{12}(x, x) = -F_{21}(x, x)$ for $x \neq 0$. However, all partial derivatives of the rational function $F$ are continuous except possibly at the origin. Thus $F_{12}(x, x) = F_{21}(x, x)$ for $x \neq 0$. Therefore, $F_{12}(x, x) = 0$ for $x \neq 0$.

b) $F_{12}$ cannot be continuous at $(0, 0)$ because its value there (which is $-2$) differs from the value of $F_{21}(0, 0)$ (which is 2). Alternatively, $F_{12}(0, 0)$ is not the limit of $F_{12}(x, x)$ as $x \to 0$.

**38.** If $w(r) = f(r) + g(s)$, where $f$ and $g$ are arbitrary twice differentiable functions, then

$$\frac{\partial^2 w}{\partial r \partial s} = \frac{\partial}{\partial r} g'(s) = 0.$$

**40.** By Exercise 39, the DE $u_t = c^2 u_{xx}$ has solution

$$u(x, t) = f(x + ct) + g(x - ct),$$

for arbitrary sufficiently smooth functions $f$ and $g$. The initial conditions imply that

$$p(x) = u(x, 0) = f(x) + g(x)$$
$$q(x) = u_t(x, 0) = cf'(x) - cg'(x).$$

Integrating the second of these equations, we get

$$f(x) - g(x) = \frac{1}{c}\int_a^x q(s)\, ds,$$

where $a$ is a constant. Solving the two equations for $f$ and $g$ we obtain

$$f(x) = \frac{1}{2}p(x) + \frac{1}{2c}\int_a^x q(s)\, ds$$
$$g(x) = \frac{1}{2}p(x) - \frac{1}{2c}\int_a^x q(s)\, ds.$$

Thus the solution to the initial-value problem is

$$u(x, t) = \frac{p(x + ct) + p(x - ct)}{2} + \frac{1}{2c}\int_{x-ct}^{x+ct} q(s)\, ds.$$

## Section 4.6 Linear Approximation, Differentiability, and Differentials (page 191)

**2.**
$$f(x, y) = \tan^{-1}\frac{y}{x} \qquad f(3, 3) = \frac{\pi}{4}$$
$$f_1(x, y) = -\frac{y}{x^2 + y^2} \qquad f_1(3, 3) = -\frac{1}{6}$$
$$f_2(x, y) = \frac{x}{x^2 + y^2} \qquad f_2(3, 3) = \frac{1}{6}$$
$$\begin{aligned} f(3.01, 2.99) &= f(3 + 0.01, 3 - 0.01) \\ &\approx f(3, 3) + 0.01 f_1(3, 3) - 0.01 f_2(3, 3) \\ &= \frac{\pi}{4} - \frac{0.01}{6} - \frac{0.01}{6} = \frac{\pi}{4} - \frac{0.01}{3} \\ &\approx 0.7820648 \end{aligned}$$

**4.**
$$f(x, y) = \frac{24}{x^2 + xy + y^2}$$
$$f_1(x, y) = \frac{-24(2x + y)}{(x^2 + xy + y^2)^2}, \qquad f_2(x, y) = \frac{-24(x + 2y)}{(x^2 + xy + y^2)^2}$$
$$f(2, 2) = 2, \qquad f_1(2, 2) = -1, \qquad f_2(2, 2) = -1$$
$$\begin{aligned} f(2.1, 1.8) &\approx f(2, 2) + 0.1 f_1(2, 2) - 0.2 f_2(2, 2) \\ &= 2 - 0.1 + 0.2 = 2.1 \end{aligned}$$

**6.**
$$f(x, y) = xe^{y+x^2} \qquad f(2, -4) = 2$$
$$f_1(x, y) = e^{y+x^2}(1 + 2x^2) \qquad f_1(2, -4) = 9$$
$$f_2(x, y) = xe^{y+x^2} \qquad f_2(2, -4) = 2$$
$$\begin{aligned} f(2.05, -3.92) &\approx f(2, -4) + 0.05 f_1(2, -4) + 0.08 f_2(2, -4) \\ &= 2 + 0.45 + 0.16 = 2.61 \end{aligned}$$

**8.** $V = \frac{1}{3}\pi r^2 h \Rightarrow dV = \frac{2}{3}\pi r h\, dr + \frac{1}{3}\pi r^2\, dh$. If $r = 25$ ft, $h = 21$ ft, and $dr = dh = 0.5/12$ ft, then

$$dV = \frac{\pi}{3}(2 \times 25 \times 21 + 25^2)\frac{0.5}{12} \approx 73.08.$$

The calculated volume can be in error by about 73 cubic feet.

**10.** If the sides and contained angle of the triangle are $x$ and $y$ m and $\theta$ radians, then its area $A$ satisfies

$$A = \frac{1}{2}xy \sin\theta$$
$$dA = \frac{1}{2}y \sin\theta\, dx + \frac{1}{2}x \sin\theta\, dy + \frac{1}{2}xy \cos\theta\, d\theta$$
$$\frac{dA}{A} = \frac{dx}{x} + \frac{dy}{y} + \cot\theta\, d\theta.$$

For $x = 224$, $y = 158$, $\theta = 64° = 64\pi/180$, $dx = dy = 0.4$, and $d\theta = 2° = 2\pi/180$, we have

$$\frac{dA}{A} = \frac{0.4}{224} + \frac{0.4}{158} + (\cot 64°)\frac{2\pi}{180} \approx 0.0213.$$

The calculated area of the plot can be in error by a little over 2%.

**12.** $w = \dfrac{x^2y^3}{z^4}$ $\qquad \dfrac{\partial w}{\partial x} = \dfrac{2xy^3}{z^4} = \dfrac{2w}{x}$

$$\frac{\partial w}{\partial y} = \frac{3x^2y^2}{z^4} = \frac{3w}{y} \qquad \frac{\partial w}{\partial z} = -\frac{4x^2y^3}{z^5} = -\frac{4w}{x}.$$

$$dw = \frac{\partial w}{\partial x}\,dx + \frac{\partial w}{\partial y}\,dy + \frac{\partial w}{\partial z}\,dz$$

$$\frac{dw}{w} = 2\frac{dx}{x} + 3\frac{dy}{y} - 4\frac{dz}{z}.$$

Since $x$ increases by 1%, then $\dfrac{dx}{x} = \dfrac{1}{100}$. Similarly, $\dfrac{dy}{y} = \dfrac{2}{100}$ and $\dfrac{dz}{z} = \dfrac{3}{100}$. Therefore

$$\frac{\Delta w}{w} \approx \frac{dw}{w} = \frac{2+6-12}{100} = -\frac{4}{100},$$

and $w$ decreases by about 4%.

**14.** $\mathbf{f}(\rho, \phi, \theta) = (\rho\sin\phi\cos\theta, \rho\sin\phi\sin\theta, \rho\cos\phi)$

$$D\mathbf{f}(\rho, \phi, \theta) = \begin{pmatrix} \sin\phi\cos\theta & \rho\cos\phi\cos\theta & -\rho\sin\phi\sin\theta \\ \sin\phi\sin\theta & \rho\cos\phi\sin\theta & \rho\sin\phi\cos\theta \\ \cos\phi & -\rho\sin\phi & 0 \end{pmatrix}$$

**16.**

$$\mathbf{g}(r, s, t) = \begin{pmatrix} r^2 s \\ r^2 t \\ s^2 - t^2 \end{pmatrix}$$

$$D\mathbf{g}(r, s, t) = \begin{pmatrix} 2rs & r^2 & 0 \\ 2rt & 0 & r^2 \\ 0 & 2s & -2t \end{pmatrix}$$

$$D\mathbf{g}(1, 3, 3) = \begin{pmatrix} 6 & 1 & 0 \\ 6 & 0 & 1 \\ 0 & 6 & -6 \end{pmatrix}$$

$$\mathbf{g}(0.99, 3.02, 2.97) \approx \mathbf{g}(1,3,3) + D\mathbf{g}(1,3,3)\begin{pmatrix} -0.01 \\ 0.02 \\ -0.03 \end{pmatrix}$$

$$= \begin{pmatrix} 3 \\ 3 \\ 0 \end{pmatrix} + \begin{pmatrix} -0.04 \\ -0.09 \\ 0.30 \end{pmatrix} = \begin{pmatrix} 2.96 \\ 2.91 \\ 0.30 \end{pmatrix}$$

**18.** Let $g(t) = f(a+th, b+tk)$. Then

$$g'(t) = hf_1(a+th, b+tk) + kf_2(a+th, b+tk).$$

If $h$ and $k$ are small enough that $(a+h, b+k)$ belongs to the disk referred to in the statement of the problem, then we can apply the (one-variable) Mean-Value Theorem to $g(t)$ on $[0, 1]$ and obtain

$$g(1) = g(0) + g'(\theta),$$

for some $\theta$ satisfying $0 < \theta < 1$, i.e.,

$$f(a+h, b+k) = f(a, b) + hf_1(a+\theta h, b+\theta k) + kf_2(a+\theta h, b+\theta k).$$

## Section 4.7 Gradients and Directional Derivatives (page 201)

**2.** $f(x, y) = \dfrac{x-y}{x+y}$, $\qquad f(1, 1) = 0$.

$\nabla f = \dfrac{2y\mathbf{i} - 2x\mathbf{j}}{(x+y)^2}$,

$\nabla f(1, 1) = \dfrac{1}{2}(\mathbf{i} - \mathbf{j})$. Tangent plane to $z = f(x, y)$ at $(1, 1, 0)$ has equation $\frac{1}{2}(x-1) - \frac{1}{2}(y-1) = z$, or $x - y - 2z = 0$.

Tangent line to $f(x, y) = 0$ at $(1, 1)$ has equation $\frac{1}{2}(x-1) - \frac{1}{2}(y-1)$, or $x = y$.

**4.** $f(x, y) = e^{xy}$, $\quad \nabla f = ye^{xy}\mathbf{i} + xe^{xy}\mathbf{j}$,

$\nabla f(2, 0) = 2\mathbf{j}$. Tangent plane to $z = f(x, y)$ at $(2, 0, 1)$ has equation $2y = z - 1$, or $2y - z = -1$.

Tangent line to $f(x, y) = 1$ at $(2, 0)$ has equation $y = 0$.

**6.**

$$f(x, y) = \frac{2xy}{x^2+y^2},$$

$$f_1(x, y) = \frac{(x^2+y^2)(2y) - 2xy(2x)}{(x^2+y^2)^2} = \frac{2y(y^2-x^2)}{(x^2+y^2)^2},$$

$$f_2(x, y) = \frac{2x(x^2-y^2)}{(x^2+y^2)^2}.$$

$$\nabla f(x, y) = \frac{2(y^2-x^2)}{(x^2+y^2)^2}(y\mathbf{i} - x\mathbf{j}),$$

$\nabla f(0, 2) = \mathbf{i}$.

Tangent plane to $z = f(x, y)$ at $(0, 2, 0)$ has equation $x = z$.

Tangent line to $f(x, y) = 0$ at $(0, 2)$ has equation $x = 0$.

**8.** $f(x, y) = \sqrt{1+xy^2}$, $\qquad f(2, -2) = 3$.

$\nabla f(x, y) = \dfrac{y^2\mathbf{i} + 2xy\mathbf{j}}{2\sqrt{1+xy^2}}$,

$\nabla f(2, -2) = \dfrac{2}{3}\mathbf{i} - \dfrac{4}{3}\mathbf{j}$.

Tangent plane to $z = f(x, y)$ at $(2, -2, 3)$ has equation $\dfrac{2}{3}(x-2) - \dfrac{4}{3}(y+2) = z - 3$, or $2x - 4y - 3z = 3$.

Tangent line to $f(x, y) = 3$ at $(2, -2)$ has equation $\dfrac{2}{3}(x-2) - \dfrac{4}{3}(y+2) = 0$, or $x - 2y = 6$.

**10.** $f(x, y, z) = \cos(x + 2y + 3z)$,

$f\left(\dfrac{\pi}{2}, \pi, \pi\right) = \cos\dfrac{11\pi}{2} = 0$.

$\nabla f(x, y, z) = -\sin(x + 2y + 3z)(\mathbf{i} + 2\mathbf{j} + 3\mathbf{k})$,

$\nabla f\left(\dfrac{\pi}{2}, \pi, \pi\right) = -\sin\dfrac{11\pi}{2}(\mathbf{i} + 2\mathbf{j} + 3\mathbf{k}) = \mathbf{i} + 2\mathbf{j} + 3\mathbf{k}$.

Tangent plane to $f(x, y, z) = 0$ at $\left(\dfrac{\pi}{2}, \pi, \pi\right)$ has equation

$$x - \frac{\pi}{2} + 2(y - \pi) + 3(z - \pi) = 0,$$

or $x + 2y + 3z = \dfrac{11\pi}{2}$.

**12.** $f(x,y)=3x-4y$, $\quad \nabla f(0,2)=\nabla f(x,y)=3\mathbf{i}-4\mathbf{j}$,
$D_{-\mathbf{i}}f(0,2)=-\mathbf{i}\bullet(3\mathbf{i}-4\mathbf{j})=-3$.

**14.** $f(x,y)=\dfrac{x}{1+y}$, $\quad \nabla f(x,y)=\dfrac{1}{1+y}\mathbf{i}-\dfrac{x}{(1+y)^2}\mathbf{j}$,
$\nabla f(0,0)=\mathbf{i}$, $\quad \mathbf{u}=\dfrac{\mathbf{i}-\mathbf{j}}{\sqrt{2}}$,

$$D_{\mathbf{u}}f(0,0)=\mathbf{i}\bullet\left(\frac{\mathbf{i}-\mathbf{j}}{\sqrt{2}}\right)=\frac{1}{\sqrt{2}}.$$

**16.** $f(x,y,z)=(y^2+\sin z)e^{-x}$.
$\nabla f(x,y,z)=-(y^2+\sin z)e^{-x}\mathbf{i}+2ye^{-x}\mathbf{j}+\cos ze^{-x}\mathbf{k}$,
$\nabla f(0,2,\pi)=-4\mathbf{i}+4\mathbf{j}-\mathbf{k}$.
A vector from $(0,2,\pi)$ towards $(1,1,0)$ is $\mathbf{v}=\mathbf{i}-\mathbf{j}-\pi\mathbf{k}$.
The rate of change of $f$ at $(0,2,\pi)$ in the direction of $\mathbf{v}$ is

$$\frac{\mathbf{v}}{|\mathbf{v}|}\bullet\nabla f(0.2.\pi)=\frac{-4-4+\pi}{\sqrt{2+\pi^2}}=-\frac{8-\pi}{\sqrt{2+\pi^2}}.$$

**18.** $f(x,y)=\ln|\mathbf{r}|$, where $\mathbf{r}=x\mathbf{i}+y\mathbf{j}$. Since $|\mathbf{r}|=\sqrt{x^2+y^2}$, we have

$$\nabla f(x,y)=\frac{1}{|\mathbf{r}|}\left(\frac{x}{|\mathbf{r}|}\mathbf{i}+\frac{y}{|\mathbf{r}|}\mathbf{j}\right)=\frac{\mathbf{r}}{|\mathbf{r}|^2}.$$

**20.** Since $x=r\cos\theta$ and $y=r\sin\theta$, we have

$$\begin{aligned}\frac{\partial f}{\partial r}&=\cos\theta\frac{\partial f}{\partial x}+\sin\theta\frac{\partial f}{\partial y}\\ \frac{\partial f}{\partial\theta}&=-r\sin\theta\frac{\partial f}{\partial x}+r\cos\theta\frac{\partial f}{\partial y}.\end{aligned}$$

Also,

$$\begin{aligned}\hat{\mathbf{r}}&=\frac{x\mathbf{i}+y\mathbf{j}}{r}=(\cos\theta)\mathbf{i}+(\sin\theta)\mathbf{j}\\ \hat{\boldsymbol{\theta}}&=\frac{-y\mathbf{i}+x\mathbf{j}}{r}=-(\sin\theta)\mathbf{i}+(\cos\theta)\mathbf{j}.\end{aligned}$$

Therefore,

$$\begin{aligned}&\frac{\partial f}{\partial r}\hat{\mathbf{r}}+\frac{1}{r}\frac{\partial f}{\partial\theta}\hat{\boldsymbol{\theta}}\\ &=\left(\cos^2\theta\frac{\partial f}{\partial x}+\sin\theta\cos\theta\frac{\partial f}{\partial y}\right)\mathbf{i}\\ &\quad+\left(\cos\theta\sin\theta\frac{\partial f}{\partial x}+\sin^2\theta\frac{\partial f}{\partial y}\right)\mathbf{j}\\ &\quad+\left(\sin^2\theta\frac{\partial f}{\partial x}-\sin\theta\cos\theta\frac{\partial f}{\partial y}\right)\mathbf{i}\\ &\quad+\left(-\cos\theta\sin\theta\frac{\partial f}{\partial x}+\cos^2\theta\frac{\partial f}{\partial y}\right)\mathbf{j}\\ &=\frac{\partial f}{\partial x}\mathbf{i}+\frac{\partial f}{\partial y}\mathbf{j}=\nabla f.\end{aligned}$$

**22.** $f(x,y,z)=x^2+y^2-z^2$.
$\nabla f(a,b,c)=2a\mathbf{i}+2b\mathbf{j}-2c\mathbf{k}$. The maximum rate of change of $f$ at $(a,b,c)$ is in the direction of $\nabla f(a,b,c)$, and is equal to $|\nabla f(a,b,c)|$.
Let $\mathbf{u}$ be a unit vector making an angle $\theta$ with $\nabla f(a,b,c)$. The rate of change of $f$ at $(a,b,c)$ in the direction of $\mathbf{u}$ will be half of the maximum rate of change of $f$ at that point provided

$$\frac{1}{2}|\nabla f(a,b,c)|=\mathbf{u}\bullet\nabla f(a,b,c)=|\nabla f(a,b,c)|\cos\theta,$$

that is, if $\cos\theta=\frac{1}{2}$, which means $\theta=60°$. At $(a,b,c)$, $f$ increases at half its maximal rate in all directions making $60°$ angles with the direction $a\mathbf{i}+b\mathbf{j}-c\mathbf{k}$.

**24.** Given the values $D_{\phi_1}f(a,b)$ and $D_{\phi_2}f(a,b)$, we can solve the equations

$$\begin{aligned}f_1(a,b)\cos\phi_1+f_2(a,b)\sin\phi_1&=D_{\phi_1}f(a,b)\\ f_1(a,b)\cos\phi_2+f_2(a,b)\sin\phi_2&=D_{\phi_2}f(a,b)\end{aligned}$$

for unique values of $f_1(a,b)$ and $f_2(a,b)$ (and hence determine $\nabla f(a,b)$ uniquely), provided the coefficients satisfy

$$0\neq\begin{vmatrix}\cos\phi_1 & \sin\phi_1\\ \cos\phi_2 & \sin\phi_2\end{vmatrix}=\sin(\phi_2-\phi_1).$$

Thus $\phi_1$ and $\phi_2$ must not differ by an integer multiple of $\pi$.

**26.** Let the curve be $y=g(x)$. At $(x,y)$ this curve has normal $\nabla\big(g(x)-y\big)=g'(x)\mathbf{i}-\mathbf{j}$.
A curve of the family $x^4+y^2=C$ has normal $\nabla(x^4+y^2)=4x^3\mathbf{i}+2y\mathbf{j}$.
These curves will intersect at right angles if their normals are perpendicular. Thus we require that

$$0=4x^3g'(x)-2y=4x^3g'(x)-2g(x),$$

or, equivalently,

$$\frac{g'(x)}{g(x)}=\frac{1}{2x^3}.$$

Integration gives $\ln|g(x)|=-\dfrac{1}{4x^2}+\ln|C|$, or $g(x)=Ce^{-(1/4x^2)}$.
Since the curve passes through $(1,1)$, we must have $1=g(1)=Ce^{-1/4}$, so $C=e^{1/4}$.
The required curve is $y=e^{(1/4)-(1/4x^2)}$.

**28.** Let $f(x,y)=e^{-(x^2+y^2)}$. Then

$$\nabla f(x,y)=-2e^{-(x^2+y^2)}(x\mathbf{i}+y\mathbf{j}).$$

The vector $\mathbf{u} = \dfrac{a\mathbf{i} + b\mathbf{j}}{\sqrt{a^2+b^2}}$ is a unit vector in the direction directly away from the origin at $(a, b)$.
The first directional derivative of $f$ at $(x, y)$ in the direction of $\mathbf{u}$ is

$$\mathbf{u} \bullet \nabla f(x, y) = -\frac{2}{\sqrt{a^2+b^2}}(ax+by)e^{-(x^2+y^2)}.$$

The second directional derivative is

$$\begin{aligned}
&\mathbf{u} \bullet \nabla \left( -\frac{2}{\sqrt{a^2+b^2}}(ax+by)e^{-(x^2+y^2)} \right) \\
&= -\frac{2}{a^2+b^2}(a\mathbf{i}+b\mathbf{j}) \bullet e^{-(x^2+y^2)} \\
&\qquad \Big[\big(a - 2x(ax+by)\big)\mathbf{i} + \big(b - 2y(ax+by)\big)\mathbf{j}\Big].
\end{aligned}$$

At $(a, b)$ this second directional derivative is

$$\begin{aligned}
&-\frac{2e^{-(a^2+b^2)}}{a^2+b^2}\big(a^2 - 2a^4 - 2a^2b^2 + b^2 - 2a^2b^2 - 2b^4\big) \\
&= \frac{2}{a^2+b^2}\big(2(a^2+b^2)^2 - a^2 - b^2\big)e^{-(a^2+b^2)} \\
&= 2\big(2(a^2+b^2) - 1\big)e^{-(a^2+b^2)}.
\end{aligned}$$

Remark: Since $f(x, y) = e^{-r^2}$ (expressed in terms of polar coordinates), the second directional derivative of $f$ at $(a, b)$ in the direction directly away from the origin (i.e., the direction of increasing $r$) can be more easily calculated as

$$\left.\frac{d^2}{dr^2}e^{-r^2}\right|_{r^2=a^2+b^2}.$$

**30.** At $(1, -1, 1)$ the surface $x^2 + y^2 = 2$ has normal

$$\mathbf{n}_1 = \nabla(x^2+y^2)\Big|_{(1,-1,1)} = 2\mathbf{i} - 2\mathbf{j},$$

and $y^2 + z^2 = 2$ has normal

$$\mathbf{n}_2 = \nabla(y^2+z^2)\Big|_{(1,-1,1)} = -2\mathbf{j} + 2\mathbf{k}.$$

A vector tangent to the curve of intersection of the two surfaces at $(1, -1, 1)$ must be perpendicular to both these normals. Since

$$(\mathbf{i} - \mathbf{j}) \times (-\mathbf{j} + \mathbf{k}) = -(\mathbf{i} + \mathbf{j} + \mathbf{k}),$$

the vector $\mathbf{i} + \mathbf{j} + \mathbf{k}$, or any scalar multiple of this vector, is tangent to the curve at the given point.

**32.** A vector tangent to the path of the fly at $(1, 1, 2)$ is given by

$$\begin{aligned}
\mathbf{v} &= \nabla(3x^2 - y^2 - z) \times \nabla(2x^2 + 2y^2 - z^2)\Big|_{(1,1,2)} \\
&= (6x\mathbf{i} - 2y\mathbf{j} - \mathbf{k}) \times (4x\mathbf{i} + 4y\mathbf{j} - 2z\mathbf{k})\Big|_{(1,1,2)} \\
&= (6\mathbf{i} - 2\mathbf{j} - \mathbf{k}) \times (4\mathbf{i} + 4\mathbf{j} - 4\mathbf{k}) \\
&= 4\begin{vmatrix} \mathbf{i} & \mathbf{j} & \mathbf{k} \\ 6 & -2 & -1 \\ 1 & 1 & -1 \end{vmatrix} = 4(3\mathbf{i} + 5\mathbf{j} + 8\mathbf{k}).
\end{aligned}$$

The temperature $T = x^2 - y^2 + z^2 + xz^2$ has gradient at $(1, 1, 2)$ given by

$$\begin{aligned}
\nabla T(1, 1, 2) &= (2x + z^2)\mathbf{i} - 2y\mathbf{j} + 2z(1+x)\mathbf{k}\Big|_{(1,1,2)} \\
&= 6\mathbf{i} - 2\mathbf{j} + 8\mathbf{k}.
\end{aligned}$$

Thus the fly, passing through $(1, 1, 2)$ with speed 7, experiences temperature changing at rate

$$\begin{aligned}
7 \times \frac{\mathbf{v}}{|\mathbf{v}|} \bullet \nabla T(1, 1, 2) &= 7\frac{3\mathbf{i} + 5\mathbf{j} + 8\mathbf{k}}{\sqrt{98}} \bullet (6\mathbf{i} - 2\mathbf{j} + 8\mathbf{k}) \\
&= \frac{1}{\sqrt{2}}(18 - 10 + 64) = \frac{72}{\sqrt{2}}.
\end{aligned}$$

We don't know which direction the fly is moving along the curve, so all we can say is that it experiences temperature changing at rate $36\sqrt{2}$ degrees per unit time.

**34.** The level surface of $f(x, y, z) = \cos(x + 2y + 3z)$ through $(\pi, \pi, \pi)$ has equation $\cos(x + 2y + 3z) = \cos(6\pi) = 1$, which simplifies to $x + 2y + 3z = 6\pi$. This level surface is a plane, and is therefore its own tangent plane. We cannot determine this plane by the method used to find the tangent plane to the level surface of $f$ through $(\pi/2, \pi, \pi)$ in Exercise 10, because $\nabla f(\pi, \pi, \pi) = \mathbf{0}$, so the gradient does not provide a usable normal vector to define the tangent plane.

**36.** Let $f(x, y) = x^3 - y^2$. Then $\nabla f(x, y) = 3x^2\mathbf{i} - 2y\mathbf{j}$ exists everywhere, but equals $\mathbf{0}$ at $(0, 0)$. The level curve of $f$ passing through $(0, 0)$ is $y^2 = x^3$, which has a cusp at $(0, 0)$, so is not smooth there.

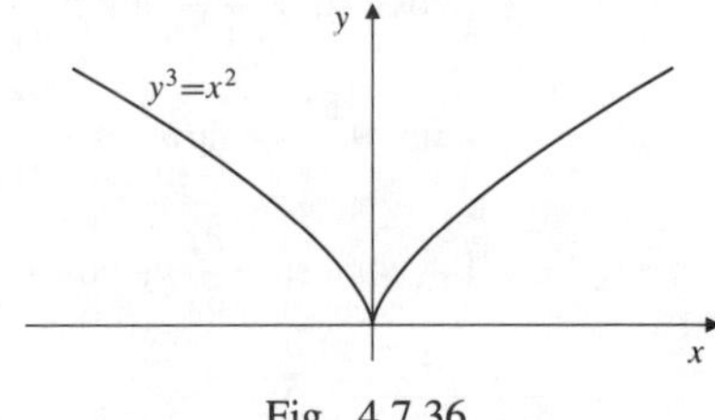


Fig. 4.7.36

**38.** $T = T(x, y, z)$. As measured by the observer,

$$\begin{aligned}\frac{dT}{dt} &= D_{\mathbf{v}(t)}T = \mathbf{v}(t) \bullet \nabla T \\ \frac{d^2T}{dt^2} &= \mathbf{a}(t) \bullet \nabla T + \mathbf{v}(t) \bullet \frac{d}{dt}\nabla T \\ &= D_{\mathbf{a}(t)}T + \left(v_1(t)\frac{d}{dt}\frac{\partial T}{\partial x} + \cdots\right) \\ &= D_{\mathbf{a}(t)}T + \left(v_1(t)\mathbf{v}(t) \bullet \nabla\frac{\partial T}{\partial x} + \cdots\right) \\ &= D_{\mathbf{a}(t)}T + \left(\left(v_1(t)\right)^2\frac{\partial^2 T}{\partial x^2} + v_1(t)v_2(t)\frac{\partial^2 T}{\partial y \partial x} + \cdots\right) \\ &= D_{\mathbf{a}(t)}T + D_{\mathbf{v}(t)}(D_{\mathbf{v}(t)}T)\end{aligned}$$

(as in Exercise 37).

**40.** $f(x, y) = \begin{cases} \dfrac{\sin(xy)}{\sqrt{x^2+y^2}} & \text{if } (x, y) \neq (0, 0) \\ 0 & \text{if } (x, y) = (0, 0) \end{cases}.$

a) $f_1(0, 0) = \lim_{h\to 0} \dfrac{0-0}{h} = 0 = f_2(0, 0)$. Thus $\nabla f(0, 0) = \mathbf{0}$.

b) If $\mathbf{u} = (\mathbf{i} + \mathbf{j})/\sqrt{2}$, then

$$D_{\mathbf{u}}f(0, 0) = \lim_{h\to 0+} \frac{1}{h}\frac{\sin(h^2/2)}{\sqrt{h^2}} = \frac{1}{2}.$$

c) $f$ cannot be differentiable at $(0, 0)$; if it were, then the directional derivative obtained in part (b) would have been $\mathbf{u} \bullet \nabla f(0, 0) = 0$.

## Section 4.8 Implicit Functions (page 211)

**2.** $xy^3 = y - z$: $\quad x = x(y, z)$

$$y^3\frac{\partial x}{\partial y} + 3xy^2 = 1$$

$$\frac{\partial x}{\partial y} = \frac{1 - 3xy^2}{y^3}.$$

The given equation has a solution $x = x(y, z)$ with this partial derivative near any point where $y \neq 0$.

**4.** $e^{yz} - x^2 z \ln y = \pi$: $\quad y = y(x, z)$

$$e^{yz}\left(z\frac{\partial y}{\partial z} + y\right) - x^2 \ln y - \frac{x^2 z}{y}\frac{\partial y}{\partial z} = 0$$

$$\frac{\partial y}{\partial z} = \frac{x^2 \ln y - ye^{yz}}{ze^{yz} - \dfrac{x^2 z}{y}} = \frac{x^2 y \ln y - y^2 e^{yz}}{yze^{yz} - x^2 z}.$$

The given equation has a solution $y = y(x, z)$ with this derivative near any point where $y > 0$, $z \neq 0$, and $ye^{yz} \neq x^2$.

**6.** $F(x, y, x^2 - y^2) = 0$: $\quad y = y(x)$

$$F_1 + F_2\frac{dy}{dx} + F_3\left(2x - 2y\frac{dy}{dx}\right) = 0$$

$$\frac{dy}{dx} = \frac{F_1(x, y, x^2 - y^2) + 2xF_3(x, y, x^2 - y^2)}{2yF_3(x, y, x^2 - y^2) - F_2(x, y, x^2 - y^2)}.$$

The given equation has a solution with this derivative near any point where $F_2(x, y, x^2 - y^2) \neq 2yF_3(x, y, x^2 - y^2)$.

**8.** $F(x^2 - z^2, y^2 + xz) = 0$: $\quad z = z(x, y)$

$$F_1\left(2x - 2z\frac{\partial z}{\partial x}\right) + F_2\left(x\frac{\partial z}{\partial x} + z\right) = 0$$

$$\frac{\partial z}{\partial x} = \frac{2xF_1(x^2 - z^2, y^2 + xz) + zF_2(x^2 - z^2, y^2 + xz)}{2zF_1(x^2 - z^2, y^2 + xz) - xF_2(x^2 - z^2, y^2 + xz)}.$$

The given equation has a solution with this derivative near any point where $xF_2(x^2 - z^2, y^2 + xz) \neq 2zF_1(x^2 - z^2, y^2 + xz)$.

**10.** $\begin{cases} xyuv = 1 \\ x + y + u + v = 0 \end{cases} \Rightarrow \begin{cases} y = y(x, u) \\ v = v(x, u) \end{cases}$

Differentiate the given equations with respect to $x$:

$$\begin{aligned} yuv + xuv\frac{\partial y}{\partial x} + xyu\frac{\partial v}{\partial x} &= 0 \\ 1 + \frac{\partial y}{\partial x} + \frac{\partial v}{\partial x} &= 0 \end{aligned}$$

Multiply the last equation by $xyu$ and subtract the two equations:

$$yuv - xyu + (xuv - xyu)\frac{\partial y}{\partial x} = 0$$

$$\left(\frac{\partial y}{\partial x}\right)_u = \frac{y(x - v)}{x(v - y)}.$$

The given equations have a solution of the indicated form with this derivative near any point where $u \neq 0$, $x \neq 0$ and $y \neq v$.

**12.** $\begin{cases} x^2 y + y^2 u - u^3 = 0 \\ x^2 + yu = 1 \end{cases} \Rightarrow \begin{cases} u = u(x) \\ y = y(x) \end{cases}$

$$\begin{aligned} 2xy + (x^2 + 2yu)\frac{dy}{dx} + (y^2 - 3u^2)\frac{du}{dx} &= 0 \\ 2x + u\frac{dy}{dx} + y\frac{du}{dx} &= 0 \end{aligned}$$

Multiply the first equation by $u$ and the second by $x^2 + 2yu$ and subtract:

$$2x(x^2 + yu) + (x^2 y + y^2 u + 3u^3)\frac{du}{dx} = 0$$

$$\frac{du}{dx} = -\frac{2x(x^2 + yu)}{3u^3 + x^2 y + y^2 u} = -\frac{x}{2u^3}.$$

The given equations have a solution with the indicated derivative near any point where $u \neq 0$.

**14.** $\begin{cases} x = r^2 + 2s \\ y = s^2 - 2r \end{cases}$

$$\frac{\partial(x,y)}{\partial(r,s)} = \begin{vmatrix} 2r & 2 \\ -2 & 2s \end{vmatrix} = 4(rs+1).$$

The given system can be solved for $r$ and $s$ as functions of $x$ and $y$ near any point $(r,s)$ where $rs \neq -1$.
We have

$$\begin{aligned} 1 &= 2r\frac{\partial r}{\partial x} + 2\frac{\partial s}{\partial x} \\ 0 &= -2\frac{\partial r}{\partial x} + 2s\frac{\partial s}{\partial x} \\ 0 &= 2r\frac{\partial r}{\partial y} + 2\frac{\partial s}{\partial y} \\ 1 &= -2\frac{\partial r}{\partial y} + 2s\frac{\partial s}{\partial y}. \end{aligned}$$

Thus

$$\frac{\partial r}{\partial x} = \frac{s}{2(rs+1)} \qquad \frac{\partial r}{\partial y} = -\frac{1}{2(rs+1)}$$
$$\frac{\partial s}{\partial x} = \frac{1}{2(rs+1)} \qquad \frac{\partial s}{\partial y} = \frac{r}{2(rs+1)}.$$

**16.** $x = \rho\sin\phi\cos\theta$, $y = \rho\sin\phi\sin\theta$, $z = \rho\cos\phi$.

$$\begin{aligned} \frac{\partial(x,y,z)}{\partial(\rho,\phi,\theta)} &= \begin{vmatrix} \sin\phi\cos\theta & \rho\cos\phi\cos\theta & -\rho\sin\phi\sin\theta \\ \sin\phi\sin\theta & \rho\cos\phi\sin\theta & \rho\sin\phi\cos\theta \\ \cos\phi & -\rho\sin\phi & 0 \end{vmatrix} \\ &= \cos\phi \begin{vmatrix} \rho\cos\phi\cos\theta & -\rho\sin\phi\sin\theta \\ \rho\cos\phi\sin\theta & \rho\sin\phi\cos\theta \end{vmatrix} \\ &\quad + \rho\sin\phi \begin{vmatrix} \sin\phi\cos\theta & -\rho\sin\phi\sin\theta \\ \sin\phi\sin\theta & \rho\sin\phi\cos\theta \end{vmatrix} \\ &= \rho^2\cos\phi\big[\cos\phi\sin\phi\cos^2\theta + \sin\phi\cos\phi\sin^2\theta\big] \\ &\quad + \rho^2\sin\phi\big[\sin^2\phi\cos^2\theta + \sin^2\phi\sin^2\theta\big] \\ &= \rho^2\cos^2\phi\sin\phi + \rho^2\sin^3\phi = \rho^2\sin\phi. \end{aligned}$$

The transformation is one-to-one (and invertible) near any point where $\rho^2\sin\phi \neq 0$, that is, near any point not on the $z$-axis.

**18.** Let $F(x,y,z,u,v) = xe^y + uz - \cos v - 2$
$G(x,y,z,u,v) = u\cos y + x^2v - yz^2 - 1.$
If $P_0$ is the point where $(x,y,z) = (2,0,1)$ and $(u,v) = (1,0)$, then

$$\begin{aligned} \left.\frac{\partial(F,G)}{\partial(u,v)}\right|_{P_0} &= \left.\begin{vmatrix} z & \sin v \\ \cos y & x^2 \end{vmatrix}\right|_{P_0} \\ &= \begin{vmatrix} 1 & 0 \\ 1 & 4 \end{vmatrix} = 4. \end{aligned}$$

Since this Jacobian is not zero, the equations $F = G = 0$ can be solved for $u$, and $v$ in terms of $x$, $y$ and $z$ near $P_0$. Also,

$$\begin{aligned} \left.\left(\frac{\partial u}{\partial z}\right)_{x,y}\right|_{(2,0,1)} &= -\frac{1}{4}\left.\frac{\partial(F,G)}{\partial(z,v)}\right|_{P_0} \\ &= -\frac{1}{4}\left.\begin{vmatrix} u & \sin v \\ -2yz & x^2 \end{vmatrix}\right|_{P_0} \\ &= -\frac{1}{4}\begin{vmatrix} 1 & 0 \\ 0 & 4 \end{vmatrix} = -1. \end{aligned}$$

**20.** $F(x,y,z,u,v) = 0$
$G(x,y,z,u,v) = 0$
$H(x,y,z,u,v) = 0$
To calculate $\dfrac{\partial x}{\partial y}$ we require that $x$ be one of three dependent variables, and $y$ be one of two independent variables. The other independent variable can be $z$ or $u$ or $v$. The possible interpretations for this partial, and their values, are

$$\left(\frac{\partial x}{\partial y}\right)_z = -\frac{\dfrac{\partial(F,G,H)}{\partial(y,u,v)}}{\dfrac{\partial(F,G,H)}{\partial(x,u,v)}}$$

$$\left(\frac{\partial x}{\partial y}\right)_u = -\frac{\dfrac{\partial(F,G,H)}{\partial(y,z,v)}}{\dfrac{\partial(F,G,H)}{\partial(x,z,v)}}$$

$$\left(\frac{\partial x}{\partial y}\right)_v = -\frac{\dfrac{\partial(F,G,H)}{\partial(y,z,u)}}{\dfrac{\partial(F,G,H)}{\partial(x,z,u)}}.$$

**22.** If $F(x,y,z) = 0 \quad\Rightarrow\quad z = z(x,y)$, then

$$F_1 + F_3\frac{\partial z}{\partial x} = 0, \qquad F_2 + F_3\frac{\partial z}{\partial y} = 0$$

$$F_{11} + F_{13}\frac{\partial z}{\partial x} + F_{31}\frac{\partial z}{\partial x} + F_{33}\left(\frac{\partial z}{\partial x}\right)^2 + F_3\frac{\partial^2 z}{\partial x^2} = 0.$$

Thus

$$\begin{aligned} \frac{\partial^2 z}{\partial x^2} &= -\frac{1}{F_3}\left[F_{11} + 2F_{13}\left(-\frac{F_1}{F_3}\right) + F_{33}\left(-\frac{F_1}{F_3}\right)^2\right] \\ &= -\frac{1}{F_3^3}\big[F_{11}F_3^2 - 2F_1F_3F_{13} + F_1^2F_{33}\big]. \end{aligned}$$

Similarly,

$$\frac{\partial^2 z}{\partial y^2} = -\frac{1}{F_3^3}\big[F_{22}F_3^2 - 2F_2F_3F_{23} + F_2^2F_{33}\big].$$

Also,

$$F_{12} + F_{13}\frac{\partial z}{\partial y} + \left(F_{32} + F_{33}\frac{\partial z}{\partial y}\right)\frac{\partial z}{\partial x} + F_3\frac{\partial^2 z}{\partial y\partial x}.$$

Therefore

$$\frac{\partial^2 z}{\partial x \partial y} = -\frac{1}{F_3}\left[F_{12} + F_{13}\left(-\frac{F_2}{F_3}\right) + F_{23}\left(-\frac{F_1}{F_3}\right) + F_{33}\left(\frac{F_1F_2}{F_3^2}\right)\right]$$

$$= -\frac{1}{F_3^2}\left[F_3^2F_{12} - F_2F_3F_{13} - F_1F_3F_{23} + F_1F_2F_{33}\right].$$

**24.** $pV = T - \dfrac{4p}{T^2}, \qquad T = T(p, V)$

a) $V = \dfrac{\partial T}{\partial p} - \dfrac{4}{T^2} + \dfrac{8p}{T^3}\dfrac{\partial T}{\partial p}$

$p = \dfrac{\partial T}{\partial V} + \dfrac{8p}{T^3}\dfrac{\partial T}{\partial V}.$

Putting $p = V = 1$ and $T = 2$, we obtain

$$2\frac{\partial T}{\partial p} = 2, \qquad 2\frac{\partial T}{\partial V} = 1,$$

so $\dfrac{\partial T}{\partial p} = 1$ and $\dfrac{\partial T}{\partial V} = \dfrac{1}{2}$.

b) $dT = \dfrac{\partial T}{\partial p}dp + \dfrac{\partial T}{\partial V}dV.$

If $p = 1$, $|dp| \le 0.001$, $V = 1$, and $|dV| \le 0.002$, then $T = 2$ and

$$|dT| \le (1)(0.001) + \frac{1}{2}(0.002) = 0.002.$$

The approximate maximum error in $T$ is 0.002.

**26.** Given $F(x, y, u, v) = 0, \qquad G(x, y, u, v) = 0$, let

$$\Delta = \frac{\partial(F, G)}{\partial(x, y)} = \frac{\partial F}{\partial x}\frac{\partial G}{\partial y} - \frac{\partial F}{\partial y}\frac{\partial G}{\partial x}.$$

Then, regarding the given equations as defining $x$ and $y$ as functions of $u$ and $v$, we have

$$\frac{\partial x}{\partial u} = -\frac{1}{\Delta}\frac{\partial(F, G)}{\partial(u, y)} \qquad \frac{\partial y}{\partial u} = -\frac{1}{\Delta}\frac{\partial(F, G)}{\partial(x, u)}$$

$$\frac{\partial x}{\partial v} = -\frac{1}{\Delta}\frac{\partial(F, G)}{\partial(v, y)} \qquad \frac{\partial y}{\partial v} = -\frac{1}{\Delta}\frac{\partial(F, G)}{\partial(x, v)}.$$

Therefore,

$$\frac{\partial(x, y)}{\partial(u, v)} = \frac{1}{\Delta^2}\left(\frac{\partial(F, G)}{\partial(u, y)}\frac{\partial(F, G)}{\partial(x, v)} - \frac{\partial(F, G)}{\partial(v, y)}\frac{\partial(F, G)}{\partial(x, u)}\right)$$

$$= \frac{1}{\Delta^2}\left[\left(\frac{\partial F}{\partial u}\frac{\partial G}{\partial y} - \frac{\partial F}{\partial y}\frac{\partial G}{\partial u}\right)\left(\frac{\partial F}{\partial x}\frac{\partial G}{\partial v} - \frac{\partial F}{\partial v}\frac{\partial G}{\partial x}\right) - \left(\frac{\partial F}{\partial v}\frac{\partial G}{\partial y} - \frac{\partial F}{\partial y}\frac{\partial G}{\partial v}\right)\left(\frac{\partial F}{\partial x}\frac{\partial G}{\partial u} - \frac{\partial F}{\partial u}\frac{\partial G}{\partial x}\right)\right]$$

$$= \frac{1}{\Delta^2}\left[\frac{\partial F}{\partial u}\frac{\partial G}{\partial y}\frac{\partial F}{\partial x}\frac{\partial G}{\partial v} - \frac{\partial F}{\partial y}\frac{\partial G}{\partial u}\frac{\partial F}{\partial x}\frac{\partial G}{\partial v} - \frac{\partial F}{\partial u}\frac{\partial G}{\partial y}\frac{\partial F}{\partial v}\frac{\partial G}{\partial x} + \frac{\partial F}{\partial y}\frac{\partial G}{\partial u}\frac{\partial F}{\partial v}\frac{\partial G}{\partial x} - \frac{\partial F}{\partial v}\frac{\partial G}{\partial y}\frac{\partial F}{\partial x}\frac{\partial G}{\partial u} + \frac{\partial F}{\partial v}\frac{\partial G}{\partial y}\frac{\partial F}{\partial u}\frac{\partial G}{\partial x} + \frac{\partial F}{\partial y}\frac{\partial G}{\partial v}\frac{\partial F}{\partial x}\frac{\partial G}{\partial u} - \frac{\partial F}{\partial y}\frac{\partial G}{\partial v}\frac{\partial F}{\partial u}\frac{\partial G}{\partial x}\right]$$

$$= \frac{1}{\Delta^2}\left[\frac{\partial F}{\partial u}\frac{\partial G}{\partial y}\frac{\partial F}{\partial x}\frac{\partial G}{\partial v} + \frac{\partial F}{\partial y}\frac{\partial G}{\partial u}\frac{\partial F}{\partial v}\frac{\partial G}{\partial x} - \frac{\partial F}{\partial v}\frac{\partial G}{\partial y}\frac{\partial F}{\partial x}\frac{\partial G}{\partial u} - \frac{\partial F}{\partial y}\frac{\partial G}{\partial v}\frac{\partial F}{\partial u}\frac{\partial G}{\partial x}\right]$$

$$= \frac{1}{\Delta^2}\left(\frac{\partial F}{\partial x}\frac{\partial G}{\partial y} - \frac{\partial F}{\partial y}\frac{\partial G}{\partial x}\right)\left(\frac{\partial F}{\partial u}\frac{\partial G}{\partial v} - \frac{\partial F}{\partial v}\frac{\partial G}{\partial u}\right)$$

$$= \frac{1}{\Delta^2}\frac{\partial(F, G)}{\partial(x, y)}\frac{\partial(F, G)}{\partial(u, v)}$$

$$= \frac{1}{\Delta}\frac{\partial(F, G)}{\partial(u, v)} = \frac{\partial(F, G)}{\partial(u, v)} \Big/ \frac{\partial(F, G)}{\partial(x, y)}.$$

**28.** By the Chain Rule,

$$\begin{pmatrix} \frac{\partial x}{\partial r} & \frac{\partial x}{\partial s} \\ \frac{\partial y}{\partial r} & \frac{\partial y}{\partial s} \end{pmatrix} = \begin{pmatrix} \frac{\partial x}{\partial u}\frac{\partial u}{\partial r} + \frac{\partial x}{\partial v}\frac{\partial v}{\partial r} & \frac{\partial x}{\partial u}\frac{\partial u}{\partial s} + \frac{\partial x}{\partial v}\frac{\partial v}{\partial s} \\ \frac{\partial y}{\partial u}\frac{\partial u}{\partial r} + \frac{\partial y}{\partial v}\frac{\partial v}{\partial r} & \frac{\partial y}{\partial u}\frac{\partial u}{\partial s} + \frac{\partial y}{\partial v}\frac{\partial v}{\partial s} \end{pmatrix} = \begin{pmatrix} \frac{\partial x}{\partial u} & \frac{\partial x}{\partial v} \\ \frac{\partial y}{\partial u} & \frac{\partial y}{\partial v} \end{pmatrix}\begin{pmatrix} \frac{\partial u}{\partial r} & \frac{\partial u}{\partial s} \\ \frac{\partial v}{\partial r} & \frac{\partial v}{\partial s} \end{pmatrix}.$$

Since the determinant of a product of matrices is the product of their determinants, we have

$$\frac{\partial(x, y)}{\partial(r, s)} = \frac{\partial(x, y)}{\partial(u, v)}\frac{\partial(u, v)}{\partial(r, s)}.$$

**30.** Let $u = f(x, y)$ and $v = g(x, y)$, and suppose that

$$\frac{\partial(u, v)}{\partial(x, y)} = \frac{\partial(f, g)}{\partial(x, y)} = 0$$

for all $(x, y)$. Thus

$$\frac{\partial f}{\partial x}\frac{\partial g}{\partial y} - \frac{\partial f}{\partial y}\frac{\partial g}{\partial x} = 0.$$

Now consider the equations $u = f(x, y)$ and $v = g(x, y)$ as defining $u$ and $y$ as functions of $x$ and $v$. Holding $v$ constant and differentiating with respect to $x$, we get

$$\frac{\partial g}{\partial x} + \frac{\partial g}{\partial y}\frac{\partial y}{\partial x} = 0,$$

and

$$\left(\frac{\partial u}{\partial x}\right)_v = \frac{\partial f}{\partial x} + \frac{\partial f}{\partial y}\frac{\partial y}{\partial x}$$
$$= \frac{1}{\frac{\partial g}{\partial y}}\left(\frac{\partial f}{\partial x}\frac{\partial g}{\partial y} - \frac{\partial f}{\partial y}\frac{\partial g}{\partial x}\right) = 0.$$

This says that $u = u(x, v)$ is independent of $x$, and so depends only on $v$: $u = k(v)$ for some function $k$ of one variable. Thus $f(x, y) = k\big(g(x, y)\big)$, so $f$ and $g$ are functionally dependent.

## Section 4.9 Taylor Series and Approximations (page 217)

**2.** Since
$$\begin{aligned} f(x, y) &= \ln(1 + x + y + xy) \\ &= \ln\big((1 + x)(1 + y)\big) \\ &= \ln(1 + x) + \ln(1 + y), \end{aligned}$$
the Taylor series for $f$ about $(0, 0)$ is

$$\sum_{n=1}^{\infty}(-1)^{n-1}\frac{x^n + y^n}{n}.$$

**4.** Let $u = x - 1$, $v = y + 1$. Thus

$$\begin{aligned} f(x, y) &= x^2 + xy + y^3 \\ &= (u + 1)^2 + (u + 1)(v - 1) + (v - 1)^3 \\ &= 1 + 2u + u^2 - 1 + v - u + uv + v^3 - 3v^2 + 3v - 1 \\ &= -1 + u + 4v + u^2 + uv - 3v^2 + v^3 \\ &= -1 + (x - 1) + 4(y + 1) + (x - 1)^2 \\ &\quad + (x - 1)(y + 1) - 3(y + 1)^2 + (y + 1)^3. \end{aligned}$$

This is the Taylor series for $f$ about $(1, -1)$.

**6.**
$$\begin{aligned} f(x, y) = \sin(2x + 3y) &= \sum_{n=0}^{\infty}(-1)^n\frac{(2x + 3y)^{2n+1}}{(2n + 1)!} \\ &= \sum_{n=0}^{\infty}\frac{(-1)^n}{(2n + 1)!}\sum_{j=0}^{2n+1}\frac{(2n + 1)!}{j!(2n + 1 - j)!}(2x)^j(3y)^{2n+1-j} \\ &= \sum_{n=0}^{\infty}\sum_{j=0}^{2n+1}\frac{(-1)^n 2^j 3^{2n+1-j}}{j!(2n + 1 - j)!}x^j y^{2n+1-j}. \end{aligned}$$
This is the Taylor series for $f$ about $(0, 0)$.

**8.** Let $u = x - 1$. Then

$$\begin{aligned} f(x, y) &= \ln(x^2 + y^2) = \ln(1 + 2u + u^2 + y^2) \\ &= (2u + u^2 + y^2) - \frac{(2u + u^2 + y^2)^2}{2} \\ &\quad + \frac{(2u + u^2 + y^2)^3}{3} - \cdots \\ &= 2u + u^2 + y^2 - 2u^2 - 2u^3 - 2uy^2 + \frac{8u^3}{3} + \cdots. \end{aligned}$$

The Taylor polynomial of degree 3 for $f$ near $(1, 0)$ is

$$\begin{aligned} &2(x - 1) - (x - 1)^2 + y^2 - 2(x - 1)^3 \\ &\quad - 2(x - 1)y^2 + \frac{8}{3}(x - 1)^3. \end{aligned}$$

**10.**
$$\begin{aligned} f(x, y) &= \cos(x + \sin y) \\ &= 1 - \frac{(x + \sin y)^2}{2!} + \frac{(x + \sin y)^4}{4!} - \cdots \\ &= 1 - \frac{\left(x + y - \frac{y^3}{6} + \cdots\right)^2}{2} + \frac{(x + y - \cdots)^4}{4} - \cdots \\ &= 1 - \frac{1}{2}\left(x^2 + y^2 + 2xy - \frac{xy^3}{3} - \frac{y^4}{3} + \cdots\right) \\ &\quad + \frac{1}{4}(x^4 + 4x^3y + 6x^2y^2 + 4xy^3 + y^4 + \cdots). \end{aligned}$$
The Taylor polynomial of degree 4 for $f$ near $(0, 0)$ is

$$\begin{aligned} &1 - \frac{x^2}{2} - xy - \frac{y^2}{2} + \frac{x^4}{4} + x^3y \\ &\quad + \frac{3x^2y^2}{2} + \frac{7xy^3}{6} + \frac{5y^4}{12}. \end{aligned}$$

**12.**
$$\begin{aligned} f(x, y) &= \frac{1 + x}{1 + x^2 + y^4} \\ &= (1 + x)\big(1 - (x^2 + y^4) + \cdots\big) \\ &= 1 + x - x^2 - \cdots. \end{aligned}$$
The Taylor polynomial of degree 2 for $f$ near $(0, 0)$ is

$$1 + x - x^2.$$

**14.** The equation can be written $F(x, y) = 0$, where $F(x, y) = e^{x+y-1} - 1 - 2y$.
Since $F(1, 0) = 0$ and $F_2(1, 0) = -1 \neq 0$, the equation has a solution of the form $y = f(x)$, with $f(1) = 0$.

To make the calculations a little easier, let $u = x - 1$ and rewrite the equation in the form

$$e^y = (1 + 2y)e^{-u}.$$

Try $y = a_1u + a_2u^2 + a_3u^3 + \cdots$. Then

$$\begin{aligned} &1 + (a_1u + a_2u^2 + a_3u^3 + \cdots) \\ &\quad + \frac{1}{2}(a_1u + a_2u^2 + \cdots)^2 + \frac{1}{6}(a_1u + \cdots)^3 + \cdots \\ &= (1 + 2a_1u + 2a_2u^2 + 2a_3u^3 + \cdots) \\ &\quad \times \left(1 - u + \frac{u^2}{2} - \frac{u^3}{6} + \cdots\right), \end{aligned}$$

that is,

$$\begin{aligned}
&1 + a_1 u + \left(a_2 + \frac{a_1^2}{2}\right)u^2 \\
&\quad + \left(a_3 + a_1 a_2 + \frac{a_1^3}{6}\right)u^3 + \cdots \\
&= 1 + (2a_1 - 1)u + \left(2a_2 - 2a_1 + \frac{1}{2}\right)u^2 \\
&\quad + \left(2a_3 - 2a_2 + a_1 - \frac{1}{6}\right)u^3 + \cdots .
\end{aligned}$$

Therefore

$$\begin{aligned}
a_1 &= 2a_1 - 1 \\
a_2 + \frac{1}{2}a_1^2 &= 2a_2 - 2a_1 + \frac{1}{2} \\
a_3 + a_1 a_2 + \frac{1}{6}a_1^3 &= 2a_3 - 2a_2 + a_1 - \frac{1}{6},
\end{aligned}$$

and $a_1 = 1$, $a_2 = 2$, $a_3 = 16/3$. The required solution is

$$y = (x-1) + 2(x-1)^2 + \frac{16}{3}(x-1)^3 + \cdots .$$

**16.** The equation $\sqrt{1+xy} = 1 + x + \ln(1+y)$ can be rewritten $F(x,y) = 0$, where $F(x,y) = \sqrt{1+xy} - 1 - x - \ln(1+y)$. Since $F(0,0) = 0$ and $F_2(0,0) = -1 \neq 0$, the given equation has a solution of the form $y = f(x)$ where $f(0) = 0$.

Try $y = a_1 x + a_2 x^2 + a_3 x^3 + a_4 x^4 + \cdots$. We have

$$\begin{aligned}
&\sqrt{1+xy} \\
&= \sqrt{1 + a_1 x^2 + a_2 x^3 + a_3 x_4 + \cdots} \\
&= 1 + \frac{1}{2}(a_1 x^2 + a_2 x^3 + a_3 x^4 + \cdots) \\
&\quad - \frac{1}{8}(a_1 x^2 + \cdots)^2 + \cdots \\
&1 + x + \ln(1+y) \\
&= 1 + x + (a_1 x + a_2 x^2 + a_3 x^3 + a_4 x^4 + \cdots) \\
&\quad - \frac{1}{2}(a_1 x + a_2 x^2 + a_3 x^3 + \cdots)^2 + \frac{1}{3}(a_1 x + a_2 x^2 \cdots)^3 - \cdots
\end{aligned}$$

Thus we must have

$$\begin{aligned}
0 &= 1 + a_1 \\
\frac{1}{2}a_1 &= a_2 - \frac{1}{2}a_1^2 \\
\frac{1}{2}a_2 &= a_3 - a_1 a_2 + \frac{1}{3}a_1^3 \\
\frac{1}{2}a_3 - \frac{1}{8}a_1^2 &= a_4 - \frac{1}{2}a_2^2 - a_1 a_3 + a_1^2 a_2,
\end{aligned}$$

and $a_1 = -1$, $a_2 = 0$, $a_3 = \frac{1}{3}$, $a_4 = -\frac{7}{24}$. The required solution is

$$y = -x + \frac{1}{3}x^3 - \frac{7}{24}x^4 + \cdots .$$

**18.** The coefficient of $x^2 y$ in the Taylor series for $f(x,y) = \tan^{-1}(x+y)$ about $(0,0)$ is

$$\frac{1}{2!1!}f_{112}(0,0) = \frac{1}{2}f_{112}(0,0).$$

But

$$\begin{aligned}
\tan^{-1}(x+y) &= x + y - \frac{1}{3}(x+y)^3 + \cdots \\
&= x + y - \frac{1}{3}(x^3 + 3x^2 y + 3xy^2 + y^3) + \cdots
\end{aligned}$$

so the coefficient of $x^2 y$ is $-1$. Hence $f_{112}(0,0) = -2$.

## Review Exercises 4 (page 218)

**2.** $$\begin{aligned}
T &= \frac{140 + 30x^2 - 60x + 120y^2}{8 + x^2 - 2x + 4y^2} \\
&= 30 - \frac{100}{(x-1)^2 + 4y^2 + 7}
\end{aligned}$$

Ellipses: centre $(1,0)$, values of $T$ between $30 - (100/7)$ (minimum) at $(1,0)$ and 30 (at infinite distance from $(1,0)$).

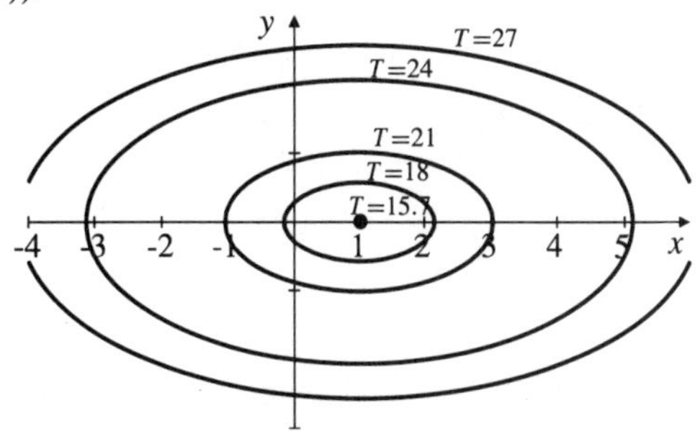


Fig. R-4.2

**4.** $f(x,y) = \begin{cases} x^3/(x^2+y^2) & \text{if } (x,y) \neq (0,0) \\ 0 & \text{if } (x,y) = (0,0) \end{cases}.$

$$\begin{aligned}
f_1(0,0) &= \lim_{h\to 0} \frac{(h^3 - 0)/h^2}{h} = 1 \\
f_2(0,0) &= \lim_{k\to 0} \frac{0-0}{k} = 0.
\end{aligned}$$

For $(x,y) \neq (0,0)$, we have

$$\begin{aligned}
f_1(x,y) &= \frac{x^4 + 3x^2y^2}{(x^2+y^2)^2} \\
f_2(x,y) &= -\frac{2x^3 y}{(x^2+y^2)^2} \\
f_{12}(0,0) &= \lim_{k\to 0} \frac{f_1(0,k) - f_1(0,0)}{k} = \lim_{k\to 0} \frac{0-1}{k} \text{ does not exist} \\
f_{21}(0,0) &= \lim_{h\to 0} \frac{f_2(h,0) - f_2(0,0)}{h} = \lim_{h\to 0} \frac{0-0}{h} = 0.
\end{aligned}$$

**6.** $f(x,y) = e^{x^2-2x-4y^2+5}$ $\qquad f(1,-1) = 1$

$f_1(x,y) = 2(x-1)e^{x^2-2x-4y^2+5}$ $\qquad f_1(1,-1) = 0$

$f_2(x,y) = -8ye^{x^2-2x-4y^2+5}$ $\qquad f_2(1,-1) = 8.$

a) The tangent plane to $z = f(x,y)$ at $(1,-1,1)$ has equation $z = 1 + 8(y+1)$, or $z = 8y + 9$.

b) $f(x,y) = C \Rightarrow (x-1)^2 - 4y^2 + 4 = \ln C$

$\Rightarrow (x-1)^2 - 4y^2 = \ln C - 4.$

These are hyperbolas with centre $(1,0)$ and asymptotes $x = 1 \pm 2y$.

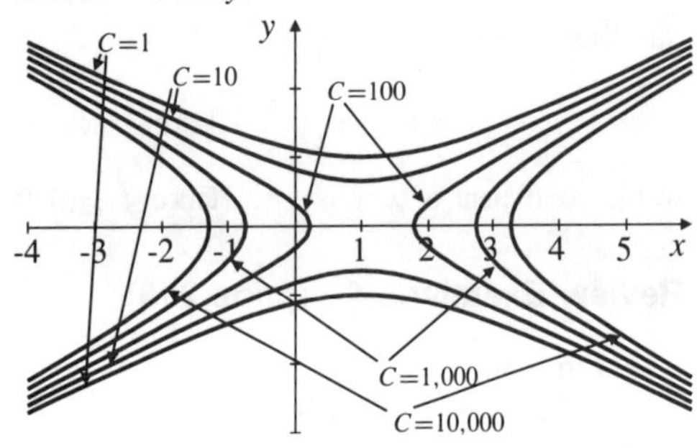


Fig. R-4.6

**8.** $\dfrac{1}{R} = \dfrac{1}{R_1} + \dfrac{1}{R_2}$

$-\dfrac{1}{R^2}\,dR = -\dfrac{1}{R_1^2}\,dR_1 - \dfrac{1}{R_2^2}\,dR_2$

If $R_1 = 100$ and $R_2 = 25$, so that $R = 20$, and if $|dR_1/R_1| = 5/100$ and $|dR_2/R_2| = 2/100$, then

$$\frac{1}{20}\left|\frac{dR}{R}\right| \le \frac{1}{100}\cdot\frac{5}{100} + \frac{1}{25}\cdot\frac{2}{100} = \frac{13}{100^2}.$$

Thus $|dR/R| \le 13/500$; $R$ can be in error by about 2.6%.

**10.** $T = x^3y + y^3z + z^3x.$

a) $\nabla T = (3x^2y + z^3)\mathbf{i} + (3y^2z + x^3)\mathbf{j} + (3z^2x + y^3)\mathbf{k}$

$\nabla T(2,-1,0) = -12\mathbf{i} + 8\mathbf{j} - \mathbf{k}.$

A unit vector in the direction from $(2,-1,0)$ towards $(1,1,2)$ is $\mathbf{u} = (-\mathbf{i} + 2\mathbf{j} + 2\mathbf{k})/3$. The directional derivative of $T$ at $(2,-1,0)$ in the direction of $\mathbf{u}$ is

$$\mathbf{u} \bullet \nabla T(2,-1,0) = \frac{12 + 16 - 2}{3} = \frac{26}{3}.$$

b) Since $\nabla(2x^2 + 3y^2 + z^2) = 4x\mathbf{i} + 6y\mathbf{j} + 2z\mathbf{k}$, at $t = 0$ the fly is at $(2,-1,0)$ and is moving in the direction $\pm(8\mathbf{i} - 6\mathbf{j})$, so its velocity is

$$\pm 5\frac{8\mathbf{i} - 6\mathbf{j}}{10} = \pm(4\mathbf{i} - 3\mathbf{j}).$$

Since the fly is moving in the direction of increasing $T$, the rate at which it experiences $T$ increasing is

$$\frac{dT}{dt} = |(4\mathbf{i} - 3\mathbf{j}) \bullet (-12\mathbf{i} + 8\mathbf{j} - \mathbf{k})| = 48 + 24 = 72.$$

**12.** $f(x,y,z) = (x^2 + z^2)\sin\dfrac{\pi xy}{2} + yz^2$, $P_0 = (1,1,-1)$.

a)
$$\nabla f = \left(2x\sin\frac{\pi xy}{2} + \frac{\pi y}{2}(x^2+z^2)\cos\frac{\pi xy}{2}\right)\mathbf{i} + \left(\frac{\pi x}{2}(x^2+z^2)\cos\frac{\pi xy}{2} + z^2\right)\mathbf{j} + 2z\left(\sin\frac{\pi xy}{2} + y\right)\mathbf{k}$$

$\nabla f(P_0) = 2\mathbf{i} + \mathbf{j} - 4\mathbf{k}.$

b) Since $f(P_0) = 2 + 1 = 3$, the linearization of $f$ at $P_0$ is

$$L(x,y,z) = 3 + 2(x-1) + (y-1) - 4(z+1).$$

c) The tangent plane at $P_0$ to the level surface of $f$ through $P_0$ has equation

$$\nabla f(P_0) \bullet \big((x-1)\mathbf{i} + (y-1)\mathbf{j} + (z+1)\mathbf{k}\big) = 0$$
$$2(x-1) + (y-1) - 4(z+1) = 0$$
$$2x + y - 4z = 7.$$

d) The bird is flying in direction

$$(2-1)\mathbf{i} + (-1-1)\mathbf{j} + (1+1)\mathbf{k} = \mathbf{i} - 2\mathbf{j} + 2\mathbf{k},$$

a vector of length 3. Since the bird's speed is 5, its velocity is

$$\mathbf{v} = \frac{5}{3}(\mathbf{i} - 2\mathbf{j} + 2\mathbf{k}).$$

The rate of change of $f$ as experienced by the bird is

$$\frac{df}{dt} = \mathbf{v} \bullet \nabla f(P_0) = \frac{5}{3}(2 - 2 - 8) = -\frac{40}{3}.$$

e) To experience the greatest rate of increase of $f$ while flying through $P_0$ at speed 5, the bird should fly in the direction of $\nabla f(P_0)$, that is, $2\mathbf{i} + \mathbf{j} - 4\mathbf{k}$.

**14.** If $F(x,y,z) = 0$, $G(x,y,z) = 0$ are solved for $x = x(y)$, $z = z(y)$, then

$$F_1\frac{dx}{dy} + F_2 + F_3\frac{dz}{dy} = 0$$
$$G_1\frac{dx}{dy} + G_2 + G_3\frac{dz}{dy} = 0.$$

Eliminating $dz/dy$ from these equations, we obtain

$$\frac{dx}{dy} = -\frac{F_2G_3 - F_3G_2}{F_1G_3 - F_3G_1}.$$

Similarly, if the equations are solved for $x = x(z)$, $y = y(z)$, then

$$\frac{dy}{dz} = -\frac{F_3G_1 - F_1G_3}{F_2G_1 - F_1G_2},$$

and if the equations are solved for $y = y(x)$, $z = z(x)$, then

$$\frac{dz}{dx} = -\frac{F_1G_2 - F_2G_1}{F_3G_2 - F_2G_3}.$$

Hence

$$\frac{dx}{dy} \cdot \frac{dy}{dz} \cdot \frac{dz}{dx} = -\frac{F_2G_3 - F_3G_2}{F_1G_3 - F_3G_1} \cdot \frac{F_3G_1 - F_1G_3}{F_2G_1 - F_1G_2} \cdot \frac{F_1G_2 - F_2G_1}{F_3G_2 - F_2G_3} = 1.$$

**16.** $u = x^2 + y^2$
$v = x^2 - 2xy^2$
Assume these equations define $x = x(u, v)$ and $y = y(u, v)$ near the point $(u, v) = (5, -7)$, with $x = 1$ and $y = 2$ at that point.

a) Differentiate the given equations with respect to $u$ to obtain

$$1 = 2x\frac{\partial x}{\partial u} + 2y\frac{\partial y}{\partial u}$$
$$0 = 2(x - y^2)\frac{\partial x}{\partial u} - 4xy\frac{\partial y}{\partial u}.$$

At $x = 1$, $y = 2$,

$$2\frac{\partial x}{\partial u} + 4\frac{\partial y}{\partial u} = 1$$
$$-6\frac{\partial x}{\partial u} - 8\frac{\partial y}{\partial u} = 0,$$

from which we obtain $\partial x/\partial u = -1$ and $\partial y/\partial u = 3/4$ at $(5, -7)$.

b) If $z = \ln(y^2 - x^2)$, then

$$\frac{\partial z}{\partial u} = \frac{1}{y^2 - x^2}\left[-2x\frac{\partial x}{\partial u} + 2y\frac{\partial y}{\partial u}\right].$$

At $(u, v) = (5, -7)$, we have $(x, y) = (1, 2)$, and so

$$\frac{\partial z}{\partial u} = \frac{1}{3}\left[-2(-1) + 4\left(\frac{3}{4}\right)\right] = \frac{5}{3}.$$

## Challenging Problems 4 (page 219)

**2.** Let the position vector of the particle at time $t$ be $\mathbf{r} = x(t)\mathbf{i} + y(t)\mathbf{j} + z(t)\mathbf{k}$. Then the velocity of the particle is

$$\mathbf{v} = \frac{dx}{dt}\mathbf{i} + \frac{dy}{dt}\mathbf{j} + \frac{dz}{dt}\mathbf{k}.$$

This velocity must be parallel to

$$\nabla f(x, y, z) = -2x\mathbf{i} - 4y\mathbf{j} + 6z\mathbf{k}$$

at every point of the path, that is,

$$\frac{dx}{dt} = -2tx, \quad \frac{dy}{dt} = -4ty, \quad \frac{dz}{dt} = 6tz,$$

so that $\dfrac{dx}{-2x} = \dfrac{dy}{-4y} = \dfrac{dz}{6z}$. Integrating these equations, we get

$$\ln|y| = 2\ln|x| + C_1, \quad \ln|z| = -3\ln|x| + C_2.$$

Since the path passes through $(1, 1, 8)$, $C_1$ and $C_2$ are determined by

$$\ln 1 = 2\ln 1 + C_1, \quad \ln 8 = -3\ln 1 + C_2.$$

Thus $C_1 = 0$ and $C_2 = \ln 8$. The path therefore has equations $y = x^2$, $z = 8/x^3$. Evidently $(2, 4, 1)$ lies on the path, and $(3, 7, 0)$ does not.

**4.** If $u(x, y, z, t) = v(\rho, t) = \dfrac{f(\rho - ct)}{\rho}$ is independent of $\theta$ and $\phi$, then

$$\frac{\partial^2 u}{\partial x^2} + \frac{\partial^2 u}{\partial y^2} + \frac{\partial^2 u}{\partial z^2} = \frac{\partial^2 v}{\partial \rho^2} + \frac{2}{\rho}\frac{\partial v}{\partial \rho}$$

by Problem 3. We have

$$\frac{\partial v}{\partial \rho} = \frac{f'(\rho - ct)}{\rho} - \frac{f(\rho - ct)}{\rho^2}$$
$$\frac{\partial^2 v}{\partial \rho^2} = \frac{f''(\rho - ct)}{\rho} - \frac{2f'(\rho - ct)}{\rho^2} + \frac{2f(\rho - ct)}{\rho^3}$$
$$\frac{\partial v}{\partial t} = -\frac{cf'(\rho - ct)}{\rho}$$
$$\frac{\partial^2 v}{\partial t^2} = \frac{c^2 f''(\rho - ct)}{\rho}$$
$$\frac{\partial^2 v}{\partial \rho^2} + \frac{2}{\rho}\frac{\partial v}{\partial \rho} = \frac{f''(\rho - ct)}{\rho} - \frac{2f'(\rho - ct)}{\rho^2} + \frac{2f(\rho - ct)}{\rho^3} + \frac{2f'(\rho - ct)}{\rho^2} - \frac{2f(\rho - ct)}{\rho^3}$$
$$= \frac{f''(\rho - ct)}{\rho}$$
$$= \frac{1}{c^2}\frac{\partial^2 v}{\partial t^2} = \frac{1}{c^2}\frac{\partial^2 u}{\partial t^2}.$$

The function $f(\rho - ct)/\rho$ represents the shape of a symmetrical wave travelling uniformly away from the origin at speed $c$. Its amplitude at distance $\rho$ from the origin decreases as $\rho$ increases; it is proportional to the reciprocal of $\rho$.

# CHAPTER 5. APPLICATIONS OF PARTIAL DERIVATIVES

## Section 5.1 Extreme Values (page 228)

**2.** $f(x, y) = xy - x + y$, $f_1 = y - 1$, $f_2 = x + 1$
$A = f_{11} = 0$, $B = f_{12} = 1$, $C = f_{22} = 0$.
Critical point $(-1, 1)$ is a saddle point since $B^2 - AC > 0$.

**4.** $f(x, y) = x^4 + y^4 - 4xy$, $f_1 = 4(x^3 - y)$, $f_2 = 4(y^3 - x)$
$A = f_{11} = 12x^2$, $B = f_{12} = -4$, $C = f_{22} = 12y^2$.
For critical points: $x^3 = y$ and $y^3 = x$. Thus $x^9 = x$, or $x(x^8 - 1) = 0$, and $x = 0$, 1, or $-1$. The critical points are $(0, 0)$, $(1, 1)$ and $(-1, -1)$.
At $(0, 0)$, $B^2 - AC = 16 - 0 > 0$, so $(0, 0)$ is a saddle point.
At $(1, 1)$ and $(-1, -1)$, $B^2 - AC = 16 - 144 < 0$, $A > 0$, so $f$ has local minima at these points.

**6.** $f(x, y) = \cos(x + y)$, $f_1 = -\sin(x + y) = f_2$.
All points on the lines $x + y = n\pi$ ($n$ is an integer) are critical points. If $n$ is even, $f = 1$ at such points; if $n$ is odd, $f = -1$ there. Since $-1 \le f(x, y) \le 1$ at all points in $\mathbb{R}^2$, $f$ must have local and absolute maximum values at points $x + y = n\pi$ with $n$ even, and local and absolute minimum values at such points with $n$ odd.

**8.** $f(x, y) = \cos x + \cos y$, $f_1 = -\sin x$, $f_2 = -\sin y$
$A = f_{11} = -\cos x$, $B = f_{12} = 0$, $C = f_{22} = -\cos y$.
The critical points are points $(m\pi, n\pi)$, where $m$ and $n$ are integers.
Here $B^2 - AC = -\cos(m\pi)\cos(n\pi) = (-1)^{m+n+1}$ which is negative if $m + n$ is even, and positive if $m + n$ is odd.
If $m + n$ is odd then $f$ has a saddle point at $(m\pi, n\pi)$.
If $m + n$ is even and $m$ is odd then $f$ has a local (and absolute) minimum value, $-2$, at $(m\pi, n\pi)$. If $m + n$ is even and $m$ is even then $f$ has a local (and absolute) maximum value, 2, at $(m\pi, n\pi)$.

**10.**
$$f(x, y) = \frac{xy}{2 + x^4 + y^4}$$
$$f_1 = \frac{(2 + x^4 + y^4)y - xy4x^3}{(2 + x^4 + y^4)^2} = \frac{y(2 + y^4 - 3x^4)}{(2 + x^4 + y^4)^2}$$
$$f_2 = \frac{x(2 + x^4 - 3y^4)}{(2 + x^4 + y^4)^2}.$$
For critical points, $y(2 + y^4 - 3x^4) = 0$ and $x(2 + x^4 - 3y^4) = 0$.
One critical point is $(0, 0)$. Since $f(0, 0) = 0$ but $f(x, y) > 0$ in the first quadrant and $f(x, y) < 0$ in the second quadrant, $(0, 0)$ must be a saddle point of $f$.

Any other critical points must satisfy $2 + y^4 - 3x^4 = 0$ and $2 + x^4 - 3y^4 = 0$, that is, $y^4 = x^4$, or $y = \pm x$. Thus $2 - 2x^4 = 0$ and $x = \pm 1$. Therefore there are four other critical points: $(1, 1)$, $(-1, -1)$, $(1, -1)$ and $(-1, 1)$. $f$ is positive at the first two of these, and negative at the other two. Since $f(x, y) \to 0$ as $x^2 + y^2 \to \infty$, $f$ must have maximum values at $(1, 1)$ and $(-1, -1)$, and minimum values at $(1, -1)$ and $(-1, 1)$.

**12.**
$$f(x, y) = \frac{1}{1 - x + y + x^2 + y^2} = \frac{1}{\left(x - \frac{1}{2}\right)^2 + \left(y + \frac{1}{2}\right)^2 + \frac{1}{2}}.$$
Evidently $f$ has absolute maximum value 2 at $\left(\frac{1}{2}, -\frac{1}{2}\right)$. Since
$$f_1(x, y) = \frac{1 - 2x}{(1 - x + y + x^2 + y^2)^2}$$
$$f_2(x, y) = -\frac{1 + 2y}{(1 - x + y + x^2 + y^2)^2},$$
$\left(\frac{1}{2}, -\frac{1}{2}\right)$ is the only critical point of $f$.

**14.** $f(x, y, z) = xyz - x^2 - y^2 - z^2$. For critical points we have
$$0 = f_1 = yz - 2x, \quad 0 = f_2 = xz - 2y, \quad 0 = f_3 = xy - 2z.$$
Thus $xyz = 2x^2 = 2y^2 = 2z^2$, so $x^2 = y^2 = z^2$. Hence $x^3 = \pm 2x^2$, and $x = \pm 2$ or 0. Similarly for $y$ and $z$. The only critical points are $(0, 0, 0)$, $(2, 2, 2)$, $(-2, -2, 2)$, $(-2, 2, -2)$, and $(2, -2, -2)$.

Let $\mathbf{u} = u\mathbf{i} + v\mathbf{j} + w\mathbf{k}$, where $u^2 + v^2 + w^2 = 1$. Then
$$D_{\mathbf{u}} f(x, y, z) = (yz - 2x)u + (xz - 2y)v + (xy - 2z)w$$
$$D_{\mathbf{u}}\big(D_{\mathbf{u}} f(x, y, z)\big) = (-2u + zv + yw)u + (zu - 2v + xw)v + (yu + xv - 2w)w.$$
At $(0, 0, 0)$, $D_{\mathbf{u}}\big(D_{\mathbf{u}} f(0, 0, 0)\big) = -2u^2 - 2v^2 - 2w^2 < 0$ for $\mathbf{u} \ne \mathbf{0}$, so $f$ has a local maximum value at $(0, 0, 0)$.

At $(2, 2, 2)$, we have
$$\begin{aligned} D_{\mathbf{u}}\big(D_{\mathbf{u}} f(2, 2, 2)\big) &= (-2u + 2v + 2w)u + (2u - 2v + 2w)v + (2u + 2v - 2w)w \\ &= -2(u^2 + v^2 + w^2) + 4(uv + vw + wu) \\ &= -2[(u - v - w)^2 - 4vw] \\ &\begin{cases} < 0 & \text{if } v = w = 0,\ u \ne 0 \\ > 0 & \text{if } v = w \ne 0,\ u - v - w = 0. \end{cases} \end{aligned}$$
Thus $(2, 2, 2)$ is a saddle point.

At $(2,-2,-2)$, we have

$$\begin{aligned}D_{\mathbf{u}}\big(D_{\mathbf{u}}f\big) &= -2(u^2+v^2+w^2+2uv+2uw-2vw)\\ &= -2[(u+v+w)^2-4vw]\\ &\begin{cases}<0 & \text{if } v=w=0,\ u\neq 0\\ >0 & \text{if } v=w\neq 0,\ u+v+w=0.\end{cases}\end{aligned}$$

Thus $(2,-2,-2)$ is a saddle point. By symmetry, so are the remaining two critical points.

**16.** $f(x,y,z)=4xyz-x^4-y^4-z^4$

$$\begin{aligned}D &= f(1+h,1+k,1+m)-f(1,1,1)\\ &= 4(1+h)(1+k)(1+m)-(1+h)^4-(1+k)^4\\ &\quad -(1+m)^4-1\\ &= 4(1+h+k+m+hk+hm+km+hkm)\\ &\quad -(1+4h+6h^2+4h^3+h^4)\\ &\quad -(1+4k+6k^2+4k^3+k^4)\\ &\quad -(1+4m+6m^2+4m^3+m^4)-1\\ &= 4(hk+hm+km)-6(h^2+k^2+m^2)+\cdots,\end{aligned}$$

where $\cdots$ stands for terms of degree 3 and 4 in the variables $h$, $k$, and $m$. Completing some squares among the quadratic terms we obtain

$$D=-2\big[(h-k)^2+(k-m)^2+(h-m)^2+h^2+k^2+m^2\big]+\cdots$$

which is negative if $|h|$, $|k|$ and $|m|$ are small and not all 0. (This is because the terms of degree 3 and 4 are smaller in size than the quadratic terms for small values of the variables.)
Hence $f$ has a local maximum value at $(1,1,1)$.

**18.**
$$\begin{aligned}f(x,y)&=\frac{x}{1+x^2+y^2}\\ f_1(x,y)&=\frac{1+y^2-x^2}{(1+x^2+y^2)^2}\\ f_2(x,y)&=\frac{-2xy}{(1+x^2+y^2)^2}.\end{aligned}$$

For critical points, $x^2-y^2=1$, and $xy=0$. The critical points are $(\pm1,0)$. $f(\pm1,0)=\pm\frac{1}{2}$.
Since $f(x,y)\to 0$ as $x^2+y^2\to\infty$, the maximum and minimum values of $f$ are $1/2$ and $-1/2$ respectively.

**20.**
$$\begin{aligned}f(x,y)&=x+8y+\frac{1}{xy}, \qquad (x>0,\quad y>0)\\ f_1(x,y)&=1-\frac{1}{x^2y}=0 \quad\Rightarrow\quad x^2y=1\\ f_2(x,y)&=8-\frac{1}{xy^2}=0 \quad\Rightarrow\quad 8xy^2=1.\end{aligned}$$

The critical points must satisfy

$$\frac{x}{y}=\frac{x^2y}{xy^2}=8,$$

that is, $x=8y$. Also, $x^2y=1$, so $64y^3=1$.
Thus $y=1/4$, and $x=2$; the critical point is $\left(2,\frac{1}{4}\right)$.
Since $f(x,y)\to\infty$ if $x\to 0+$, $y\to 0+$, or $x^2+y^2\to\infty$, the critical point must give a minimum value for $f$. The minimum value is $f\left(2,\frac{1}{4}\right)=2+2+2=6$.

**22.** Let the length, width, and height of the box be $x$, $y$, and $z$, respectively. Then $V=xyz$. If the top and side walls cost $\$k$ per unit area, then the total cost of materials for the box is

$$\begin{aligned}C&=2kxy+kxy+2kxz+2kyz\\ &=k\left[3xy+2(x+y)\frac{V}{xy}\right]=k\left[3xy+\frac{2V}{x}+\frac{2V}{y}\right],\end{aligned}$$

where $x>0$ and $y>0$. Since $C\to\infty$ as $x\to 0+$ or $y\to 0+$ or $x^2+y^2\to\infty$, $C$ must have a minimum value at a critical point in the first quadrant. For CP:

$$\begin{aligned}0&=\frac{\partial C}{\partial x}=k\left(3y-\frac{2V}{x^2}\right)\\ 0&=\frac{\partial C}{\partial y}=k\left(3x-\frac{2V}{y^2}\right).\end{aligned}$$

Thus $3x^2y=2V=3xy^2$, so that $x=y=(2V/3)^{1/3}$ and $z=V/(2V/3)^{2/3}=(9V/4)^{1/3}$.

**25.** Differentiate the given equation

$$e^{2zx-x^2}-3e^{2zy+y^2}=2$$

with respect to $x$ and $y$, regarding $z$ as a function of $x$ and $y$:

$$e^{2zx-x^2}\left(2x\frac{\partial z}{\partial x}+2z-2x\right)-3e^{2zy+y^2}\left(2y\frac{\partial z}{\partial x}\right)=0 \qquad (*)$$

$$e^{2zx-x^2}\left(2x\frac{\partial z}{\partial y}\right)-3e^{2zy+y^2}\left(2y\frac{\partial z}{\partial y}+2z+2y\right)=0 \qquad (**)$$

For a critical point we have $\dfrac{\partial z}{\partial x}=0$ and $\dfrac{\partial z}{\partial y}=0$, and it follows from the equations above that $z=x$ and $z=-y$. Substituting these into the given equation, we get

$$\begin{aligned}e^{z^2}-3e^{-z^2}&=2\\ (e^{z^2})^2-2e^{z^2}-3&=0\\ (e^{z^2}-3)(e^{z^2}+1)&=0.\end{aligned}$$

Thus $e^{z^2}=3$ or $e^{z^2}=-1$. Since $e^{z^2}=-1$ is not possible, we have $e^{z^2}=3$, so $z=\pm\sqrt{\ln 3}$.
The critical points are $(\sqrt{\ln 3},-\sqrt{\ln 3})$, and $(-\sqrt{\ln 3},\sqrt{\ln 3})$.

**26.** We will use the second derivative test to classify the two critical points calculated in Exercise 25. To calculate the second partials

$$A = \frac{\partial^2 z}{\partial x^2}, \qquad B = \frac{\partial^2 z}{\partial x \partial y}, \qquad C = \frac{\partial^2 z}{\partial y^2},$$

we differentiate the expressions (*), and (**) obtained in Exercise 25.
Differentiating (*) with respect to $x$, we obtain

$$\begin{aligned} e^{2zx-x^2}\bigg[&\left(2x\frac{\partial z}{\partial x} + 2z - 2x\right)^2 \\ &+ 4\frac{\partial z}{\partial x} + 2x\frac{\partial^2 z}{\partial x^2} - 2\bigg] \\ - 3e^{2zy+y^2}&\left[\left(2y\frac{\partial z}{\partial x}\right)^2 + 2y\frac{\partial^2 z}{\partial x^2}\right] = 0. \end{aligned}$$

At a critical point, $\dfrac{\partial z}{\partial x} = 0$, $z = x$, $z = -y$, and $z^2 = \ln 3$, so

$$3\left(2x\frac{\partial^2 z}{\partial x^2} - 2\right) - \frac{3}{3}\left(2y\frac{\partial^2 z}{\partial x^2}\right) = 0,$$

$$A = \frac{\partial^2 z}{\partial x^2} = \frac{6}{6x - 2y}.$$

Differentiating (**) with respect to $y$ gives

$$\begin{aligned} &e^{2zx-x^2}\left[\left(2x\frac{\partial z}{\partial y}\right)^2 + 2x\frac{\partial^2 z}{\partial y^2}\right] \\ &- 3e^{2zy+y^2}\left[\left(2y\frac{\partial z}{\partial y} + 2z + 2y\right)^2 + 4\frac{\partial z}{\partial y} + 2y\frac{\partial^2 z}{\partial y^2} + 2\right] = 0, \end{aligned}$$

and evaluation at a critical point gives

$$3\left(2x\frac{\partial^2 z}{\partial y^2}\right) - \frac{3}{3}\left(2y\frac{\partial^2 z}{\partial y^2} + 2\right) = 0,$$

$$C = \frac{\partial^2 z}{\partial y^2} = \frac{2}{6x - 2y}.$$

Finally, differentiating (*) with respect to $y$ gives

$$\begin{aligned} e^{2zx-x^2}\bigg[&\left(2x\frac{\partial z}{\partial x} + 2z - 2x\right)\left(2x\frac{\partial z}{\partial y}\right) \\ &+ 2x\frac{\partial^2 z}{\partial x \partial y} + 2\frac{\partial z}{\partial y}\bigg] \\ - 3e^{2zy+y^2}\bigg[&\left(2y\frac{\partial z}{\partial y} + 2z + 2y\right)\left(2y\frac{\partial z}{\partial x}\right) \\ &+ 2\frac{\partial z}{\partial x} + 2y\frac{\partial^2 z}{\partial x \partial y}\bigg] = 0, \end{aligned}$$

and, evaluating at a critical point,

$$(6x - 2y)\frac{\partial^2 z}{\partial x \partial y} = 0,$$

so that

$$B = \frac{\partial^2 z}{\partial x \partial y} = 0.$$

At the critical point $(\sqrt{\ln 3}, -\sqrt{\ln 3})$ we have

$$A = \frac{6}{8\ln 3}, \qquad B = 0, \qquad C = \frac{2}{8\ln 3},$$

so $B^2 - AC < 0$, and $f$ has a local minimum at that critical point.
At the critical point $(-\sqrt{\ln 3}, \sqrt{\ln 3})$ we have

$$A = -\frac{6}{8\ln 3}, \qquad B = 0, \qquad C = -\frac{2}{8\ln 3},$$

so $B^2 - AC < 0$, and $f$ has a local maximum at that critical point.

**28.** Given that

$$Q(u, v, w) = Au^2 + Bv^2 + Cw^2 + 2Duv + 2Euw + 2Fvw$$

and that

$$A > 0, \qquad \begin{vmatrix} A & D \\ D & B \end{vmatrix} = AB - D^2 > 0$$

$$\begin{vmatrix} A & D & E \\ D & B & F \\ E & F & C \end{vmatrix} = ABC + 2DEF - BE^2 - CD^2 - AF^2 > 0,$$

we want to show that $Q(u, v, w) > 0$ if $(u, v, w) \neq (0, 0, 0)$. We have

$$\begin{aligned} &Q(u, v, w) \\ &= A\left[u^2 + 2u\frac{Dv + Ew}{A} + \left(\frac{Dv + Ew}{A}\right)^2\right] \\ &\quad + \frac{AB - D^2}{A}v^2 + \frac{AC - E^2}{A}w^2 + \frac{2(AF - DE)}{A}vw \\ &= A\left(u + \frac{Dv + Ew}{A}\right)^2 \\ &\quad + \frac{AB - D^2}{A}\left(v^2 + \frac{2(AF - DE)}{AB - D^2}vw + \left(\frac{AF - DE}{AB - D^2}\right)^2 w^2\right) \\ &\quad + \left[\frac{AC - E^2}{A} - \frac{(AF - DE)^2}{A(AB - D^2)}\right]w^2 \\ &= A\left(u + \frac{Dv + Ew}{A}\right)^2 + \frac{AB - D^2}{A}\left(v + \frac{AF - DE}{AB - D^2}w\right)^2 \\ &\quad + \frac{A(ABC - BE^2 - AF^2 - CD^2 + 2DEF)}{A(AB - D^2)}w^2. \end{aligned}$$

Under the given conditions, this expression is a sum of squares which is $\geq 0$, and $\neq 0$ unless $w = 0$, $v = 0$ and $u = 0$.

## Section 5.2 Extreme Values of Functions Defined on Restricted Domains (page 234)

**2.** $f(x, y) = xy - 2x$ on
$R = \{(x, y) : -1 \le x \le 1,\ 0 \le y \le 1\}$.
For critical points:

$$0 = f_1(x, y) = y - 2, \qquad 0 = f_2(x, y) = x.$$

The only CP is $(0, 2)$, which lies outside $R$. Therefore the maximum and minimum values of $f$ on $R$ lie on one of the four boundary segments of $R$.
On $x = -1$ we have $f(-1, y) = 2 - y$ for $0 \le y \le 1$, which has maximum value 2 and minimum value 1.
On $x = 1$ we have $f(1, y) = y - 2$ for $0 \le y \le 1$, which has maximum value $-1$ and minimum value $-2$.
On $y = 0$ we have $f(x, 0) = -2x$ for $-1 \le x \le 1$, which has maximum value 2 and minimum value $-2$.
On $y = 1$ we have $f(x, 1) = -x$ for $-1 \le x \le 1$, which has maximum value 1 and minimum value $-1$.
Thus the maximum and minimum values of $f$ on the rectangle $R$ are 2 and $-2$ respectively.

**4.** $f(x, y) = x + 2y$ on the closed disk $x^2 + y^2 \le 1$. Since $f_1 = 1$ and $f_2 = 2$, $f$ has no critical points, and the maximum and minimum values of $f$, which must exist because $f$ is continuous on a closed, bounded set in the plane, must occur at boundary points of the domain, that is, points of the circle $x^2 + y^2 = 1$. This circle can be parametrized $x = \cos t$, $y = \sin t$, so that

$$f(x, y) = f(\cos t, \sin t) = \cos t + 2\sin t = g(t), \text{ say.}$$

For critical points of $g$: $0 = g'(t) = -\sin t + 2\cos t$. Thus $\tan t = 2$, and $x = \pm 1/\sqrt{5}$, $y = \pm 2/\sqrt{5}$. The critical points are $(-1/\sqrt{5}, -2/\sqrt{5})$, where $f$ has value $-\sqrt{5}$, and $(1/\sqrt{5}, 2/\sqrt{5})$, where $f$ has value $\sqrt{5}$. Thus the maximum and minimum values of $f(x, y)$ on the disk are $\sqrt{5}$ and $-\sqrt{5}$ respectively.

**6.**

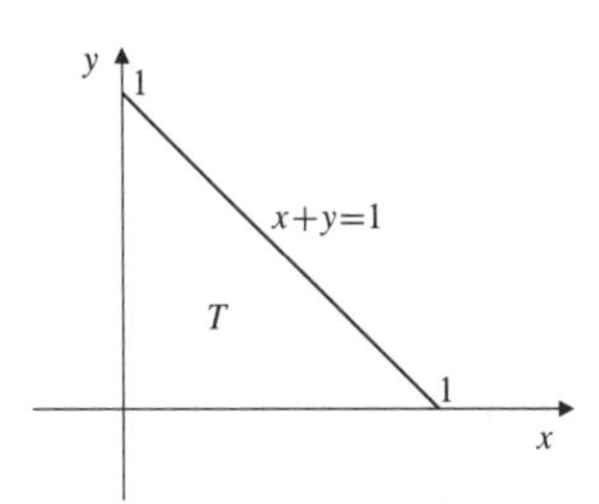


Fig. 5.2.6

$f(x, y) = xy(1 - x - y)$ on the triangle $T$ shown in the figure. Evidently $f(x, y) = 0$ on all three boundary segements of $T$, and $f(x, y) > 0$ inside $T$. Thus the minimum value of $f$ on $T$ is 0, and the maximum value must occur at an interior critical point. For critical points:

$$0 = f_1(x, y) = y(1-2x-y), \qquad 0 = f_2(x, y) = x(1-x-2y).$$

The only critical points are $(0, 0)$, $(1, 0)$ and $(0, 1)$, which are on the boundary of $T$, and $(1/3, 1/3)$, which is inside $T$. The maximum value of $f$ over $T$ is $f(1/3, 1/3) = 1/27$.

**8.**

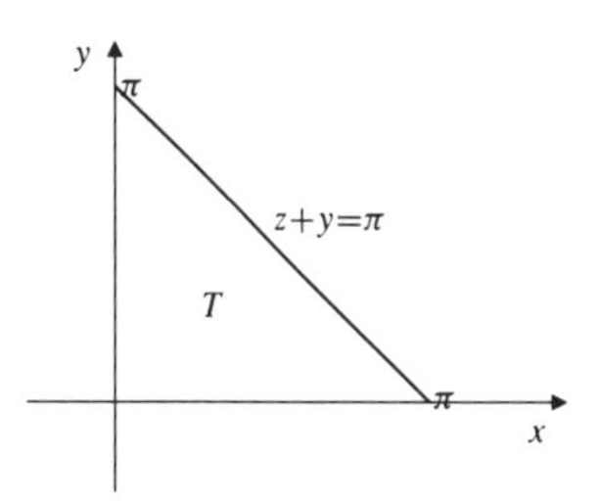


Fig. 5.2.8

$f(x, y) = \sin x \sin y \sin(x + y)$ on the triangle $T$ shown in the figure. Evidently $f(x, y) = 0$ on the boundary of $T$, and $f(x, y) > 0$ at all points inside $T$. Thus the minimum value of $f$ on $T$ is zero, and the maximum value must occur at an interior critical point. For critical points inside $T$ we must have

$$0 = f_1(x, y) = \cos x \sin y \sin(x + y) + \sin x \sin y \cos(x + y)$$
$$0 = f_2(x, y) = \sin x \cos y \sin(x + y) + \sin x \sin y \cos(x + y).$$

Therefore $\cos x \sin y = \cos y \sin x$, which implies $x = y$ for points inside $T$, and

$$\cos x \sin x \sin 2x + \sin^2 x \cos 2x = 0$$
$$2\sin^2 x \cos^2 x + 2\sin^2 x \cos^2 x - \sin^2 x = 0$$
$$4\cos^2 x = 1.$$

Thus $\cos x = \pm 1/2$, and $x = \pm\pi/3$. The interior critical point is $(\pi/3, \pi/3)$, where $f$ has the value $3\sqrt{3}/8$. This is the maximum value of $f$ on $T$.

**10.** $f(x, y) = \dfrac{x - y}{1 + x^2 + y^2}$ on the half-plane $y \ge 0$.
For critical points:

$$0 = f_1(x, y) = \frac{1 - x^2 + y^2 + 2xy}{(1 + x^2 + y^2)^2}$$
$$0 = f_2(x, y) = \frac{-1 - x^2 + y^2 - 2xy}{(1 + x^2 + y^2)^2}.$$

Any critical points must satisfy $1 - x^2 + y^2 + 2xy = 0$ and $-1 - x^2 + y^2 - 2xy = 0$, and hence $x^2 = y^2$ and $2xy = -1$. Therefore $y = -x = \pm 1/\sqrt{2}$. The only critical point in the region $y \ge 0$ is $(-1/\sqrt{2}, 1/\sqrt{2})$, where $f$ has the value $-1/\sqrt{2}$.

On the boundary $y = 0$ we have

$$f(x, 0) = \frac{x}{1 + x^2} = g(x), \qquad (-\infty < x < \infty).$$

Evidently, $g(x) \to 0$ as $x \to \pm\infty$.
Since $g'(x) = \dfrac{1-x^2}{(1+x^2)^2}$, the critical points of $g$ are $x = \pm 1$. We have $g(\pm 1) = \pm\dfrac{1}{2}$.
The maximum and minimum values of $f$ on the upper half-plane $y \ge 0$ are $1/2$ and $-1/\sqrt{2}$ respectively.

**12.** Let $f(x, y, z) = xz + yz$ on the ball $x^2 + y^2 + z^2 \le 1$. First look for interior critical points:

$$0 = f_1 = z, \quad 0 = f_2 = z, \quad 0 = f_3 = x + y.$$

All points on the line $z = 0$, $x + y = 0$ are CPs, and $f = 0$ at all such points.

Now consider the boundary sphere $x^2 + y^2 + z^2 = 1$. On it

$$f(x, y, z) = (x + y)z = \pm(x + y)\sqrt{1 - x^2 - y^2} = g(x, y),$$

where $g$ has domain $x^2 + y^2 \le 1$. On the boundary of its domain, $g$ is identically 0, although $g$ takes both positive and negative values at some points inside its domain. Therefore, we need consider only critical points of $g$ in $x^2 + y^2 < 1$. For such CPs:

$$\begin{aligned}
0 = g_1 &= \sqrt{1 - x^2 - y^2} + \frac{(x + y)(-2x)}{2\sqrt{1 - x^2 - y^2}} \\
&= \frac{1 - x^2 - y^2 - x^2 - xy}{\sqrt{1 - x^2 - y^2}} \\
0 = g_2 &= \frac{1 - x^2 - y^2 - xy - y^2}{\sqrt{1 - x^2 - y^2}}.
\end{aligned}$$

Therefore $2x^2 + y^2 + xy = 1 = x^2 + 2y^2 + xy$, from which $x^2 = y^2$.
Case I: $x = -y$. Then $g = 0$, so $f = 0$.
Case II: $x = y$. Then $2x^2 + x^2 + x^2 = 1$, so $x^2 = 1/4$ and $x = \pm 1/2$. $g$ (which is really two functions depending on our choice of the "+" or "−" sign) has four CPs, two corresponding to $x = y = 1/2$ and two to $x = y = -1/2$. The values of $g$ at these four points are $\pm 1/\sqrt{2}$.

Since we have considered all points where $f$ can have extreme values, we conclude that the maximum value of $f$ on the ball is $1/\sqrt{2}$ (which occurs at the boundary points $\pm(\frac{1}{2}, \frac{1}{2}, \frac{1}{\sqrt{2}})$) and minimum value $-1/\sqrt{2}$ (which occurs at the boundary points $\pm(\frac{1}{2}, \frac{1}{2}, -\frac{1}{\sqrt{2}})$).

**14.** $f(x, y) = xy^2 e^{-xy}$ on $Q = \{(x, y) : x \ge 0,\ y \ge 0\}$.
Note that $f(x, 0) = f(0, y) = 0$ and
$\lim_{x\to\infty} f(x, kx) = k^2 x^3 e^{-x^2} = 0$.
Also, $f(0, y) = 0$ while $f\left(\dfrac{1}{y}, y\right) = \dfrac{y}{e} \to \infty$ as $y \to \infty$, so that $f$ has no limit as $x^2 + y^2 \to \infty$ in $Q$, and $f$ has no maximum value on $Q$.

**16.** Let the dimensions be as shown in the figure. Then $2x + y = 100$, the length of the fence. For maximum area $A$ of the enclosure we will have $x > 0$ and $0 < \theta < \pi/2$. Since $h = x\cos\theta$, the area $A$ is

$$\begin{aligned}
A &= xy\cos\theta + 2 \times \frac{1}{2}(x\sin\theta)(x\cos\theta) \\
&= x(100 - 2x)\cos\theta + x^2\sin\theta\cos\theta \\
&= (100x - 2x^2)\cos\theta + \frac{1}{2}x^2\sin 2\theta.
\end{aligned}$$

We look for a critical point of $A$ satisfying $x > 0$ and $0 < \theta < \pi/2$.

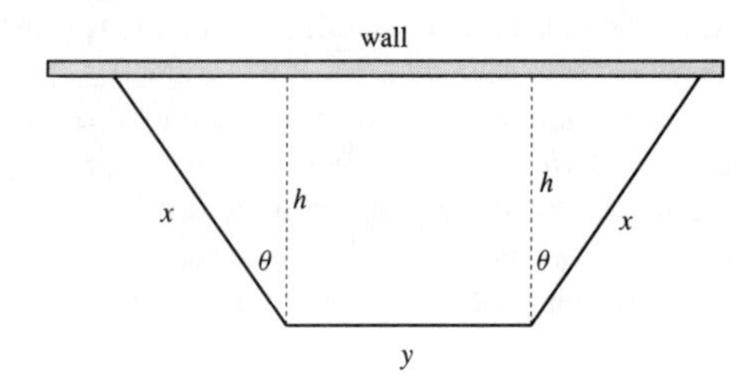


Fig. 5.2.16

$$\begin{aligned}
0 &= \frac{\partial A}{\partial x} = (100 - 4x)\cos\theta + x\sin 2\theta \\
&\Rightarrow \cos\theta(100 - 4x + 2x\sin\theta) = 0 \\
&\Rightarrow 4x - 2x\sin\theta = 100 \Rightarrow x = \frac{50}{2 - \sin\theta} \\
0 &= \frac{\partial A}{\partial \theta} = -(100x - 2x^2)\sin\theta + x^2\cos 2\theta \\
&\Rightarrow x(1 - 2\sin^2\theta) + 2x\sin\theta - 100\sin\theta = 0.
\end{aligned}$$

Substituting the first equation into the second we obtain

$$\begin{aligned}
&\frac{50}{2 - \sin\theta}\big(1 - 2\sin^2\theta + 2\sin\theta\big) - 100\sin\theta = 0 \\
&50(1 - 2\sin^2\theta + 2\sin\theta) = 100(2\sin\theta - \sin^2\theta) \\
&50 = 100\sin\theta.
\end{aligned}$$

Thus $\sin\theta = 1/2$, and $\theta = \pi/6$.
Therefore $x = \dfrac{50}{2 - (1/2)} = \dfrac{100}{3}$, and $y = 100 - 2x = \dfrac{100}{3}$.
The maximum area for the enclosure is

$$A = \left(\frac{100}{3}\right)^2 \frac{\sqrt{3}}{2} + \left(\frac{100}{3}\right)^2 \frac{1}{2}\frac{\sqrt{3}}{2} = \frac{2500}{\sqrt{3}}$$

square units. All three segments of the fence will be the same length, and the bend angles will be 120°.

**18.** Minimize $F(x, y, z) = 2x + 3y + 4z$ subject to

$$\begin{aligned}
&x \ge 0, \qquad y \ge 0, \qquad z \ge 0, \\
&x + y \ge 2, \qquad y + z \ge 2, \qquad x + z \ge 2.
\end{aligned}$$

Here the constraint region has vertices $(1, 1, 1)$, $(2, 2, 0)$, $(2, 0, 2)$, and $(0, 2, 2)$. Since $F(1, 1, 1) = 9$, $F(2, 2, 0) = 10$, $F(2, 0, 2) = 12$, and $F(0, 2, 2) = 14$, the minimum value of $F$ subject to the constraints is 9.

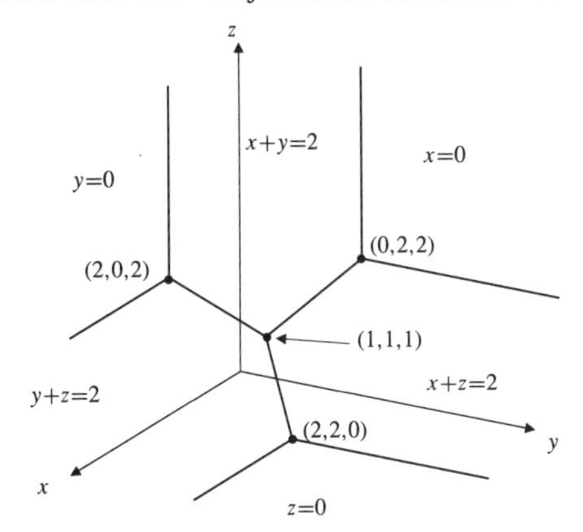


Fig. 5.2.18

**20.** If the developer builds $x$ houses, $y$ duplex units, and $z$ apartments, his profit will be

$$P = 40,000x + 20,000y + 16,000z.$$

The legal constraints imposed require that

$$\frac{x}{6} + \frac{y}{8} + \frac{z}{12} \le 10, \quad \text{that is } 4x + 3y + 2z \le 240,$$

and also

$$z \ge x + y.$$

Evidently we must also have $x \ge 0$, $y \ge 0$, and $z \ge 0$. The planes $4x + 3y + 2z = 240$ and $z = x + y$ intersect where $6x + 5y = 240$. Thus the constraint region has vertices $(0, 0, 0)$, $(40, 0, 40)$, $(0, 48, 48)$, and $(0, 0, 120)$, which yield revenues of \$0, \$2,240,000, \$1,728,000, and \$1,920,000 respectively.
For maximum profit, the developer should build 40 houses, no duplex units, and 40 apartments.

## Section 5.3 Lagrange Multipliers (page 242)

**2.** a) Let $D$ be the distance from $(3, 0)$ to the point $(x, y)$ on the curve $y = x^2$. Then

$$D^2 = (x-3)^2 + y^2 = (x-3)^2 + x^4.$$

For a minimum, $0 = \dfrac{dD^2}{dx} = 2(x-3) + 4x^3$. Thus $2x^3 + x - 3 = 0$. Clearly $x = 1$ is a root of this cubic equation. Since

$$\frac{2x^3 + x - 3}{x - 1} = 2x^2 + 2x + 3,$$

and $2x^2 + 2x + 3$ has negative discriminant, $x = 1$ is the only critical point. Thus the minimum distance from $(3, 0)$ to $y = x^2$ is $D = \sqrt{(-2)^2 + 1^4} = \sqrt{5}$ units.

b) We want to minimize $D^2 = (x-3)^2 + y^2$ subject to the constraint $y = x^2$. Let $L = (x-3)^2 + y^2 + \lambda(x^2 - y)$. For critical points of $L$ we want

$$0 = \frac{\partial L}{\partial x} = 2(x-3) + 2\lambda x$$
$$\Rightarrow (1+\lambda)x - 3 = 0 \qquad (A)$$
$$0 = \frac{\partial L}{\partial y} = 2y - \lambda \qquad (B)$$
$$0 = \frac{\partial L}{\partial \lambda} = x^2 - y. \qquad (C)$$

Eliminating $\lambda$ from (A) and (B), we get $x + 2xy - 3 = 0$.
Substituting (C) then leads to $2x^3 + x - 3 = 0$, or $(x-1)(2x^2 + 2x + 3) = 0$. The only real solution is $x = 1$, so the point on $y = x^2$ closest to $(3, 0)$ is $(1, 1)$.
Thus the minimum distance from $(3, 0)$ to $y = x^2$ is $D = \sqrt{(1-3)^2 + 1^2} = \sqrt{5}$ units.

**4.** Let $f(x, y, z) = x + y - z$, and define the Lagrangian

$$L = x + y - z + \lambda(x^2 + y^2 + z^2 - 1).$$

Solutions to the constrained problem will be found among the critical points of $L$. To find these we have

$$0 = \frac{\partial L}{\partial x} = 1 + 2\lambda x,$$
$$0 = \frac{\partial L}{\partial y} = 1 + 2\lambda y,$$
$$0 = \frac{\partial L}{\partial z} = -1 + 2\lambda z,$$
$$0 = \frac{\partial L}{\partial \lambda} = x^2 + y^2 + z^2 - 1.$$

Therefore $2\lambda x = 2\lambda y = -2\lambda z$. Either $\lambda = 0$ or $x = y = -z$. $\lambda = 0$ is not possible. (It implies $0 = 1$ from the first equation.) From $x = y = -z$ we obtain $1 = x^2 + y^2 + z^2 = 3x^2$, so $x = \pm\dfrac{1}{\sqrt{3}}$. $L$ has critical points at $\left(\dfrac{1}{\sqrt{3}}, \dfrac{1}{\sqrt{3}}, -\dfrac{1}{\sqrt{3}}\right)$ and $\left(\dfrac{1}{-\sqrt{3}}, -\dfrac{1}{\sqrt{3}}, \dfrac{1}{\sqrt{3}}\right)$.
At the first $f = \sqrt{3}$, which is the maximum value of $f$ on the sphere; at the second $f = -\sqrt{3}$, which is the minimum value.

**6.** Let $L = x^2 + y^2 + z^2 + \lambda(xyz^2 - 2)$. For critical points:

$$0 = \frac{\partial L}{\partial x} = 2x + \lambda yz^2 \quad \Leftrightarrow \quad -\lambda xyz^2 = 2x^2$$
$$0 = \frac{\partial L}{\partial y} = 2y + \lambda xz^2 \quad \Leftrightarrow \quad -\lambda xyz^2 = 2y^2$$
$$0 = \frac{\partial L}{\partial z} = 2z + 2\lambda xyz \quad \Leftrightarrow \quad -\lambda xyz^2 = z^2$$
$$0 = \frac{\partial L}{\partial \lambda} = xyz^2 - 2.$$

From the first three equations, $x^2 = y^2$ and $z^2 = 2x^2$. The fourth equation then gives $x^2y^24z^4 = 4$, or $x^8 = 1$. Thus $x^2 = y^2 = 1$ and $z^2 = 2$.
The shortest distance from the origin to the surface $xyz^2 = 2$ is

$$\sqrt{1+1+2} = 2 \text{ units.}$$

**8.** Let $L = x^2 + y^2 + \lambda(3x^2 + 2xy + 3y^2 - 16)$. We have

$$0 = \frac{\partial L}{\partial x} = 2x + 6\lambda x + 2\lambda y \qquad (A)$$
$$0 = \frac{\partial L}{\partial y} = 2y + 6\lambda y + 2\lambda x. \qquad (B)$$

Multiplying $(A)$ by $y$ and $(B)$ by $x$ and subtracting we get

$$2\lambda(y^2 - x^2) = 0.$$

Thus, either $\lambda = 0$, or $y = x$, or $y = -x$.
$\lambda = 0$ is not possible, since it implies $x = 0$ and $y = 0$, and the point $(0, 0)$ does not lie on the given ellipse.
If $y = x$, then $8x^2 = 16$, so $x = y = \pm\sqrt{2}$.
If $y = -x$, then $4x^2 = 16$, so $x = -y = \pm 2$.
The points on the ellipse nearest the origin are $(\sqrt{2}, \sqrt{2})$ and $(-\sqrt{2}, -\sqrt{2})$. The points farthest from the origin are $(2, -2)$ and $(-2, 2)$. The major axis of the ellipse lies along $y = -x$ and has length $4\sqrt{2}$. The minor axis lies along $y = x$ and has length 4.

**10.** Let $L = x + 2y - 3z + \lambda(x^2 + 4y^2 + 9z^2 - 108)$. For CPs of $L$:

$$0 = \frac{\partial L}{\partial x} = 1 + 2\lambda x \qquad (A)$$
$$0 = \frac{\partial L}{\partial y} = 2 + 8\lambda y \qquad (B)$$
$$0 = \frac{\partial L}{\partial z} = -3 + 18\lambda z \qquad (C)$$
$$0 = \frac{\partial L}{\partial \lambda} = x^2 + 4y^2 + 9z^2 - 108. \qquad (D)$$

From (A), (B), and (C),

$$\lambda = -\frac{1}{2x} = -\frac{2}{8y} = \frac{3}{18z},$$

so $x = 2y = -3z$. From (D):

$$x^2 + 4\left(\frac{x^2}{4}\right) + 9\left(\frac{x^2}{9}\right) = 108,$$

so $x^2 = 36$, and $x = \pm 6$. There are two CPs: $(6, 3, -2)$ and $(-6, -3, 2)$. At the first, $x + 2y - 3z = 18$, the maximum value, and at the second, $x + 2y - 3z = -18$, the minimum value.

**12.** Let $L = x^2 + y^2 + z^2 + \lambda(x^2 + y^2 - z^2) + \mu(x - 2z - 3)$. For critical points of $L$:

$$0 = \frac{\partial L}{\partial x} = 2x(1 + \lambda) + \mu \qquad (A)$$
$$0 = \frac{\partial L}{\partial y} = 2y(1 + \lambda) \qquad (B)$$
$$0 = \frac{\partial L}{\partial z} = 2z(1 - \lambda) - 2\mu \qquad (C)$$
$$0 = \frac{\partial L}{\partial \lambda} = x^2 + y^2 - z^2 \qquad (D)$$
$$0 = \frac{\partial L}{\partial \mu} = x - 2z - 3. \qquad (E)$$

From (B), either $y = 0$ or $\lambda = -1$.

CASE I. $y = 0$. Then (D) implies $x = \pm z$.
If $x = z$ then (E) implies $z = -3$, so we get the point $(-3, 0, -3)$.
If $x = -z$ then (E) implies $z = -1$, so we get the point $(1, 0, -1)$.

CASE II. $\lambda = -1$. Then (A) implies $\mu = 0$ and (C) implies $z = 0$. By (D), $x = y = 0$, and this contradicts (E), so this case is not possible.

If $f(x, y, z) = x^2 + y^2 + z^2$, then $f(-3, 0, -3) = 18$ is the maximum value of $f$ on the ellipse $x^2 + y^2 = z^2$, $x - 2z = 3$, and $f(1, 0, -1) = 2$ is the minimum value.

**14.** The max and min values of $f(x, y, z) = x + y^2 z$ subject to the constraints $y^2 + z^2 = 2$ and $z = x$ will be found among the critical points of

$$L = x + y^2 z + \lambda(y^2 + z^2 - 2) + \mu(z - x).$$

Thus

$$0 = \frac{\partial L}{\partial x} = 1 - \mu = 0,$$
$$0 = \frac{\partial L}{\partial y} = 2yz + 2\lambda y = 0,$$
$$0 = \frac{\partial L}{\partial z} = y^2 + 2\lambda z + \mu = 0,$$
$$0 = \frac{\partial L}{\partial \lambda} = y^2 + z^2 - 2,$$
$$0 = \frac{\partial L}{\partial \mu} = z - x.$$

From the first equation $\mu = 1$. From the second, either $y = 0$ or $z = -\lambda$.
If $y = 0$ then $z^2 = 2$, $z = x$, so critical points are $(\sqrt{2}, 0, \sqrt{2})$ and $(-\sqrt{2}, 0, -\sqrt{2})$. $f$ has the values $\pm\sqrt{2}$ at these points. If $z = -\lambda$ then $y^2 - 2z^2 + 1 = 0$. Thus $2z^2 - 1 = 2 - z^2$, or $z^2 = 1$, $z = \pm 1$. This leads to critical points $(1, \pm 1, 1)$ and $(-1, \pm 1, -1)$ where $f$ has values $\pm 2$. The maximum value of $f$ subject to the constraints is 2; the minimum value is $-2$.

**16.** Let $L = x_1 + x_2 + \cdots + x_n + \lambda(x_1^2 + x_2^2 + \cdots + x_n^2 - 1)$. For critical points of $L$ we have

$$0 = \frac{\partial L}{\partial x_1} = 1 + 2\lambda x_1, \quad \ldots \quad 0 = \frac{\partial L}{\partial x_n} = 1 + 2\lambda x_n$$
$$0 = \frac{\partial L}{\partial \lambda} = x_1^2 + x_2^2 + \cdots + x_n^2 - 1.$$

The first $n$ equations give

$$x_1 = x_2 = \cdots = x_n = -\frac{1}{2\lambda},$$

and the final equation gives

$$\frac{1}{4\lambda^2} + \frac{1}{4\lambda^2} + \cdots + \frac{1}{4\lambda^2} = 1,$$

so that $4\lambda^2 = n$, and $\lambda = \pm\sqrt{n}/2$.
The maximum and minimum values of $x_1 + x_2 + \cdots + x_n$ subject to $x_1^2 + \cdots + x_n^2 = 1$ are $\pm\dfrac{n}{2\lambda}$, that is, $\sqrt{n}$ and $-\sqrt{n}$ respectively.

**18.** Let the width, depth, and height of the box be $x$, $y$ and $z$ respectively. We want to minimize the surface area

$$S = xy + 2xz + 2yz$$

subject to the constraint that $xyz = V$, where $V$ is a given positive volume. Let

$$L = xy + 2xz + 2yz + \lambda(xyz - V).$$

For critical points of $L$,

$$0 = \frac{\partial L}{\partial x} = y + 2z + \lambda yz \quad \Leftrightarrow \quad -\lambda xyz = xy + 2xz$$
$$0 = \frac{\partial L}{\partial y} = x + 2z + \lambda xz \quad \Leftrightarrow \quad -\lambda xyz = xy + 2yz$$
$$0 = \frac{\partial L}{\partial z} = 2x + 2y + \lambda xy \quad \Leftrightarrow \quad -\lambda xyz = 2xz + 2yz$$
$$0 = \frac{\partial L}{\partial \lambda} = xyz - V.$$

From the first three equations, $xy = 2xz = 2yz$. Since $x$, $y$, and $z$ are all necessarily positive, we must therefore have $x = y = 2z$. Thus the most economical box with no top has width and depth equal to twice the height.

**20.** We want to maximize $xyz$ subject to $xy + 2yz + 3xz = 18$. Let

$$L = xyz + \lambda(xy + 2yz + 3xz - 18).$$

For critical points of $L$,

$$0 = \frac{\partial L}{\partial x} = yz + \lambda(y + 3z) \quad \Leftrightarrow \quad -xyz = \lambda(xy + 3xz)$$
$$0 = \frac{\partial L}{\partial y} = xz + \lambda(x + 2z) \quad \Leftrightarrow \quad -xyz = \lambda(xy + 2yz)$$
$$0 = \frac{\partial L}{\partial z} = xy + \lambda(2y + 3x) \quad \Leftrightarrow \quad -xyz = \lambda(2yz + 3xz)$$
$$0 = \frac{\partial L}{\partial \lambda} = xy + 2yz + 3xz - 18.$$

From the first three equations $xy = 2yz = 3xz$. From the fourth equation, the sum of these expressions is 18. Thus

$$xy = 2yz = 3xz = 6.$$

Thus the maximum volume of the box is

$$V = xyz = \sqrt{(xy)(yz)(xz)} = \sqrt{6 \times 3 \times 2} = 6 \text{ cubic units.}$$

**22.** $f(x, y, z) = xy + z^2$ on $B = \{(x, y, z) : x^2 + y^2 + z^2 \le 1\}$. For critical points of $f$,

$$0 = f_1(x, y, z) = y, \qquad 0 = f_2(x, y, z) = x,$$
$$0 = f_3(x, y, z) = 2z.$$

Thus the only critical point is the interior point $(0, 0, 0)$, where $f$ has the value 0, evidently neither a maximum nor a minimum. The maximum and minimum must therefore occur on the boundary of $B$, that is, on the sphere $x^2 + y^2 + z^2 = 1$. Let

$$L = xy + z^2 + \lambda(x^2 + y^2 + z^2 - 1).$$

For critical points of $L$,

$$0 = \frac{\partial L}{\partial x} = y + 2\lambda x \qquad (A)$$
$$0 = \frac{\partial L}{\partial y} = x + 2\lambda y \qquad (B)$$
$$0 = \frac{\partial L}{\partial z} = 2z(1 + \lambda) \qquad (C)$$
$$0 = \frac{\partial L}{\partial \lambda} = x^2 + y^2 + z^2 - 1. \qquad (D)$$

From (C) either $z = 0$ or $\lambda = -1$.

CASE I. $z = 0$. (A) and (B) imply that $y^2 = x^2$ and (D) then implies that $x^2 = y^2 = 1/2$. At the four points

$$\left(\frac{1}{\sqrt{2}}, \pm\frac{1}{\sqrt{2}}, 0\right) \quad \text{and} \quad \left(-\frac{1}{\sqrt{2}}, \pm\frac{1}{\sqrt{2}}, 0\right)$$

$f$ takes the values $\frac{1}{2}$ and $-\frac{1}{2}$.

CASE II. $\lambda = -1$. (A) and (B) imply that $x = y = 0$, and so by (D), $z = \pm 1$. $f$ has the value 1 at the points $(0, 0, \pm 1)$.

Thus the maximum and minimum values of $f$ on $B$ are 1 and $-1/2$ respectively.

**24.** Let $L = \sin\frac{x}{2}\sin\frac{y}{2}\sin\frac{z}{2} + \lambda(x + y + z - \pi)$. Then

$$0 = \frac{\partial L}{\partial x} = \frac{1}{2}\cos\frac{x}{2}\sin\frac{y}{2}\sin\frac{z}{2} + \lambda \qquad (A)$$
$$0 = \frac{\partial L}{\partial y} = \frac{1}{2}\sin\frac{x}{2}\cos\frac{y}{2}\sin\frac{z}{2} + \lambda \qquad (B)$$
$$0 = \frac{\partial L}{\partial z} = \frac{1}{2}\sin\frac{x}{2}\sin\frac{y}{2}\cos\frac{z}{2} + \lambda. \qquad (C)$$

For any triangle we must have $0 \le x \le \pi$, $0 \le y \le \pi$ and $0 \le z \le \pi$. Also

$$P = \sin\frac{x}{2}\sin\frac{y}{2}\sin\frac{z}{2}$$

is 0 if any of $x$, $y$ or $z$ is 0 or $\pi$. Subtracting equations $(A)$ and $(B)$ gives

$$\frac{1}{2}\sin\frac{z}{2}\sin\frac{x-y}{2} = 0.$$

It follows that we must have $x = y$; all other possibilities lead to a zero value for $P$. Similarly, $y = z$. Thus the triangle for which $P$ is maximum must be equilateral: $x = y = z = \pi/3$. Since $\sin(\pi/3) = 1/2$, the maximum value of $P$ is 1/8.

**26.**

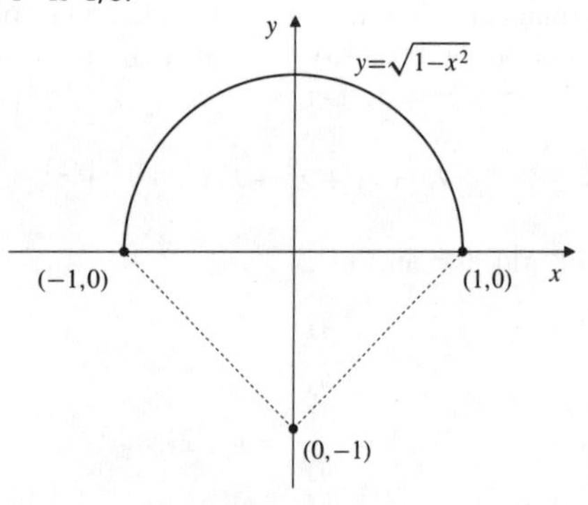


Fig. 5.3.26

As can be seen in the figure, the minimum distance from $(0, -1)$ to points of the semicircle $y = \sqrt{1 - x^2}$ is $\sqrt{2}$, the closest points to $(0, -1)$ on the semicircle being $(\pm 1, 0)$. These points will not be found by the method of Lagrange multipliers because the level curve $f(x, y) = 2$ of the function $f$ giving the square of the distance from $(x, y)$ to $(0, -1)$ is not tangent to the semicircle at $(\pm 1, 0)$. This could only have happened because $(\pm 1, 0)$ are *endpoints* of the semicircle.

## Section 5.4 The Method of Least Squares (page 249)

**2.** We want to minimize $S = \sum_{i=1}^{n}(ax_i^2 - y_i)^2$. Thus

$$0 = \frac{dS}{da} = \sum_{i=1}^{n} 2(ax_i^2 - y_i)x_i^2 = 2\sum_{i=1}^{n}(ax_i^4 - x_i^2 y_i),$$

and $a = \left(\sum_{i=1}^{n} x_i^2 y_i\right) \Big/ \left(\sum_{i=1}^{n} x_i^4\right)$.

**4.** We choose $a$, $b$, and $c$ to minimize

$$S = \sum_{i=1}^{n}\left(ax_i + by_i + c - z_i\right)^2.$$

Thus

$$0 = \frac{\partial S}{\partial a} = 2\sum_{i=1}^{n}(ax_i + by_i + c - z_i)x_i$$
$$0 = \frac{\partial S}{\partial b} = 2\sum_{i=1}^{n}(ax_i + by_i + c - z_i)y_i$$
$$0 = \frac{\partial S}{\partial c} = 2\sum_{i=1}^{n}(ax_i + by_i + c - z_i).$$

Let $A = \sum x_i^2$, $B = \sum x_i y_i$, $C = \sum x_i$, $D = \sum y_i^2$, $E = \sum y_i$, $F = \sum x_i z_i$, $G = \sum y_i z_i$, and $H = \sum z_i$. In terms of these quantities the above equations become

$$\begin{array}{ccccccc} Aa & + & Bb & + & Cc & = & F \\ Ba & + & Db & + & Ec & = & G \\ Ca & + & Eb & + & nc & = & H. \end{array}$$

By Cramer's Rule (Theorem 5 of Section 3.6) the solution is

$$a = \frac{1}{\Delta}\begin{vmatrix} F & B & C \\ G & D & E \\ H & E & n \end{vmatrix}, \qquad b = \frac{1}{\Delta}\begin{vmatrix} A & F & C \\ B & G & E \\ C & H & n \end{vmatrix},$$
$$c = \frac{1}{\Delta}\begin{vmatrix} A & B & F \\ B & D & G \\ C & E & H \end{vmatrix}, \qquad \text{where } \Delta = \begin{vmatrix} A & B & C \\ B & D & E \\ C & E & n \end{vmatrix}.$$

**6.** The relationship $y = p + qx^2$ is linear in $p$ and $q$, so we choose $p$ and $q$ to minimize

$$S = \sum_{i=1}^{n}(p + qx_i^2 - y_i)^2.$$

Thus

$$0 = \frac{\partial S}{\partial p} = 2\sum_{i=1}^{n}(p + qx_i^2 - y_i)$$
$$0 = \frac{\partial S}{\partial q} = 2\sum_{i=1}^{n}(p + qx_i^2 - y_i)x_i^2,$$

that is,

$$\begin{aligned} np \quad &+ \quad \left(\sum x_i^2\right) q \quad = \quad \sum y_i \\ \left(\sum x_i^2\right) p \quad &+ \quad \left(\sum x_i^4\right) q \quad = \quad \sum x_i^2 y_i, \end{aligned}$$

so

$$p = \frac{\left(\sum y_i\right)\left(\sum x_i^4\right) - \left(\sum x_i^2 y_i\right)\left(\sum x_i^2\right)}{n\left(\sum x_i^4\right) - \left(\sum x_i^2\right)^2}$$
$$q = \frac{n\left(\sum x_i^2 y_i\right) - \left(\sum y_i\right)\left(\sum x_i^2\right)}{n\left(\sum x_i^4\right) - \left(\sum x_i^2\right)^2}.$$

This is the result obtained by direct linear regression. (No transformation of variables was necessary.)

**8.** We transform $y = \ln(p + qx)$ into the form $e^y = p + qx$, which is linear in $p$ and $q$. We let $\eta_i = e^{y_i}$ and use the regression line $\eta = ax + b$ obtained from the data $(x_i, \eta_i)$, with $a = q$ and $b = p$.
Using the formulas for $a$ and $b$ obtained in the text, we have

$$q = a = \frac{n\left(\sum x_i e^{y_i}\right) - \left(\sum x_i\right)\left(\sum e^{y_i}\right)}{n\left(\sum x_i^2\right) - \left(\sum x_i\right)^2}$$
$$p = b = \frac{\left(\sum x_i^2\right)\left(\sum e^{y_i}\right) - \left(\sum x_i\right)\left(\sum x_i e^{y_i}\right)}{n\left(\sum x_i^2\right) - \left(\sum x_i\right)^2}.$$

These values of $p$ and $q$ are not the same values that minimize the expression

$$S = \sum_{i=1}^{n}\left(\ln(p + qx_i) - y_i\right)^2.$$

**10.** We transform $y = \sqrt{(px + q)}$ into the form $y^2 = px + q$, which is linear in $p$ and $q$. We let $\eta_i = y_i^2$ and use the regression line $\eta = ax + b$ obtained from the data $(x_i, \eta_i)$, with $a = p$ and $b = q$.
Using the formulas for $a$ and $b$ obtained in the text, we have

$$p = a = \frac{n\left(\sum x_i y_i^2\right) - \left(\sum x_i\right)\left(\sum y_i^2\right)}{n\left(\sum x_i^2\right) - \left(\sum x_i\right)^2}$$
$$q = b = \frac{\left(\sum x_i^2\right)\left(\sum y_i^2\right) - \left(\sum x_i\right)\left(\sum x_i y_i^2\right)}{n\left(\sum x_i^2\right) - \left(\sum x_i\right)^2}.$$

These values of $p$ and $q$ are not the same values that minimize the expression

$$S = \sum_{i=1}^{n}\left(\sqrt{px_i + q} - y_i\right)^2.$$

**12.** We use the result of Exercise 6. We have $n = 6$ and

$$\sum x_i^2 = 115, \qquad \sum x_i^4 = 4051,$$
$$\sum y_i = 55.18, \qquad \sum x_i^2 y_i = 1984.50.$$

Therefore

$$\begin{aligned} p &= \frac{\left(\sum y_i\right)\left(\sum x_i^4\right) - \left(\sum x_i^2 y_i\right)\left(\sum x_i^2\right)}{n\left(\sum x_i^4\right) - \left(\sum x_i^2\right)^2} \\ &= \frac{55.18 \times 4051 - 1984.50 \times 115}{6 \times 4051 - 115^2} \approx -0.42 \\ q &= \frac{n\left(\sum x_i^2 y_i\right) - \left(\sum y_i\right)\left(\sum x_i^2\right)}{n\left(\sum x_i^4\right) - \left(\sum x_i^2\right)^2} \\ &= \frac{6 \times 1984.50 - 55.18 \times 115}{6 \times 4051 - 115^2} \approx 0.50. \end{aligned}$$

We have (approximately) $y = -0.42 + 0.50x^2$. The predicted value of $y$ at $x = 5$ is $-0.42 + 0.50 \times 25 \approx 12.1$.

**14.** Since $y = pe^x + q + re^{-x}$ is equivalent to

$$e^x y = p(e^x)^2 + qe^x + r,$$

we let $\xi_i = e^{x_i}$ and $\eta_i = e^{x_i} y_i$ for $i = 1, 2, \ldots, n$. We then have $p = a$, $q = b$, and $r = c$, where $a$, $b$, and $c$ are the values calculated by the formulas in Exercise 13, but for the data $(\xi_i, \eta_i)$ instead of $(x_i, y_i)$.

**16.** To maximize $I = \displaystyle\int_0^\pi \left(ax(\pi - x) - \sin x\right)^2 dx$, we choose $a$ so that

$$\begin{aligned} 0 = \frac{dI}{da} &= \int_0^\pi 2\left(ax(\pi - x) - \sin x\right)x(\pi - x)\,dx \\ &= 2a\int_0^\pi x^2(\pi - x)^2\,dx - 2\int_0^\pi x(\pi - x)\sin x\,dx \\ &= \frac{\pi^5 a}{15} - 8. \end{aligned}$$

(We have omited the details of evaluation of these integrals.) Hence $a = 120/\pi^5$. The minimum value of $I$ is

$$\int_0^\pi \left(\frac{120}{\pi^5}x(\pi - x) - \sin x\right)^2 dx = \frac{\pi}{2} - \frac{480}{\pi^5} \approx 0.00227.$$

**18.** To minimize $\int_0^1 (x^3 - ax^2 - bx - c)^2\,dx$, choose $a$, $b$ and $c$ so that

$$0 = 2\int_0^1 (x^3 - ax^2 - bx - c)(-x^2)\,dx$$
$$0 = 2\int_0^1 (x^3 - ax^2 - bx - c)(-x)\,dx$$
$$0 = 2\int_0^1 (x^3 - ax^2 - bx - c)(-1)\,dx,$$

that is,

$$\begin{aligned} \frac{a}{5} + \frac{b}{4} + \frac{c}{3} &= \frac{1}{6} \\ \frac{a}{4} + \frac{b}{3} + \frac{c}{2} &= \frac{1}{5} \\ \frac{a}{3} + \frac{b}{2} + c &= \frac{1}{4} \end{aligned}$$

for which the solution is $a = \frac{3}{2}$, $b = -\frac{3}{5}$, and $c = \frac{1}{20}$.

**20.** $J = \int_{-1}^1 (x - a\sin\pi x - b\sin 2\pi x - c\sin 3\pi x)^2\,dx.$

To minimize $J$, choose $a$, $b$, and $c$ to satisfy

$$\begin{aligned} 0 &= \frac{\partial J}{\partial a} \\ &= -2\int_{-1}^1 (x - a\sin\pi x - b\sin 2\pi x - c\sin 3\pi x)\sin\pi x\,dx \\ &= \frac{2}{\pi}(\pi a - 2) \\ 0 &= \frac{\partial J}{\partial b} \\ &= -2\int_{-1}^1 (x - a\sin\pi x - b\sin 2\pi x - c\sin 3\pi x)\sin 2\pi x\,dx \\ &= \frac{2}{\pi}(\pi b + 1) \\ 0 &= \frac{\partial J}{\partial c} \\ &= -2\int_{-1}^1 (x - a\sin\pi x - b\sin 2\pi x - c\sin 3\pi x)\sin 3\pi x\,dx \\ &= \frac{2}{3\pi}(3\pi c - 2). \end{aligned}$$

We have omitted the details of evaluation of these integrals, but note that

$$\int_{-1}^1 \sin m\pi x \sin n\pi x\,dx = 0$$

if $m$ and $n$ are different integers.

The equations above imply that $a = 2/\pi$, $b = -1/\pi$, and $c = 2/(3\pi)$. These are the values that minimize $J$.

**22.** The Fourier sine series coefficients for $f(x) = x$ on $(0, \pi)$ are

$$b_n = \frac{2}{\pi}\int_0^\pi x\sin(nx)\,dx = (-1)^{n-1}\frac{2}{n}$$

for $n = 1, 2, \ldots$. Thus the series is

$$\sum_{n=0}^\infty (-1)^{n-1}\frac{2}{n}\sin nx.$$

Since $x$ and the functions $\sin nx$ are all odd functions, we would also expect the series to converge to $x$ on $(-\pi, 0)$.

**24.** We are given that $x_1 \le x_2 \le x_3 \le \ldots \le x_n$.
To motivate the method, look at a special case, $n = 5$ say.

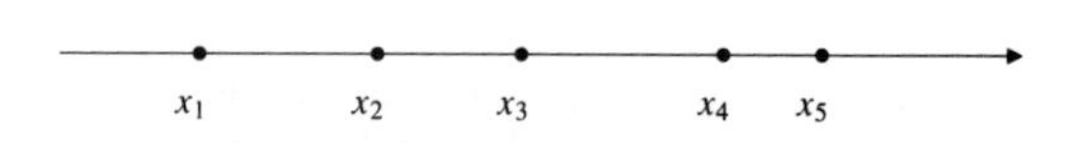


Fig. 5.4.24

If $x = x_3$, then

$$\begin{aligned} &\sum_{i=1}^5 |x - x_i| \\ &= (x_3 - x_1) + (x_3 - x_2) + 0 + (x_4 - x_3) + (x_5 - x_3) \\ &= (x_5 - x_1) + (x_4 - x_2). \end{aligned}$$

If $x$ moves away from $x_3$ in either direction, then

$$\sum_{i=1}^5 |x - x_i| = (x_5 - x_1) + (x_4 - x_2) + |x - x_3|.$$

Thus the minimum sum occurs if $x = x_3$.
In general, if $n$ is odd, then $\sum_{i=1}^n |x - x_i|$ is minimum if $x = x_{(n+1)/2}$, the middle point of the set of points $\{x_1, x_2, \ldots, x_n\}$. The value of $x$ is unique in this case.
If $n$ is even and $x$ satisfies $x_{n/2} \le x \le x_{(n/2)+1}$, then

$$\sum_{i=1}^n |x - x_i| = \sum_{i=1}^{n/2} |x_{n+1-i} - x_i|,$$

and the sum will increase if $x$ is outside that interval. In this case the value of $x$ which minimizes the sum is not unique unless it happens that $x_{n/2} = x_{(n/2)+1}$.

## Section 5.5 Parametric Problems (page 258)

**2.** $\int_{-\infty}^{\infty} e^{-u^2}\,du = \sqrt{\pi}$ Let $u = xt$, $du = x\,dt$

$\int_{-\infty}^{\infty} e^{-x^2t^2}\,dt = \frac{\sqrt{\pi}}{x}.$

Differentiate with respect to $x$:

$$\int_{-\infty}^{\infty} -2xt^2 e^{-t^2x^2}\,dt = -\frac{\sqrt{\pi}}{x^2}$$

$$\int_{-\infty}^{\infty} t^2 e^{-x^2t^2}\,dt = \frac{\sqrt{\pi}}{2x^3}. \qquad (*)$$

If $x = 1$ we get $\int_{-\infty}^{\infty} t^2 e^{-t^2}\,dt = \frac{\sqrt{\pi}}{2}.$

Differentiate $(*)$ with respect to $x$ again:

$$\int_{-\infty}^{\infty} -2xt^4 e^{-x^2t^2}\,dt = -\frac{3\sqrt{\pi}}{2x^4}.$$

Divide by $-2$ and let $x = 1$:

$$\int_{-\infty}^{\infty} t^4 e^{-t^2}\,dt = \frac{3\sqrt{\pi}}{4}.$$

**4.** Let $I(x,y) = \int_0^1 \frac{t^x - t^y}{\ln t}\,dt$, where $x > -1$ and $y > -1$. Then

$$\frac{\partial I}{\partial x} = \int_0^1 t^x\,dt = \frac{1}{x+1}$$

$$\frac{\partial I}{\partial y} = -\frac{1}{y+1}.$$

Thus

$$I(x,y) = \int \frac{dx}{x+1} = \ln(x+1) + C_1(y)$$

$$\frac{-1}{y+1} = \frac{\partial I}{\partial y} = \frac{\partial C_1}{\partial y} \Rightarrow C_1(y) = -\ln(y+1) + C_2$$

$$I(x,y) = \ln\left(\frac{x+1}{y+1}\right) + C_2.$$

But $I(x,x) = 0$, so $C_2 = 0$. Thus

$$I(x,y) = \int_0^1 \frac{t^x - t^y}{\ln t}\,dt = \ln\left(\frac{x+1}{y+1}\right)$$

for $x > -1$ and $y > -1$.

**6.** $F(x) = \int_0^{\infty} e^{-xt}\frac{\sin t}{t}\,dt$

$F'(x) = \int_0^{\infty} -e^{-xt}\sin t\,dt = -\frac{1}{1+x^2} \quad (x > 0).$

Therefore $F(x) = -\int \frac{dx}{1+x^2} = -\tan^{-1}x + C.$

Now, make the change of variable $xt = s$ in the integral defining $F(x)$, and obtain

$$F(x) = \int_0^{\infty} e^{-s}\frac{\sin(s/x)}{s/x}\frac{ds}{x} = \int_0^{\infty}\frac{e^{-s}}{s}\sin\frac{s}{x}\,ds.$$

Since $|\sin(s/x)| \le s/x$ if $s > 0$, $x > 0$, we have

$$|F(x)| \le \frac{1}{|x|}\int_0^{\infty} e^{-s}\,ds = \frac{1}{|x|} \to 0 \quad \text{as } x \to \infty.$$

Hence $-\frac{\pi}{2} + C = 0$, and $C = \frac{\pi}{2}$. Therefore

$$F(x) = \int_0^{\infty} e^{-xt}\frac{\sin t}{t}\,dt = \frac{\pi}{2} - \tan^{-1}x.$$

In particular, $\int_0^{\infty}\frac{\sin t}{t}\,dt = \lim_{x\to 0} F(x) = \frac{\pi}{2}.$

**8.** $\int_0^x \frac{dt}{x^2+t^2} = \frac{1}{x}\tan^{-1}\frac{t}{x}\Big|_0^x = \frac{\pi}{4x}$ for $x > 0$.

Differentiate with respect to $x$:

$$\frac{1}{2x^2} + \int_0^x \frac{-2x\,dt}{(x^2+t^2)^2} = -\frac{\pi}{4x^2}$$

$$\int_0^x \frac{dt}{(x^2+t^2)^2} = -\frac{1}{2x}\left[-\frac{\pi}{4x^2} - \frac{1}{2x^2}\right] = \frac{\pi}{8x^3} + \frac{1}{4x^3}.$$

Differentiate with respect to $x$ again:

$$\frac{1}{4x^4} + \int_0^x \frac{-4x\,dt}{(x^2+t^2)^3} = -\frac{3}{x^4}\left[\frac{\pi}{8} + \frac{1}{4}\right]$$

$$\int_0^x \frac{dt}{(x^2+t^2)^3} = -\frac{1}{4x}\left[-\frac{3\pi}{8x^4} - \frac{3}{4x^4} - \frac{1}{4x^4}\right] = \frac{3\pi}{32x^5} + \frac{1}{4x^5}.$$

**10.** $f(x) = Cx + D + \int_0^x (x-t)f(t)\,dt \quad \Rightarrow \quad f(0) = D$

$f'(x) = C + \int_0^x f(t)\,dt \quad \Rightarrow \quad f'(0) = C$

$f''(x) = f(x) \quad \Rightarrow \quad f(x) = A\cosh x + B\sinh x$

$D = f(0) = A, \qquad C = f'(0) = B$

$\Rightarrow f(x) = D\cosh x + C\sinh x.$

**12.** $f(x) = 1 + \int_0^1 (x+t)f(t)\,dt$

$f'(x) = \int_0^1 f(t)\,dt = C, \quad \text{say},$

since the integral giving $f'(x)$ does not depend on $x$. Thus $f(x) = A + Cx$, where $A = f(0)$. Substituting this expression into the given equation, we obtain

$$\begin{aligned} A + Cx &= 1 + \int_0^1 (x+t)(A+Ct)\,dt \\ &= 1 + Ax + \frac{A}{2} + \frac{Cx}{2} + \frac{C}{3}. \end{aligned}$$

Therefore

$$\frac{A}{2} - 1 - \frac{C}{3} + x\left(\frac{C}{2} - A\right) = 0.$$

This can hold for all $x$ only if

$$\frac{A}{2} - 1 - \frac{C}{3} = 0 \quad \text{and} \quad \frac{C}{2} - A = 0.$$

Thus $C = 2A$ and $\frac{A}{2} - \frac{2A}{3} = 1$, so that $A = -6$ and $C = -12$. Therefore $f(x) = -6 - 12x$.

**14.** We eliminate $c$ from the pair of equations

$$\begin{aligned} f(x,y,c) &= y - (x-c)\cos c - \sin c = 0 \\ \frac{\partial}{\partial c} f(x,y,c) &= \cos c + (x-c)\sin c - \cos c = 0. \end{aligned}$$

Thus $c = x$ and $y - 0 - \sin x = 0$.
The envelope is $y = \sin x$.

**16.** We eliminate $c$ from the pair of equations

$$\begin{aligned} f(x,y,c) &= \frac{x}{\cos c} + \frac{y}{\sin c} - 1 = 0 \\ \frac{\partial}{\partial c} f(x,y,c) &= \frac{x\sin c}{\cos^2 c} - \frac{y\cos c}{\sin^2 c} = 0. \end{aligned}$$

From the second equation, $y = x\tan^3 c$. Thus

$$\frac{x}{\cos c}(1 + \tan^2 c) = 1$$

which implies that $x = \cos^3 c$, and hence $y = \sin^3 c$. The envelope is the astroid $x^{2/3} + y^{2/3} = 1$.

**18.** We eliminate $c$ from the pair of equations

$$\begin{aligned} f(x,y,c) &= (x-c)^2 + (y-c)^2 - 1 = 0 \\ \frac{\partial}{\partial c} f(x,y,c) &= 2(c-x) + 2(c-y) = 0. \end{aligned}$$

Thus $c = (x+y)/2$, and

$$\left(\frac{x-y}{2}\right)^2 + \left(\frac{y-x}{2}\right)^2 = 1$$

or $x - y = \pm\sqrt{2}$. These two parallel lines constitute the envelope of the given family which consists of circles of radius 1 with centres along the line $y = x$

**20.** The curve $x^2 + (y-c)^2 = kc^2$ is a circle with centre $(0,c)$ and radius $\sqrt{k}c$, provided $k > 0$. Consider the system:

$$\begin{aligned} f(x,y,c) &= x^2 + (y-c)^2 - kc^2 = 0 \\ \frac{\partial}{\partial c} f(x,y,c) &= -2(y-c) - 2kc = 0. \end{aligned}$$

The second equation implies that $y - c = -kc$, and the first equation then says that $x^2 = k(1-k)c^2$. This is only possible if $0 \le k \le 1$.
The cases $k = 0$ and $k = 1$ are degenerate. If $k = 0$ the "curves" are just points on the $y$-axis. If $k = 1$ the curves are circles, all of which are tangent to the $x$-axis at the origin. There is no reasonable envelope in either case. If $0 < k < 1$, the envelope is the pair of lines given by $x^2 = \frac{k}{1-k}y^2$, that is, the lines $\sqrt{1-k}\,x = \pm\sqrt{k}\,y$. These lines make angle $\sin^{-1}\sqrt{k}$ with the $y$-axis.

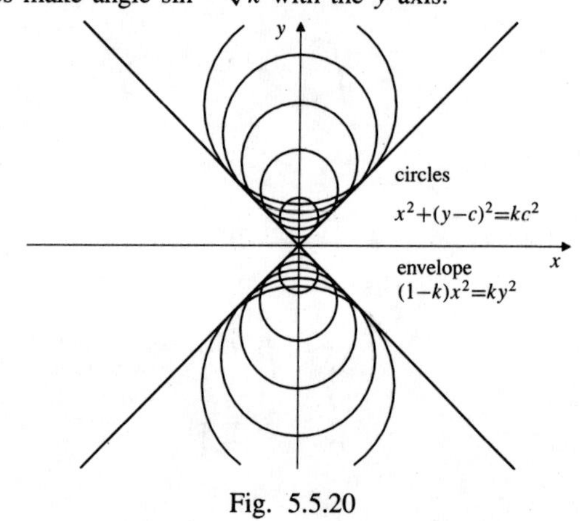


Fig. 5.5.20

**22.** If the family of surfaces $f(x,y,z,\lambda,\mu) = 0$ has an envelope, that envelope will have parametric equations

$$x = x(\lambda,\mu), \qquad y = y(\lambda,\mu), \qquad z = z(\lambda,\mu),$$

giving the point on the envelope where the envelope is tangent to the particular surface in the family having parameter values $\lambda$ and $\mu$. Thus

$$f\big(x(\lambda,\mu), y(\lambda,\mu), z(\lambda,\mu), \lambda, \mu\big) = 0.$$

Differentiating with respect to $\lambda$, we obtain

$$f_1\frac{\partial x}{\partial \lambda} + f_2\frac{\partial y}{\partial \lambda} + f_3\frac{\partial z}{\partial \lambda} + f_4 = 0.$$

However, since for fixed $\mu$, the parametric curve

$$x = x(t, \mu), \qquad y = y(t, \mu), \qquad z = z(t, \mu)$$

is tangent to the surface $f(x, y, z, \lambda, \mu) = 0$ at $t = \lambda$, its tangent vector there,

$$\mathbf{T} = \frac{\partial x}{\partial \lambda}\mathbf{i} + \frac{\partial y}{\partial \lambda}\mathbf{j} + \frac{\partial z}{\partial \lambda}\mathbf{k},$$

is perpendicular to the normal

$$\mathbf{N} = \nabla f = f_1\mathbf{i} + f_2\mathbf{j} + f_3\mathbf{k},$$

so

$$f_1\frac{\partial x}{\partial \lambda} + f_2\frac{\partial y}{\partial \lambda} + f_3\frac{\partial z}{\partial \lambda} = 0.$$

Hence we must also have $\dfrac{\partial f}{\partial \lambda} = f_4(x, y, z, \lambda, \mu) = 0.$
Similarly, $\dfrac{\partial f}{\partial \mu} = 0.$
The parametric equations of the envelope must therefore satisfy the three equations

$$\begin{aligned} f(x, y, z, \lambda, \mu) &= 0 \\ \frac{\partial}{\partial \lambda} f(x, y, z, \lambda, \mu) &= 0 \\ \frac{\partial}{\partial \mu} f(x, y, z, \lambda, \mu) &= 0. \end{aligned}$$

The envelope can be found by eliminating $\lambda$ and $\mu$ from these three equations.

**24.** $(x - \lambda)^2 + (y - \mu)^2 + z^2 = \dfrac{\lambda^2 + \mu^2}{2}.$
Differentiate with respect to $\lambda$ and $\mu$:

$$-2(x - \lambda) = \lambda, \qquad -2(y - \mu) = \mu.$$

Thus $\lambda = 2x$, $\mu = 2y$, and

$$x^2 + y^2 + z^2 = 2x^2 + 2y^2.$$

The envelope is the cone $z^2 = x^2 + y^2$.

**26.**
$$\begin{aligned} &y^2 + \epsilon e^{-y^2} = 1 + x^2 \\ &2yy_\epsilon + e^{-y^2} - 2y\epsilon e^{-y^2} y_\epsilon = 0 \\ &2y\left(1 - \epsilon e^{-y^2}\right) y_\epsilon + e^{-y^2} = 0 \\ &2y_\epsilon\left(1 - \epsilon e^{-y^2}\right) y_\epsilon - 2ye^{-y^2} y_\epsilon + 2y\left(2y\epsilon e^{-y^2} y_\epsilon\right) y_\epsilon \\ &\qquad + 2y\left(1 - \epsilon e^{-y^2}\right) y_{\epsilon\epsilon} - 2ye^{-y^2} y_\epsilon = 0. \end{aligned}$$

At $\epsilon = 0$ we have $y(x, 0) = \sqrt{1 + x^2}$, and

$$\begin{aligned} &2\sqrt{1 + x^2}\, y_\epsilon(x, 0) + e^{-(1+x^2)} = 0 \\ &y_\epsilon(x, 0) = -\frac{1}{2\sqrt{1 + x^2}} e^{-(1+x^2)} \\ &2y_\epsilon^2 - 4ye^{-y^2} y_\epsilon + 2yy_{\epsilon\epsilon} = 0 \\ &yy_{\epsilon\epsilon} = 2yy_\epsilon e^{-y^2} - y_\epsilon^2 \\ &y_{\epsilon\epsilon}(x, 0) = -\left(\frac{1}{\sqrt{1 + x^2}} + \frac{1}{4(1 + x^2)^{3/2}}\right) e^{-2(1+x^2)}. \end{aligned}$$

Thus

$$\begin{aligned} y = y(x, \epsilon) &= y(x, 0) + \epsilon y_\epsilon(x, 0) + \frac{\epsilon^2}{2!} y_{\epsilon\epsilon}(x, 0) + \cdots \\ &= \sqrt{1 + x^2} - \frac{\epsilon}{2\sqrt{1 + x^2}} e^{-(1+x^2)} \\ &\quad - \frac{\epsilon^2}{2}\left(\frac{1}{\sqrt{1 + x^2}} + \frac{1}{4(1 + x^2)^{3/2}}\right) e^{-2(1+x^2)} + \cdots. \end{aligned}$$

**28.** Let $y(x, \epsilon)$ be the solution of $y + \epsilon y^5 = \dfrac{1}{2}$. Then we have

$$\begin{aligned} &y_\epsilon\left(1 + 5\epsilon y^4\right) + y^5 = 0 \\ &y_{\epsilon\epsilon}\left(1 + 5\epsilon y^4\right) + 20\epsilon y^3 y_\epsilon^2 + 10y^4 y_\epsilon = 0 \\ &y_{\epsilon\epsilon\epsilon}\left(1 + 5\epsilon y^4\right) + y_{\epsilon\epsilon}\left(60\epsilon y^3 y_\epsilon + 15y^4\right) \\ &\qquad + 60\epsilon y_\epsilon^3 y^2 + 60y^3 y_\epsilon^2 = 0. \end{aligned}$$

At $\epsilon = 0$ we have

$$\begin{aligned} y(x, 0) &= \frac{1}{2} \\ y_\epsilon(x, 0) &= -\frac{1}{32} \\ y_{\epsilon\epsilon}(x, 0) &= -\frac{10}{16}\left(-\frac{1}{32}\right) = \frac{5}{16^2} \\ y_{\epsilon\epsilon\epsilon}(x, 0) &= -\frac{5}{16^2}\left(\frac{15}{16}\right) - \frac{60}{8}\left(-\frac{1}{32}\right)^2 = -\frac{105}{4096}. \end{aligned}$$

For $\epsilon = \dfrac{1}{100}$ we have

$$\begin{aligned} y &= \frac{1}{2} - \frac{1}{32} \times \frac{1}{100} + \frac{5}{256} \times \frac{1}{2 \times 100^2} \\ &\quad - \frac{105}{4096} \times \frac{1}{6 \times 100^3} + \cdots \\ &\approx 0.49968847 \end{aligned}$$

with error less than $10^{-8}$ in magnitude.

## Section 5.6 Newton's Method (page 263)

For each of Exercises 1–6, we sketch the graphs of the two given equations, $f(x, y) = 0$ and $g(x, y) = 0$, and use their intersections to make initial guesses $x_0$ and $y_0$ for the solutions. These guesses are then refined using the formulas

$$x_{n+1} = x_n - \left.\frac{fg_2 - gf_2}{f_1g_2 - g_1f_2}\right|_{(x_n,y_n)}, \qquad y_{n+1} = y_n - \left.\frac{f_1g - g_1f}{f_1g_2 - g_1f_2}\right|_{(x_n,y_n)}.$$

NOTE: The numerical values in the tables below were obtained by programming a microcomputer to calculate the iterations of the above formulas. In most cases the computer was using more significant digits than appear in the tables, and did not truncate the values obtained at one step before using them to calculate the next step. If you use a calculator, and use the numbers as quoted on one line of a table to calculate the numbers on the next line, your results may differ slightly (in the last one or two decimal places).

**2.** $f(x, y) = x^2 + y^2 - 1 \qquad f_1(x, y) = 2x \qquad g_1(x, y) = -e^x.$
$g(x, y) = y - e^x \qquad f_2(x, y) = 2y \qquad g_2(x, y) = 1$

Evidently one solution is $x = 0$, $y = 1$. The second solution is near $(-1, 0)$. We try $x_0 = -0.9$, $y_0 = 0.2$.

| $n$ | $x_n$ | $y_n$ | $f(x_n, y_n)$ | $g(x_n, y_n)$ |
|---|---|---|---|---|
| 0 | −0.9000000 | 0.2000000 | −0.1500000 | −0.2065697 |
| 1 | −0.9411465 | 0.3898407 | 0.0377325 | −0.0003395 |
| 2 | −0.9170683 | 0.3995751 | 0.0006745 | −0.0001140 |
| 3 | −0.9165628 | 0.3998911 | 0.0000004 | −0.0000001 |
| 4 | −0.9165626 | 0.3998913 | 0.0000000 | 0.0000000 |

The second solution is $x = -0.9165626$, $y = 0.3998913$.

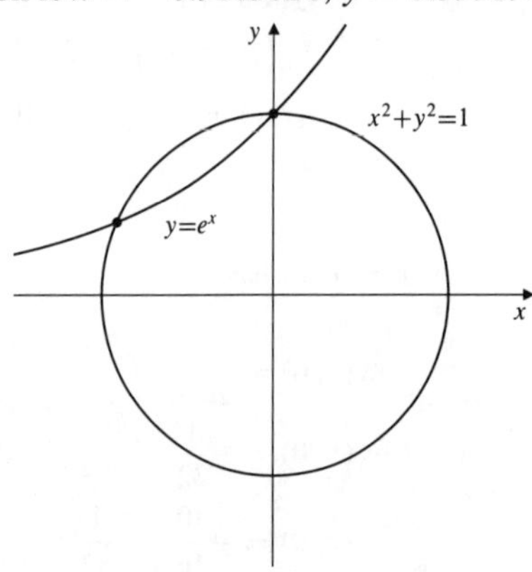


Fig. 5.6.2

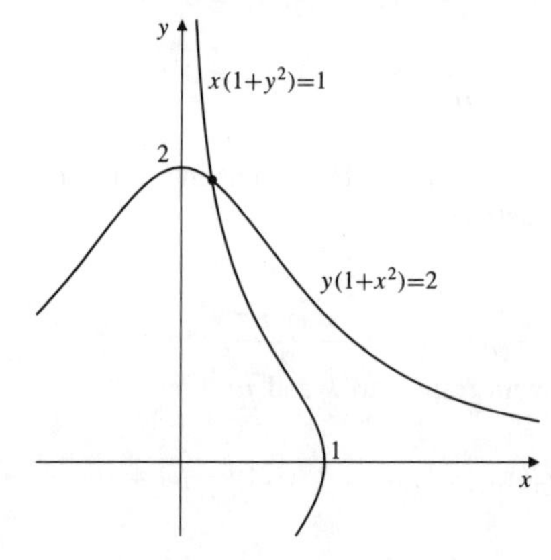


Fig. 5.6.4

**4.** $f(x, y) = x(1 + y^2) - 1 \qquad f_1(x, y) = 1 + y^2 \qquad g_1(x, y) = 2xy \quad .$
$g(x, y) = y(1 + x^2) - 2 \qquad f_2(x, y) = 2xy \qquad g_2(x, y) = 1 + x^2$

The solution appears to be near $x = 0.2$, $y = 1.8$.

| $n$ | $x_n$ | $y_n$ | $f(x_n, y_n)$ | $g(x_n, y_n)$ |
|---|---|---|---|---|
| 0 | 0.2000000 | 1.8000000 | −0.1520000 | −0.1280000 |
| 1 | 0.2169408 | 1.9113487 | 0.0094806 | 0.0013031 |
| 2 | 0.2148268 | 1.9117785 | −0.0000034 | 0.0000081 |
| 3 | 0.2148292 | 1.9117688 | 0.0000000 | 0.0000000 |

The solution is $x = 0.2148292$, $y = 1.9117688$.

**6.**

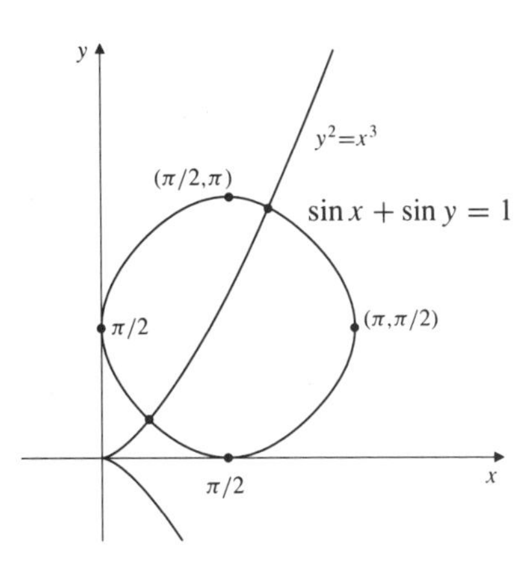


Fig. 5.6.6

$$f(x,y)=\sin x+\sin y-1 \qquad f_1(x,y)=\cos x \qquad g_1(x,y)=-3x^2.$$
$$g(x,y)=y^2-x^3 \qquad f_2(x,y)=\cos y \qquad g_2(x,y)=2y$$

There are infinitely many solutions for the given pair of equations, since the level curve of $f(x,y)=0$ is repeated periodically throughout the plane. We will find the two solutions closest to the origin in the first quadrant. From the figure, it appears that these solutions are near (0.6, 0.4) and (2, 3).

| $n$ | $x_n$ | $y_n$ | $f(x_n,y_n)$ | $g(x_n,y_n)$ |
|---|---|---|---|---|
| 0 | 0.6000000 | 0.4000000 | −0.0459392 | −0.0560000 |
| 1 | 0.5910405 | 0.4579047 | −0.0007050 | 0.0032092 |
| 2 | 0.5931130 | 0.4567721 | −0.0000015 | −0.0000063 |
| 3 | 0.5931105 | 0.4567761 | 0.0000000 | 0.0000000 |
| 4 | 0.5931105 | 0.4567761 | 0.0000000 | 0.0000000 |

| $n$ | $x_n$ | $y_n$ | $f(x_n,y_n)$ | $g(x_n,y_n)$ |
|---|---|---|---|---|
| 0 | 2.0000000 | 3.0000000 | 0.0504174 | 1.0000000 |
| 1 | 2.0899016 | 3.0131366 | −0.0036336 | −0.0490479 |
| 2 | 2.0854887 | 3.0116804 | −0.0000086 | −0.0001199 |
| 3 | 2.0854779 | 3.0116770 | 0.0000000 | 0.0000000 |
| 4 | 2.0854779 | 3.0116770 | 0.0000000 | 0.0000000 |

The solutions are $x=0.5931105$, $y=0.4567761$, and $x=2.0854779$, $y=3.0116770$.

**8.**

$$f(x,y,z)=y^2+z^2-3 \qquad g(x,y,z)=x^2+z^2-2 \qquad h(x,y,z)=x^2-z$$
$$f_1(x,y,z)=0 \qquad g_1(x,y,z)=2x \qquad h_1(x,y,z)=2x$$
$$f_2(x,y,z)=2y \qquad g_2(x,y,z)=0 \qquad h_2(x,y,z)=0$$
$$f_3(x,y,z)=2z \qquad g_3(x,y,z)=2z \qquad h_3(x,y,z)=-1$$

It is easily seen that the system

$$f(x,y,z)=0, \qquad g(x,y,z)=0, \qquad h(x,y,z)=0$$

has first-quadrant solution $x=z=1$, $y=\sqrt{2}$. Let us start at the "guess" $x_0=y_0=z_0=2$.

| $n$ | $x_n$ | $y_n$ | $z_n$ | $f(x_n, y_n, z_n)$ | $g(x_n, y_n, z_n)$ | $h(x_n, y_n, z_n)$ |
|---|---|---|---|---|---|---|
| 0 | 2.0000000 | 2.0000000 | 2.0000000 | 5.0000000 | 6.0000000 | 2.0000000 |
| 1 | 1.3000000 | 1.5500000 | 1.2000000 | 0.8425000 | 1.1300000 | 0.4900000 |
| 2 | 1.0391403 | 1.4239564 | 1.0117647 | 0.0513195 | 0.1034803 | 0.0680478 |
| 3 | 1.0007592 | 1.4142630 | 1.0000458 | 0.0002313 | 0.0016104 | 0.0014731 |
| 4 | 1.0000003 | 1.4142136 | 1.0000000 | 0.0000000 | 0.0000006 | 0.0000006 |
| 5 | 1.0000000 | 1.4142136 | 1.0000000 | 0.0000000 | 0.0000000 | 0.0000000 |

## Review Exercises 5 (page 263)

**2.** $f(x, y) = x^2y - 2xy^2 + 2xy$
$f_1(x, y) = 2xy - 2y^2 + 2y = 2y(x - y + 1)$
$f_2(x, y) = x^2 - 4xy + 2x = x(x - 4y + 2)$
$A = f_{11} = 2y$
$B = f_{12} = 2x - 4y + 2$
$C = f_{22} = -4x.$
For CP: either $y = 0$ or $x - y + 1 = 0$, and either $x = 0$ or $x - 4y + 2 = 0$. The CPs are $(0, 0)$, $(0, 1)$, $(-2, 0)$, and $(-2/3, 1/3)$.

| CP | $A$ | $B$ | $C$ | $AC - B^2$ | class |
|---|---|---|---|---|---|
| $(0, 0)$ | 0 | 2 | 0 | $-4$ | saddle |
| $(0, 1)$ | 2 | $-2$ | 0 | $-4$ | saddle |
| $(-2, 0)$ | 0 | $-2$ | 8 | $-4$ | saddle |
| $(-\frac{2}{3}, \frac{1}{3})$ | $\frac{2}{3}$ | $-\frac{2}{3}$ | $\frac{8}{3}$ | $\frac{4}{3}$ | loc. min |

**4.** $f(x, y) = x^2y(2 - x - y) = 2x^2y - x^3y - x^2y^2$
$f_1(x, y) = 4xy - 3x^2y - 2xy^2 = xy(4 - 3x - 2y)$
$f_2(x, y) = 2x^2 - x^3 - 2x^2y = x^2(2 - x - 2y)$
$A = f_{11} = 4y - 6xy - 2y^2$
$B = f_{12} = 4x - 3x^2 - 4xy$
$C = f_{22} = -2x^2.$
$(0, y)$ is a CP for any $y$. If $x \neq 0$ but $y = 0$, then $x = 2$ from the second equation. Thus $(2, 0)$ is a CP. If neither $x$ nor $y$ is 0, then $x + 2y = 2$ and $3x + 2y = 4$, so that $x = 1$ and $y = 1/2$. The third CP is $(1, 1/2)$.

| CP | $A$ | $B$ | $C$ | $AC - B^2$ | class |
|---|---|---|---|---|---|
| $(0, y)$ | $4y - 2y^2$ | 0 | 0 | 0 | ? |
| $(2, 0)$ | 0 | $-4$ | $-8$ | $-16$ | saddle |
| $(1, \frac{1}{2})$ | $-\frac{3}{2}$ | $-1$ | $-2$ | 2 | loc. max |

The second derivative test is unable to classify the line of critical points along the $y$-axis. However, direct inspection of $f(x, y)$ shows that these are local minima if $y(2 - y) > 0$ (that is, if $0 < y < 2$) and local maxima if $y(2 - y) < 0$ (that is, if $y < 0$ or $y > 2$). The points $(0, 0)$ and $(0, 2)$ are neither maxima nor minima, so they are saddle points.

**6.**
$$x^2 + y^2 + z^2 - xy - xz - yz$$
$$= \frac{1}{2}\big[(x^2 - 2xy + y^2) + (x^2 - 2xz + z^2) + (y^2 - 2yz + z^2)\big]$$
$$= \frac{1}{2}\big[(x - y)^2 + (x - z)^2 + (y - z)^2\big] \geq 0.$$
The minimum value, 0, is assumed at the origin and at all points of the line $x = y = z$.

**8.** $f(x, y) = (4x^2 - y^2)e^{-x^2+y^2}$
$f_1(x, y) = e^{-x^2+y^2}2x(4 - 4x^2 + y^2)$
$f_2(x, y) = e^{-x^2+y^2}(-2y)(1 - 4x^2 + y^2).$
$f$ has CPs $(0, 0)$, $(\pm 1, 0)$. $f(0, 0) = 0$.
$f(\pm 1, 0) = 4/e$.

a) Since $f(0, y) = -y^2e^{y^2} \to -\infty$ as $y \to \pm\infty$, and since $f(x, x) = 3x^2e^0 = 3x^2 \to \infty$ as $x \to \pm\infty$, $f$ does not have a minimum or a maximum value on the $xy$-plane.

b) On $y = 3x$, $f(x, 3x) = -5x^2e^{8x^2} \to -\infty$ as $x \to \infty$. Thus $f$ can have no minimum value on the wedge $0 \leq y \leq 3x$. However, as noted in (a), $f(x, x) \to \infty$ as $x \to \infty$. Since $(x, x)$ is in the wedge for $x > 0$, $f$ cannot have a maximum value on the wedge either.

**10.** Let the length, width, and height of the box be $x$, $y$, and $z$ in, respectively. Then the girth is $g = 2x + 2y$. We require $g + z \leq 120$ in. The volume $V = xyz$ of the box will be maximized under the constraint $2x + 2y + z = 120$, so we look for CPs of

$$L = xyz + \lambda(2x + 2y + z - 120).$$

For CPs:

$$\begin{aligned}
0 &= \frac{\partial L}{\partial x} = yz + 2\lambda && \text{(A)}\\
0 &= \frac{\partial L}{\partial y} = xz + 2\lambda && \text{(B)}\\
0 &= \frac{\partial L}{\partial z} = xy + \lambda && \text{(C)}\\
0 &= \frac{\partial L}{\partial \lambda} = 2x + 2y + z - 120. && \text{(D)}
\end{aligned}$$

Comparing (A), (B), and (C), we see that $x = y = z/2$. Then (D) implies that $3z = 120$, so $z = 40$ and $x = y = 20$ in. The largest box has volume

$$V = (20)(20)(40) = 16,000 \text{ in}^3,$$

or, about 9.26 cubic feet.

**12.** The ellipsoid $(x/a)^2 + (y/b)^2 + (z/c)^2 = 1$ contains the rectangle $-1 \le x \le 1$, $-2 \le y \le 2$, $-3 \le z \le 3$, provided $(1/a^2) + (4/b^2) + (9/c)^2 = 1$. The volume of the ellipsoid is $V = 4\pi abc/3$. We minimize $V$ by looking for critical points of

$$L = \frac{4\pi}{3}abc + \lambda\left(\frac{1}{a^2} + \frac{4}{b^2} + \frac{9}{c^2} - 1\right).$$

For CPs:

$$\begin{aligned}
0 &= \frac{\partial L}{\partial a} = \frac{4\pi}{3}bc - \frac{2\lambda}{a^3} && \text{(A)}\\
0 &= \frac{\partial L}{\partial b} = \frac{4\pi}{3}ac - \frac{8\lambda}{b^3} && \text{(B)}\\
0 &= \frac{\partial L}{\partial c} = \frac{4\pi}{3}ab - \frac{18\lambda}{c^3} && \text{(C)}\\
0 &= \frac{\partial L}{\partial \lambda} = \frac{1}{a^2} + \frac{4}{b^2} + \frac{9}{c^2} - 1. && \text{(D)}
\end{aligned}$$

Multiplying (A) by $a$, (B) by $b$, and (C) by $c$, we obtain $2\lambda/a^2 = 8\lambda/b^2 = 18\lambda/c^2$, so that either $\lambda = 0$ or $b = 2a$, $c = 3a$. Now $\lambda = 0$ implies $bc = 0$, which is inconsistent with (D). If $b = 2a$ and $c = 3a$, then (D) implies that $3/a^2 = 1$, so $a = \sqrt{3}$. The smallest volume of the ellipsoid is

$$V = \frac{4\pi}{3}(\sqrt{3})(2\sqrt{3})(3\sqrt{3}) = 24\sqrt{3}\pi \text{ cubic units.}$$

**14.**

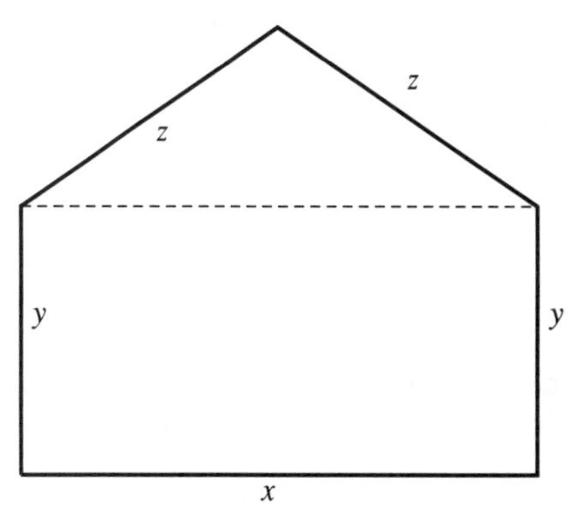


Fig. R-5.14

The area of the window is

$$A = xy + \frac{x}{2}\sqrt{z^2 - \frac{x^2}{4}},$$

or, since $x + 2y + 2z = L$,

$$A = \frac{x}{2}\left(L - x - 2z + \sqrt{z^2 - \frac{x^2}{4}}\right).$$

For maximum $A$, we look for critical points:

$$\begin{aligned}
0 = \frac{\partial A}{\partial x} &= \frac{1}{2}\left(L - x - 2z + \sqrt{z^2 - \frac{x^2}{4}}\right)\\
&\quad + \frac{x}{2}\left(-1 - \frac{x}{4\sqrt{z^2 - \frac{x^2}{4}}}\right)\\
&= \frac{L}{2} - x - z + \frac{2z^2 - x^2}{4\sqrt{z^2 - \frac{x^2}{4}}} && \text{(A)}\\
0 = \frac{\partial A}{\partial z} &= -x + \frac{xz}{2\sqrt{z^2 - \frac{x^2}{4}}}. && \text{(B)}
\end{aligned}$$

Now (B) implies that either $x = 0$ or $z = 2\sqrt{z^2 - (x^2/4)}$. But $x = 0$ gives zero area rather than maximum area, so the second alternative must hold, and it implies that $z = x/\sqrt{3}$. Then (A) gives

$$\frac{L}{2} = \left(1 + \frac{1}{\sqrt{3}}\right)x + \frac{x}{2\sqrt{3}},$$

from which we obtain $x = L/(2+\sqrt{3})$. The maximum area of the window is, therefore,

$$A\bigg|_{x=\frac{L}{2+\sqrt{3}},\ z=\frac{L/\sqrt{3}}{2+\sqrt{3}}} = \frac{1}{4}\frac{L^2}{2+\sqrt{3}}$$
$$\approx 0.0670L^2 \text{ sq. units.}$$

**16.** The envelope of $y = (x-c)^3 + 3c$ is found by eliminating $c$ from that equation and

$$0 = \frac{\partial}{\partial c}[(x-c)^3 + 3c] = -3(x-c)^2 + 3.$$

This later equation implies that $(x-c)^2 = 1$, so $x - c = \pm 1$.
The envelope is $y = (\pm 1)^3 + 3(x \mp 1)$, or $y = 3x \pm 2$.

**18.** Oops, there must be a mistake in the question. As it stands, the answer is easily seen to be $x = 1$, $y = 0$, or vice versa $x = 0$, $y = 1$. Starting at other points nearby and using Newton's method would lead to these solutions.

## Challenging Problems 5 (page 264)

**2.** If $f(x) = \begin{cases} 0 & \text{for } -\pi \le x < 0 \\ x & \text{for } 0 \le x \le \pi \end{cases}$, then

$$a_0 = \frac{1}{\pi}\int_0^\pi x\,dx = \frac{\pi}{2}$$
$$a_k = \frac{1}{\pi}\int_0^\pi x\cos kx\,dx$$
$$\begin{array}{ll} U = x & dV = \cos kx\,dx \\ dU = dx & V = \frac{1}{k}\sin kx \end{array}$$
$$= \frac{1}{\pi k}\left(x\sin kx\bigg|_0^\pi - \int_0^\pi \sin kx\,dx\right)$$
$$= \frac{\cos k\pi - 1}{\pi k^2} = \begin{cases} 0 & \text{if } k \text{ is even} \\ -\frac{2}{\pi k^2} & \text{if } k \text{ is odd} \end{cases}$$
$$b_k = \frac{1}{\pi}\int_0^\pi x\sin kx\,dx$$
$$\begin{array}{ll} U = x & dV = \sin kx\,dx \\ dU = dx & V = -\frac{1}{k}\cos kx \end{array}$$
$$= -\frac{1}{\pi k}\left(x\cos kx\bigg|_0^\pi - \int_0^\pi \cos kx\,dx\right)$$
$$= \frac{(-1)^{k+1}}{k}.$$

Using properties of trigonometric integrals,

$$\int_{-\pi}^{\pi}\left(\frac{a_0}{2} + \sum_{k=1}^{n}(a_k\cos kx + b_k\sin kx)\right)^2 dx$$
$$= \frac{\pi a_0^2}{2} + \pi\sum_{k=0}^{n}(a_k^2 + b_k^2)$$
$$\int_{-\pi}^{\pi} f(x)\left(\frac{a_0}{2} + \sum_{k=1}^{n}(a_k\cos kx + b_k\sin kx)\right) dx$$
$$= \frac{\pi a_0^2}{2} + \pi\sum_{k=0}^{n}(a_k^2 + b_k^2).$$

Therefore

$$I_n = \int_{-\pi}^{\pi}\left[f(x) - \left(\frac{a_0}{2} + \sum_{k=1}^{n}(a_k\cos kx + b_k\sin kx)\right)\right]^2 d$$
$$= \int_{-\pi}^{\pi}(f(x))^2\,dx - 2\left(\frac{\pi a_0^2}{2} + \pi\sum_{k=0}^{n}(a_k^2 + b_k^2)\right)$$
$$+ \frac{\pi a_0^2}{2} + \pi\sum_{k=0}^{n}(a_k^2 + b_k^2)$$
$$= \int_{-\pi}^{\pi}(f(x))^2\,dx - \left(\frac{\pi a_0^2}{2} + \pi\sum_{k=0}^{n}(a_k^2 + b_k^2)\right).$$

In fact, it can be shown that $I_n \to 0$ as $n \to \infty$.

**4.**

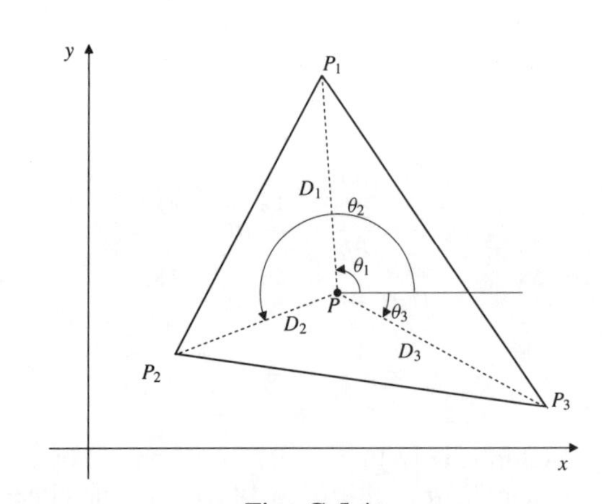


Fig. C-5.4

If $D_i = |PP_i|$ for $i = 1, 2, 3$, then

$$D_i^2 = (x - x_i)^2 + (y - y_i)^2$$
$$2D_i\frac{\partial D_i}{\partial x} = 2(x - x_i)$$
$$\frac{\partial D_i}{\partial x} = \frac{x - x_i}{D_i} = \cos\theta_i$$

where $\theta_i$ is the angle between $\overrightarrow{PP_i}$ and $\mathbf{i}$. Similarly $\partial D_i/\partial y = \sin\theta_i$. To minimize $S = D_1 + D_2 + D_3$ we look for critical points:

$$0 = \frac{\partial S}{\partial x} = \cos\theta_1 + \cos\theta_2 + \cos\theta_3$$

$$0 = \frac{\partial S}{\partial y} = \sin\theta_1 + \sin\theta_2 + \sin\theta_3.$$

Thus $\cos\theta_1 + \cos\theta_2 = -\cos\theta_3$ and $\sin\theta_1 + \sin\theta_2 = -\sin\theta_3$. Squaring and adding these two equations we get

$$2 + 2(\cos\theta_1\cos\theta_2 + \sin\theta_1\sin\theta_2) = 1,$$

or $\cos(\theta_1 - \theta_2) = -1/2$. Thus $\theta_1 - \theta_2 = \pm 2\pi/3$. Similarly $\theta_1 - \theta_3 = \theta_2 - \theta_3 = \pm 2\pi/3$. Thus $P$ should be chosen so that $\overrightarrow{PP_1}$, $\overrightarrow{PP_2}$,and $\overrightarrow{PP_3}$ make $120°$ angles with each other. This is possible only if all three angles of the triangle are less than $120°$. If the triangle has an angle of $120°$ or more (say at $P_1$), then $P$ should be that point on the side $P_2P_3$ such that $PP_1 \perp P_2P_3$.

## CHAPTER 6. MULTIPLE INTEGRATION

### Section 6.1 Double Integrals (page 270)

**2.** $$R = 1 \times \big[f(1,1) + f(1,2) + f(2,1) + f(2,2) + f(3,1) + f(3,2)\big] = 3 + 2 + 2 + 1 + 1 + 0 = 9$$

**4.** $$R = 1 \times \big[f(1,0) + f(1,1) + f(2,0) + f(2,1) + f(3,0) + f(3,1)\big] = 4 + 3 + 3 + 2 + 2 + 1 = 15$$

**6.** $I = \iint_D (5 - x - y)\, dA$ is the volume of the solid in the figure.

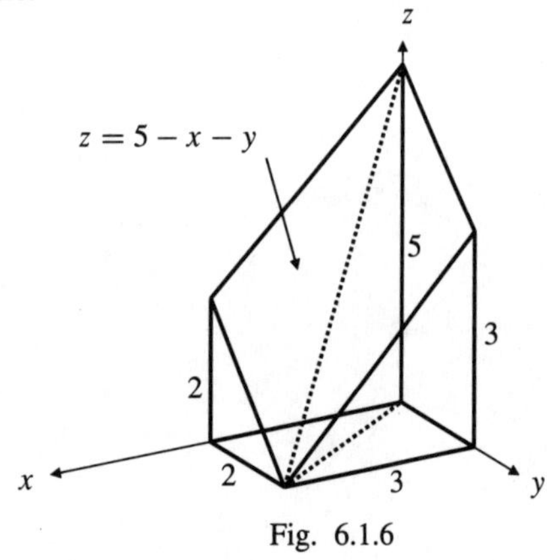


Fig. 6.1.6

The solid is split by the vertical plane through the $z$-axis and the point $(3, 2, 0)$ into two pyramids, each with a trapezoidal base; one pyramid's base is in the plane $y = 0$ and the other's is in the plane $z = 0$. $I$ is the sum of the volumes of these pyramids:

$$I = \frac{1}{3}\left(\frac{5+2}{2}(3)(2)\right) + \frac{1}{3}\left(\frac{5+3}{2}(2)(3)\right) = 15.$$

**8.** $R = 4 \times 1 \times [4 + 4 + 4 + 3 + 0] = 60$

**10.** $J = \text{area of disk} = \pi(5^2) \approx 78.54$

**12.** $f(x, y) = x^2 + y^2$

$$\begin{aligned} R = 4 \times 1 \times \big[ & f(\tfrac{1}{2},\tfrac{1}{2}) + f(\tfrac{3}{2},\tfrac{1}{2}) + f(\tfrac{5}{2},\tfrac{1}{2}) + f(\tfrac{7}{2},\tfrac{1}{2}) \\ & + f(\tfrac{9}{2},\tfrac{1}{2}) + f(\tfrac{1}{2},\tfrac{3}{2}) + f(\tfrac{3}{2},\tfrac{3}{2}) + f(\tfrac{5}{2},\tfrac{3}{2}) \\ & + f(\tfrac{7}{2},\tfrac{3}{2}) + f(\tfrac{9}{2},\tfrac{3}{2}) \\ & + f(\tfrac{1}{2},\tfrac{5}{2}) + f(\tfrac{3}{2},\tfrac{5}{2}) + f(\tfrac{5}{2},\tfrac{5}{2}) + f(\tfrac{7}{2},\tfrac{5}{2}) \\ & + f(\tfrac{1}{2},\tfrac{7}{2}) + f(\tfrac{3}{2},\tfrac{7}{2}) + f(\tfrac{5}{2},\tfrac{7}{2}) + f(\tfrac{1}{2},\tfrac{9}{2}) + f(\tfrac{3}{2},\tfrac{9}{2})\big] \\ = 918 & \end{aligned}$$

**14.** $$\begin{aligned} \iint_D (x+3)\, dA &= \iint_D x\, dA + 3\iint_D dA \\ &= 0 + 3(\text{area of } D) \\ &= 3 \times \frac{\pi 2^2}{2} = 6\pi. \end{aligned}$$

The integral of $x$ over $D$ is zero because $D$ is symmetrical about $x = 0$.

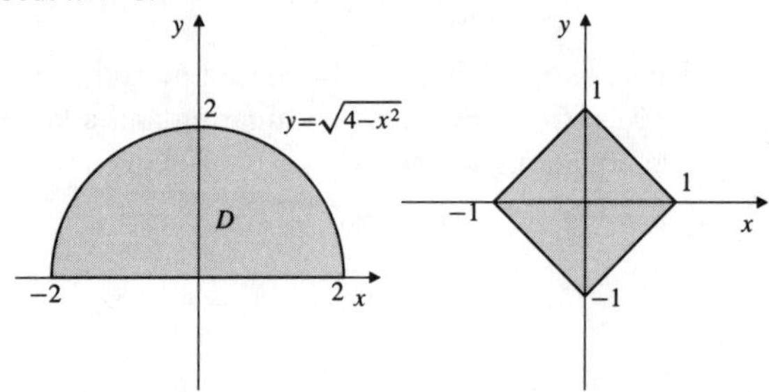


Fig. 6.1.14 Fig. 6.1.16

**16.** $$\begin{aligned} &\iint_{|x|+|y|\le 1} \left(x^3\cos(y^2) + 3\sin y - \pi\right) dA \\ &= 0 + 0 - \pi\left(\text{area bounded by } |x| + |y| = 1\right. \\ &= -\pi \times 4 \times \frac{1}{2}(1)(1) = -2\pi. \end{aligned}$$

(Each of the first two terms in the integrand is an odd function of one of the variables, and the square is symmetrical about each coordinate axis.)

**18.** $$\begin{aligned} &\iint_{x^2+y^2\le a^2} \sqrt{a^2 - x^2 - y^2}\, dA \\ &= \text{volume of hemisphere shown in the figure} \\ &= \frac{1}{2}\left(\frac{4}{3}\pi a^3\right) = \frac{2}{3}\pi a^3. \end{aligned}$$

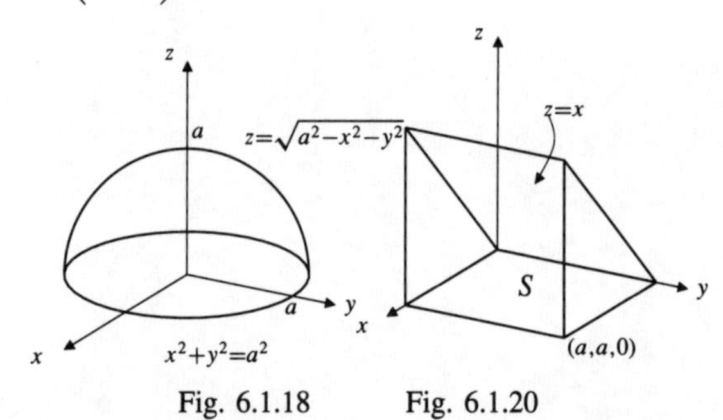


Fig. 6.1.18 Fig. 6.1.20

**20.** By the symmetry of $S$ with respect to $x$ and $y$ we have

$$\begin{aligned} \iint_S (x+y)\, dA &= 2\iint_S x\, dA \\ &= 2 \times (\text{volume of wedge shown in the figure}) \\ &= 2 \times \frac{1}{2}(a^2)a = a^3. \end{aligned}$$

**22.** $\displaystyle\iint_R \sqrt{b^2-y^2}\,dA$

= volume of the quarter cylinder shown in the figure

$\displaystyle = \frac{1}{4}(\pi b^2)a = \frac{1}{4}\pi ab^2.$

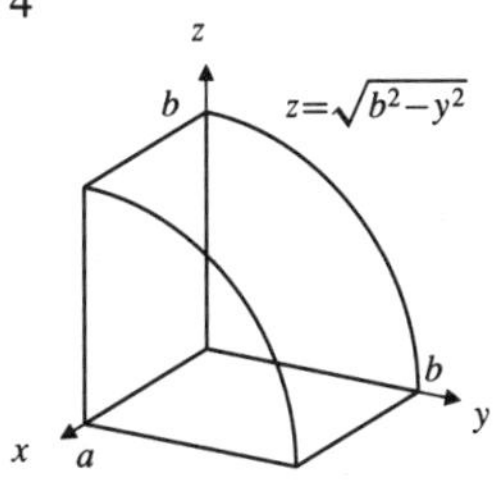


Fig. 6.1.22

## Section 6.2 Iteration of Double Integrals in Cartesian Coordinates (page 277)

**2.** $\displaystyle\int_0^1\int_0^y (xy+y^2)\,dx\,dy$

$\displaystyle = \int_0^1 \left(\frac{x^2y}{2}+xy^2\right)\bigg|_{x=0}^{x=y} dy$

$\displaystyle = \frac{3}{2}\int_0^1 y^3\,dy = \frac{3}{8}.$

**4.** $\displaystyle\int_0^2 dy\int_0^y y^2e^{xy}\,dx$

$\displaystyle = \int_0^2 y^2\,dy\left(\frac{1}{y}e^{xy}\bigg|_{x=0}^{x=y}\right)$

$\displaystyle = \int_0^2 y(e^{y^2}-1)\,dy = \frac{e^{y^2}-y^2}{2}\bigg|_0^2 = \frac{e^4-5}{2}.$

**6.** $\displaystyle\iint_R x^2y^2\,dA = \int_0^a x^2\,dx\int_0^b y^2\,dy$

$\displaystyle = \frac{a^3}{3}\,\frac{b^3}{3} = \frac{a^3b^3}{9}.$

**8.** $\displaystyle\iint_T (x-3y)\,dA = \int_0^a dx\int_0^{b(1-(x/a))}(x-3y)\,dy$

$\displaystyle = \int_0^a dx\left(xy-\frac{3}{2}y^2\right)\bigg|_{y=0}^{y=b(1-(x/a))}$

$\displaystyle = \int_0^a \left[b\left(x-\frac{x^2}{a}\right)-\frac{3}{2}b^2\left(1-\frac{2x}{a}+\frac{x^2}{a^2}\right)\right]dx$

$\displaystyle = \left(b\frac{x^2}{2}-\frac{b}{a}\frac{x^3}{3}-\frac{3}{2}b^2x+\frac{3}{2}\frac{b^2x^2}{a}-\frac{1}{2}\frac{b^2x^3}{a^2}\right)\bigg|_0^a$

$\displaystyle = \frac{a^2b}{6}-\frac{ab^2}{2}.$

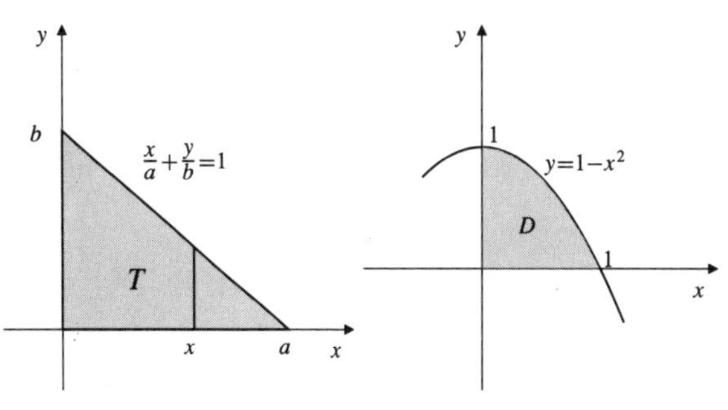


Fig. 6.2.8 Fig. 6.2.10

**10.** $\displaystyle\iint_D x\cos y\,dA$

$\displaystyle = \int_0^1 x\,dx\int_0^{1-x^2}\cos y\,dy$

$\displaystyle = \int_0^1 x\,dx\,(\sin y)\bigg|_{y=0}^{y=1-x^2}$

$\displaystyle = \int_0^1 x\sin(1-x^2)\,dx$ Let $u=1-x^2$, $du=-2x\,dx$

$\displaystyle = -\frac{1}{2}\int_1^0 \sin u\,du = \frac{1}{2}\cos u\bigg|_1^0 = \frac{1-\cos(1)}{2}.$

**12.** $\displaystyle\iint_T \sqrt{a^2-y^2}\,dA = \int_0^a \sqrt{a^2-y^2}\,dy\int_y^a dx$

$\displaystyle = \int_0^a (a-y)\sqrt{a^2-y^2}\,dy$

$\displaystyle = a\int_0^a \sqrt{a^2-y^2}\,dy - \int_0^a y\sqrt{a^2-y^2}\,dy$

Let $u=a^2-y^2$, $du=-2y\,dy$

$\displaystyle = a\frac{\pi a^2}{4}+\frac{1}{2}\int_{a^2}^0 u^{1/2}\,du$

$\displaystyle = \frac{\pi a^3}{4}-\frac{1}{3}u^{3/2}\bigg|_0^{a^2} = \left(\frac{\pi}{4}-\frac{1}{3}\right)a^3.$

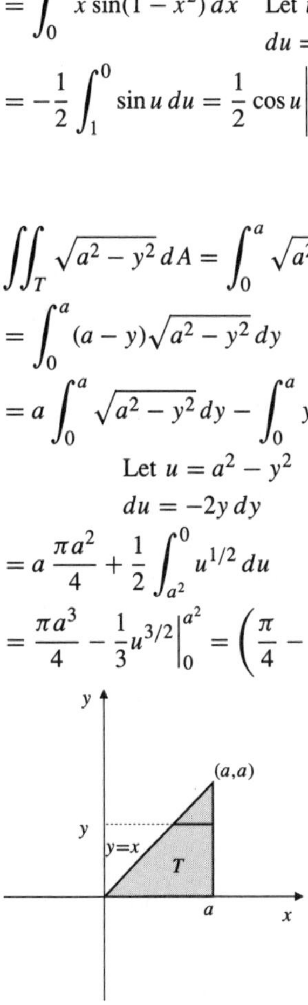

Fig. 6.2.12 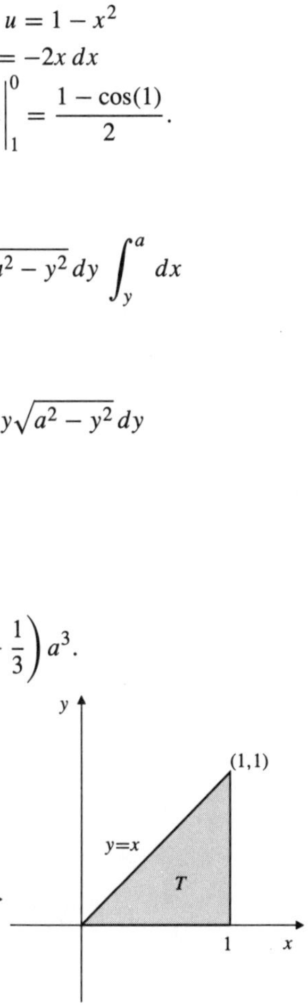 Fig. 6.2.14

**14.** $$\iint_T \frac{xy}{1+x^4}\,dA = \int_0^1 \frac{x}{1+x^4}\,dx \int_0^x y\,dy$$
$$= \frac{1}{2}\int_0^1 \frac{x^3}{1+x^4}\,dx$$
$$= \frac{1}{8}\ln(1+x^4)\Big|_0^1 = \frac{\ln 2}{8}.$$

**16.** $$\int_0^{\pi/2} dy \int_y^{\pi/2} \frac{\sin x}{x}\,dx = \iint_R \frac{\sin x}{x}\,dA \quad (R \text{ as shown})$$
$$= \int_0^{\pi/2} \frac{\sin x}{x}\,dx \int_0^x dy = \int_0^{\pi/2} \sin x\,dx = 1.$$

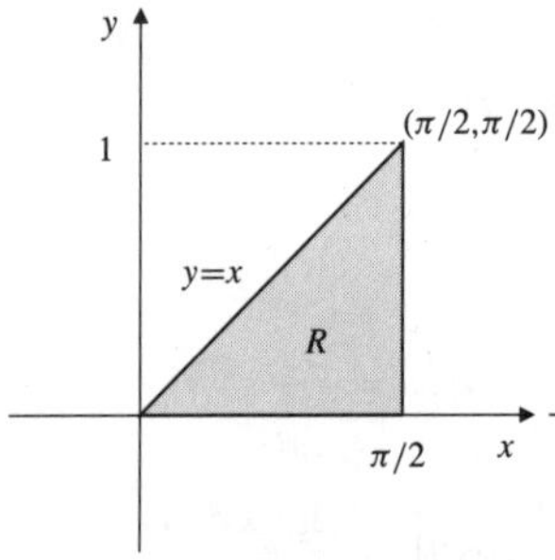

Fig. 6.2.16 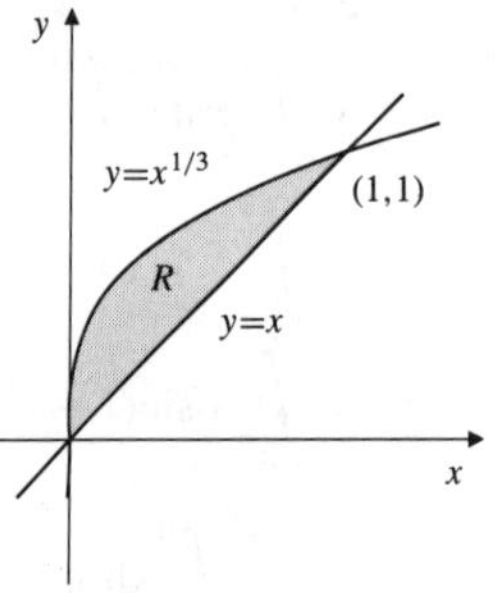 Fig. 6.2.18

**18.** $$\int_0^1 dx \int_x^{x^{1/3}} \sqrt{1-y^4}\,dy$$
$$= \iint_R \sqrt{1-y^4}\,dA \quad (R \text{ as shown})$$
$$= \int_0^1 y\sqrt{1-y^4}\,dy - \int_0^1 y^3\sqrt{1-y^4}\,dy$$
Let $u = y^2$, $du = 2y\,dy$; Let $v = 1-y^4$, $dv = -4y^3\,dy$
$$= \frac{1}{2}\int_0^1 \sqrt{1-u^2}\,du + \frac{1}{4}\int_1^0 v^{1/2}\,dv$$
$$= \frac{1}{2}\left(\frac{\pi}{4}\times 1^2\right) + \frac{1}{6}v^{3/2}\Big|_1^0 = \frac{\pi}{8} - \frac{1}{6}.$$

**20.** $$V = \int_0^1 dy \int_0^y (1-x^2)\,dx$$
$$= \int_0^1 \left(y - \frac{y^3}{3}\right) dy = \frac{1}{2} - \frac{1}{12} = \frac{5}{12} \text{ cu. units.}$$

**22.** $z = 1-y^2$ and $z = x^2$ intersect on the cylinder $x^2+y^2 = 1$. The volume lying below $z = 1-y^2$ and above $z = x^2$ is

$$V = \iint_{x^2+y^2\le 1} (1-y^2-x^2)\,dA$$
$$= 4\int_0^1 dx \int_0^{\sqrt{1-x^2}} (1-x^2-y^2)\,dy$$
$$= 4\int_0^1 dx \left((1-x^2)y - \frac{y^3}{3}\right)\Bigg|_{y=0}^{y=\sqrt{1-x^2}}$$
$$= \frac{8}{3}\int_0^1 (1-x^2)^{3/2}\,dx \quad \text{Let } x = \sin u,\ dx = \cos u\,du$$
$$= \frac{8}{3}\int_0^{\pi/2} \cos^4 u\,du = \frac{2}{3}\int_0^{\pi/2} (1+\cos 2u)^2\,du$$
$$= \frac{2}{3}\int_0^{\pi/2} \left(1 + 2\cos 2u + \frac{1+\cos 4u}{2}\right) du$$
$$= \frac{2}{3}\frac{3}{2}\frac{\pi}{2} = \frac{\pi}{2} \text{ cu. units.}$$

**24.** $$V = \int_0^{\pi^{1/4}} dy \int_0^y x^2 \sin(y^4)\,dx$$
$$= \frac{1}{3}\int_0^{\pi^{1/4}} y^3 \sin(y^4)\,dy \quad \text{Let } u = y^4,\ du = 4y^3\,dy$$
$$= \frac{1}{12}\int_0^{\pi} \sin u\,du = \frac{1}{6} \text{ cu. units.}$$

**26.** $$\text{Vol} = \iint_T \left(2 - \frac{x}{a} - \frac{y}{b}\right) dA$$
$$= \int_0^a dx \int_0^{b(1-(x/a))} \left(2 - \frac{x}{a} - \frac{y}{b}\right) dy$$
$$= \int_0^a \left[\left(2-\frac{x}{a}\right) b\left(1-\frac{x}{a}\right) - \frac{1}{2b}b^2\left(1-\frac{x}{a}\right)^2\right] dx$$
$$= \frac{b}{2}\int_0^a \left(3 - \frac{4x}{a} + \frac{x^2}{a^2}\right) dx$$
$$= \frac{b}{2}\left(3x - \frac{2x^2}{a} + \frac{x^3}{3a^2}\right)\Bigg|_0^a = \frac{2}{3}ab \text{ cu. units.}$$

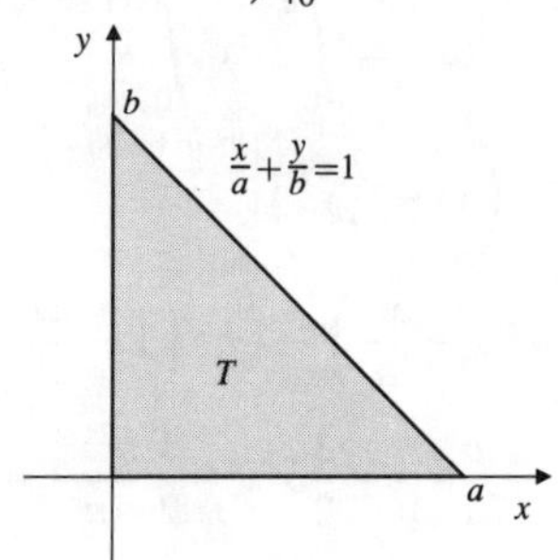


Fig. 6.2.26

**28.** The part of the plane $z = 8 - x$ lying inside the elliptic cylinder $x^2 = 2y^2 = 8$ lies above $z = 0$. The part of the plane $z = y - 4$ inside the cylinder lies below $z = 0$. Thus the required volume is

$$\begin{aligned}
\text{Vol} &= \iint_{x^2+2y^2\le 8} \big(8 - x - (y-4)\big)\,dA \\
&= \iint_{x^2+2y^2\le 8} 12\,dA \qquad \text{(by symmetry)} \\
&= 12 \times \text{area of ellipse } \frac{x^2}{8} + \frac{y^2}{4} = 1 \\
&= 12 \times \pi(2\sqrt{2})(2) = 48\sqrt{2}\pi \text{ cu. units.}
\end{aligned}$$

**30.** Since $F'(x) = f(x)$ and $G'(x) = g(x)$ on $a \le x \le b$, we have

$$\begin{aligned}
\iint_T f(x)g(x)\,dA &= \int_a^b f(x)\,dx \int_a^x G'(y)\,dy \\
&= \int_a^b f(x)\big(G(x) - G(a)\big)\,dx \\
&= \int_a^b f(x)G(x)\,dx - G(a)F(b) + G(a)F(a)
\end{aligned}$$

$$\begin{aligned}
\iint_T f(x)g(x)\,dA &= \int_a^b g(y)\,dy \int_y^b F'(x)\,dx \\
&= \int_a^b g(y)\big(F(b) - F(y)\big)\,dy \\
&= F(b)G(b) - F(b)G(a) - \int_a^b F(y)g(y)\,dx.
\end{aligned}$$

Thus

$$\int_a^b f(x)G(x)\,dx = F(b)G(b) - F(a)G(a) - \int_a^b g(y)F(y)\,dy.$$

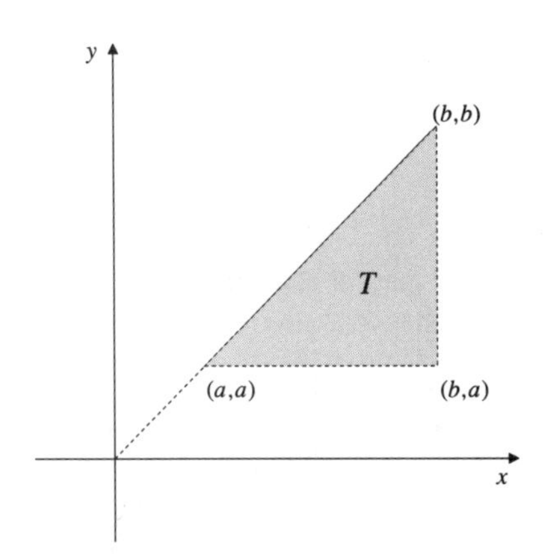


Fig. 6.2.30

## Section 6.3 Improper Integrals and a Mean-Value Theorem (page 283)

**2.**
$$\begin{aligned}
\iint_Q \frac{dA}{(1+x^2)(1+y^2)} &= \int_0^\infty \frac{dx}{1+x^2} \int_0^\infty \frac{dy}{1+y^2} \\
&= \left( \lim_{R\to\infty} (\tan^{-1} x)\Big|_0^R \right)^2 = \frac{\pi^2}{4}
\end{aligned}$$

(converges)

**4.**
$$\begin{aligned}
\iint_T \frac{1}{x\sqrt{y}}\,dA &= \int_0^1 \frac{dx}{x} \int_x^{2x} \frac{dy}{\sqrt{y}} \\
&= \int_0^1 \frac{2(\sqrt{2x} - \sqrt{x})}{x}\,dx \\
&= 2(\sqrt{2} - 1) \int_0^1 \frac{dx}{\sqrt{x}} = 4(\sqrt{2} - 1) \text{ (converges)}
\end{aligned}$$

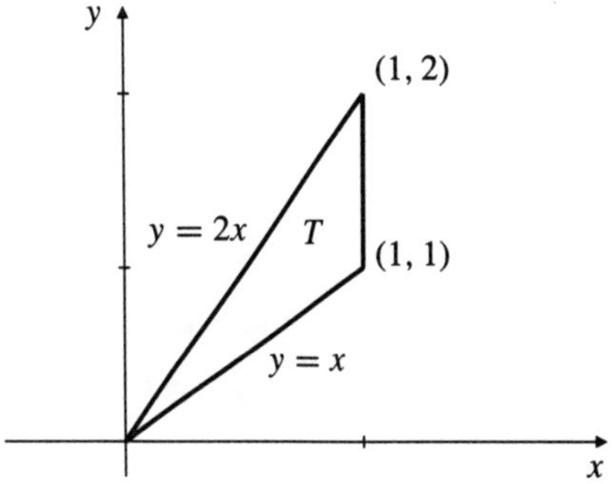

Fig. 6.3.4

**6.**
$$\begin{aligned}
\iint_H \frac{dA}{1+x+y} &= \int_0^\infty dx \int_0^1 \frac{1}{1+x+y}\,dy \\
&= \int_0^\infty \left( \ln(1+x+y)\Big|_{y=0}^{y=1} \right) dx \\
&= \int_0^\infty \ln\left(\frac{2+x}{1+x}\right) dx = \int_0^\infty \ln\left(1 + \frac{1}{1+x}\right) dx.
\end{aligned}$$

Since $\lim_{u\to 0+} \dfrac{\ln(1+u)}{u} = 1$, we have $\ln(1+u) \ge u/2$ on some interval $(0, u_0)$. Therefore

$$\ln\left(1 + \frac{1}{1+x}\right) \ge \frac{1}{2(1+x)}$$

on some interval $(x_0, \infty)$, and

$$\int_0^\infty \ln\left(1+\frac{1}{1+x}\right)dx \ge \int_{x_0}^\infty \frac{1}{2(1+x)}\,dx,$$

which diverges to infinity. Thus the given double integral diverges to infinity by comparison.

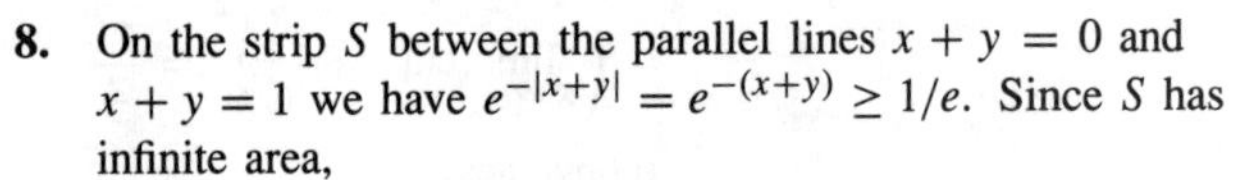

**8.** On the strip $S$ between the parallel lines $x+y=0$ and $x+y=1$ we have $e^{-|x+y|} = e^{-(x+y)} \ge 1/e$. Since $S$ has infinite area,

$$\iint_S e^{-|x+y|}\,dA = \infty.$$

Since $e^{-|x+y|} > 0$ for all $(x,y)$ in $\mathbb{R}^2$, we have

$$\iint_{\mathbb{R}^2} e^{-|x+y|}\,dA > \iint_S e^{-|x+y|}\,dA,$$

and the given integral diverges to infinity.

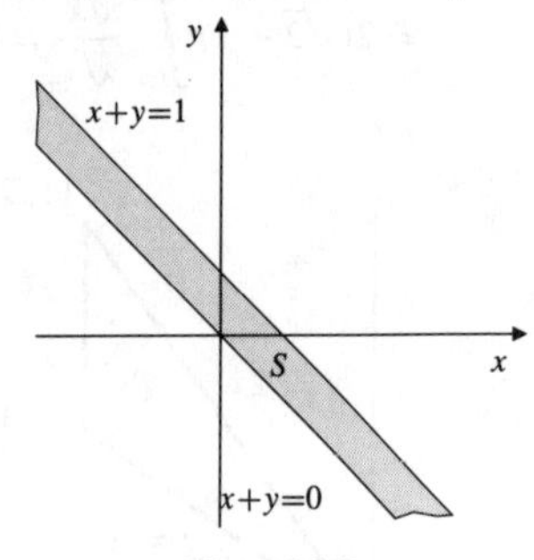


Fig. 6.3.8

**10.**
$$\iint_T \frac{dA}{x^2+y^2} = \int_1^\infty dx \int_0^x \frac{dy}{x^2+y^2}$$
$$= \int_1^\infty dx \left(\frac{1}{x}\tan^{-1}\frac{y}{x}\Big|_{y=0}^{y=x}\right)$$
$$= \frac{\pi}{4}\int_1^\infty \frac{dx}{x} = \infty$$
(The integral diverges to infinity.)

**12.**
$$\iint_R \frac{dA}{x^4+y^2} = \int_1^\infty dx \int_0^{x^2} \frac{dy}{x^4+y^2}$$
$$= \int_1^\infty dx \left(\frac{1}{x^2}\tan^{-1}\frac{y}{x^2}\Big|_{y=0}^{y=x^2}\right)$$
$$= \frac{\pi}{4}\int_1^\infty \frac{dx}{x^2} = \frac{\pi}{4}$$
(The integral converges.)

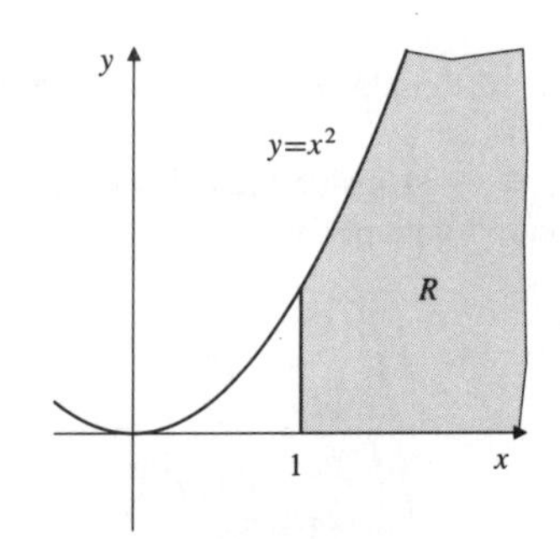


Fig. 6.3.12

**14.**
$$\iint_R \frac{1}{x}\sin\frac{1}{x}\,dA = \int_{2/\pi}^\infty \frac{1}{x}\sin\frac{1}{x}\,dx \int_0^{1/x} dy$$
$$= \int_{2/\pi}^\infty \frac{1}{x^2}\sin\frac{1}{x}\,dx \qquad \text{Let } u = 1/x,\ du = -1/x^2\,dx$$
$$= -\int_{\pi/2}^0 \sin u\,du = \cos u\Big|_{\pi/2}^0 = 1$$
(The integral converges.)

**16.**
$$\text{Vol} = \iint_S \frac{2xy}{x^2+y^2}\,dA$$
$$= 4\iint_T \frac{2xy}{x^2+y^2}\,dA \quad (T \text{ as in \#9(b)})$$
$$= 4\int_0^1 x\,dx \int_0^x \frac{y\,dy}{x^2+y^2} \qquad \text{Let } u = x^2+y^2,\ du = 2y\,dy$$
$$= 2\int_0^1 x\,dx \int_{x^2}^{2x^2} \frac{du}{u}$$
$$= 2\ln 2 \int_0^1 x\,dx = \ln 2 \text{ cu. units.}$$

**18.** $\displaystyle\iint_{D_k} y^b\,dA = \int_0^1 dx \int_0^{x^k} y^b\,dy = \int_0^1 \frac{x^{k(b+1)}}{b+1}\,dx$ if $b > -1$. This latter integral converges if $k(b+1) > -1$. Thus, the given integral converges if $b > -1$ and $k > -1/(b+1)$.

**20.** $\displaystyle\iint_{R_k} \frac{dA}{y^b} = \int_1^\infty dx \int_0^{x^k} \frac{dy}{y^b} = \int_1^\infty \frac{x^{k(1-b)}}{1-b}\,dx$ if $b < 1$. This latter integral converges if $k(1-b) < -1$. Thus, the given integral converges if $b < 1$ and $k < -1/(1-b)$.

**22.** $\displaystyle\iint_{R_k} x^a y^b\,dA = \int_1^\infty x^a\,dx \int_0^{x^k} y^b\,dy = \int_1^\infty \frac{x^{a+(b+1)k}}{b+1}\,dx,$ if $b > -1$. This latter integral converges if $a + (b+1)k < -1$. Thus, the given integral converges if $b > -1$ and $k < -(a+1)/(b+1)$.

**24.** The average value of $x^2$ over the rectangle $R$ is

$$\begin{aligned}&\frac{1}{(b-a)(d-c)}\iint_R x^2\,dA\\&=\frac{1}{(b-a)(d-c)}\int_a^b x^2\,dx\int_c^d dy\\&=\frac{1}{b-a}\,\frac{b^3-a^3}{3}=\frac{a^2+ab+b^2}{3}.\end{aligned}$$

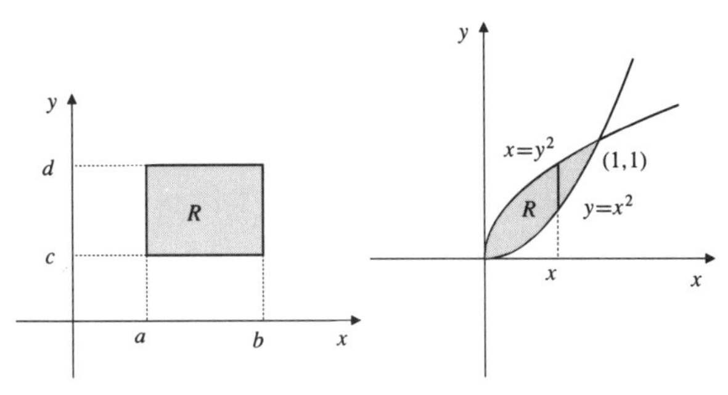


Fig. 6.3.24 Fig. 6.3.26

**26.** The area of region $R$ is

$$\int_0^1(\sqrt{x}-x^2)\,dx=\frac{1}{3}\text{ sq. units.}$$

The average value of $1/x$ over $R$ is

$$\begin{aligned}3\iint_R\frac{dA}{x}&=3\int_0^1\frac{dx}{x}\int_{x^2}^{\sqrt{x}}dy\\&=3\int_0^1\left(x^{-1/2}-x\right)dx=\frac{9}{2}.\end{aligned}$$

**28.** Let $R$ be the region $0\le x<\infty$, $0\le y\le 1/(1+x^2)$. If $f(x,y)=x$, then

$$\int_R f(x,y)\,dA=\int_0^\infty x\,dx\int_0^{1/(1+x^2)}dy=\int_0^\infty\frac{x\,dx}{1+x^2}$$

which diverges to infinity. Thus $f$ has no average value on $R$.

**30.** The integral in Example 2 reduced to

$$\int_1^\infty \ln\left(1+\frac{1}{x^2}\right)dx$$

$$U=\ln\left(1+\frac{1}{x^2}\right)\qquad dV=dx$$
$$dU=-\frac{2\,dx}{x(x^2+1)}\qquad V=x$$

$$\begin{aligned}&=\lim_{R\to\infty}\left[x\ln\left(1+\frac{1}{x^2}\right)\Big|_1^R+2\int_1^R\frac{dx}{1+x^2}\right]\\&=2\left(\frac{\pi}{2}-\frac{\pi}{4}\right)-\ln 2+\lim_{R\to\infty}\frac{\ln\big(1+(1/R^2)\big)}{1/R}\\&=\frac{\pi}{2}-\ln 2+\lim_{R\to\infty}\frac{-(2/R^3)}{\big(1+(1/R^2)\big)(-1/R^2)}\\&=\frac{\pi}{2}-\ln 2.\end{aligned}$$

**32.** If $R=\{(x,y)\,:\,a\le x\le a+h,\ b\le y\le b+k\}$, then

$$\begin{aligned}&\iint_R f_{12}(x,y)\,dA=\int_a^{a+h}dx\int_b^{b+k}f_{12}(x,y)\,dy\\&=\int_a^{a+h}\big[f_1(x,b+k)-f_1(x,b)\big]\,dx\\&=f(a+h,b+k)-f(a,b+k)-f(a+h,b)+f(a,b)\\&\iint_R f_{21}(x,y)\,dA=\int_b^{b+k}dy\int_a^{a+h}f_{21}(x,y)\,dx\\&=\int_b^{b+k}\big[f_2(a+h,y)-f_2(a,y)\big]\,dy\\&=f(a+h,b+k)-f(a+h,b)-f(a,b+k)+f(a,b).\end{aligned}$$

Thus

$$\iint_R f_{12}(x,y)\,dA=\iint_R f_{21}(x,y)\,dA.$$

Divide both sides of this identity by $hk$ and let $(h,k)\to(0,0)$ to obtain, using the result of Exercise 31,

$$f_{12}(a,b)=f_{21}(a,b).$$

## Section 6.4 Double Integrals in Polar Coordinates (page 292)

**2.** $$\iint_D\sqrt{x^2+y^2}\,dA=\int_0^{2\pi}d\theta\int_0^a r\,r\,dr=\frac{2\pi a^3}{3}$$

**4.** $$\begin{aligned}\iint_D|x|\,dA&=4\int_0^{\pi/2}d\theta\int_0^a r\cos\theta\,r\,dr\\&=4\sin\theta\Big|_0^{\pi/2}\frac{a^3}{3}=\frac{4a^3}{3}\end{aligned}$$

**6.** $$\iint_D x^2y^2\,dA = 4\int_0^{\pi/2} d\theta \int_0^a r^4\cos^2\theta\sin^2\theta r\,dr$$
$$= \frac{a^6}{6}\int_0^{\pi/2}\sin^2(2\theta)\,d\theta$$
$$= \frac{a^6}{12}\int_0^{\pi/2}\big(1-\cos(4\theta)\big)\,d\theta = \frac{\pi a^6}{24}$$

**8.** $\iint_Q (x+y)\,dA = \frac{2a^3}{3}$; by symmetry, the value is twice that obtained in the previous exercise.

**10.** $$\iint_Q \frac{2xy}{x^2+y^2}\,dA = \int_0^{\pi/2} d\theta\int_0^a \frac{2r^2\sin\theta\cos\theta}{r^2}\,r\,dr$$
$$= \frac{a^2}{2}\int_0^{\pi/2}\sin(2\theta)\,d\theta = -\frac{a^2\cos(2\theta)}{4}\bigg|_0^{\pi/2} = \frac{a^2}{2}$$

**12.** $$\iint_S x\,dA = 2\int_0^{\pi/4} d\theta\int_{\sec\theta}^{\sqrt{2}} r\cos\theta\, r\,dr$$
$$= \frac{2}{3}\int_0^{\pi/4}\cos\theta\big(2\sqrt{2}-\sec^3\theta\big)\,d\theta$$
$$= \frac{4\sqrt{2}}{3}\sin\theta\bigg|_0^{\pi/4} - \frac{2}{3}\tan\theta\bigg|_0^{\pi/4}$$
$$= \frac{4}{3}-\frac{2}{3} = \frac{2}{3}$$

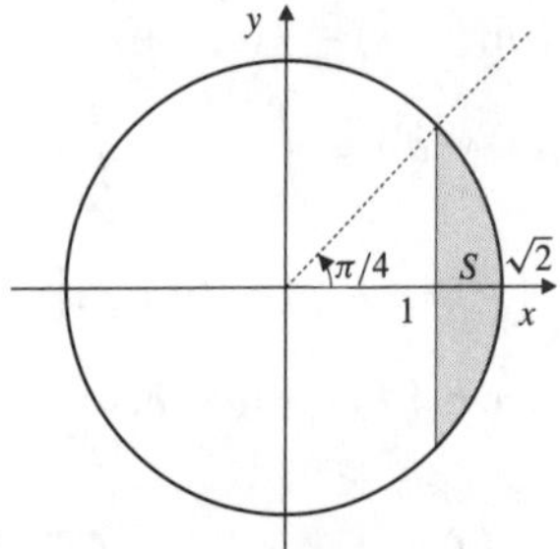


Fig. 6.4.12

**14.** $$\iint_{x^2+y^2\le 1}\ln(x^2+y^2)\,dA = \int_0^{2\pi} d\theta\int_0^1 (\ln r^2)r\,dr$$
$$= 4\pi\int_0^1 r\ln r\,dr$$
$$U = \ln r \quad dV = r\,dr$$
$$dU = \frac{dr}{r} \quad V = \frac{r^2}{2}$$
$$= 4\pi\left[\frac{r^2}{2}\ln r\bigg|_0^1 - \frac{1}{2}\int_0^1 r\,dr\right]$$
$$= 4\pi\left[0-0-\frac{1}{4}\right] = -\pi$$

(Note that the integral is improper, but converges since $\lim_{r\to 0+} r^2\ln r = 0$.)

**16.** The annular region $R$: $0 < a \le \sqrt{x^2+y^2} \le b$ has area $\pi(b^2-a^2)$. The average value of $e^{-(x^2+y^2)}$ over the region is

$$\frac{1}{\pi(b^2-a^2)}\iint_R e^{-(x^2+y^2)}\,dA$$
$$= \frac{1}{\pi(b^2-a^2)}\int_0^{2\pi} d\theta\int_a^b e^{-r^2} r\,dr \quad \text{Let } u=r^2,\ du = 2r\,dr$$
$$= \frac{1}{\pi(b^2-a^2)}(2\pi)\frac{1}{2}\int_{a^2}^{b^2} e^{-u}\,du$$
$$= \frac{1}{b^2-a^2}\big(e^{-a^2}-e^{-b^2}\big).$$

**18.** $$\iint_{\mathbb{R}^2}\frac{dA}{(1+x^2+y^2)^k}$$
$$= \int_0^{2\pi} d\theta\int_0^\infty \frac{r\,dr}{(1+r^2)^k} \quad \text{Let } u = 1+r^2,\ du = 2r\,dr$$
$$= \pi\int_1^\infty u^{-k}\,du = \frac{-\pi}{1-k} \text{ if } k>1.$$

The integral converges to $\frac{\pi}{k-1}$ if $k>1$.

**20.** $$\iint_C y\,dA = \int_0^\pi d\theta\int_0^{1+\cos\theta} r\sin\theta r\,dr$$
$$= \frac{1}{3}\int_0^\pi \sin\theta(1+\cos\theta)^3\,d\theta \quad \text{Let } u = 1+\cos\theta,\ du = -\sin\theta\,d\theta$$
$$= \frac{1}{3}\int_0^2 u^3\,du = \frac{u^4}{12}\bigg|_0^2 = \frac{4}{3}$$

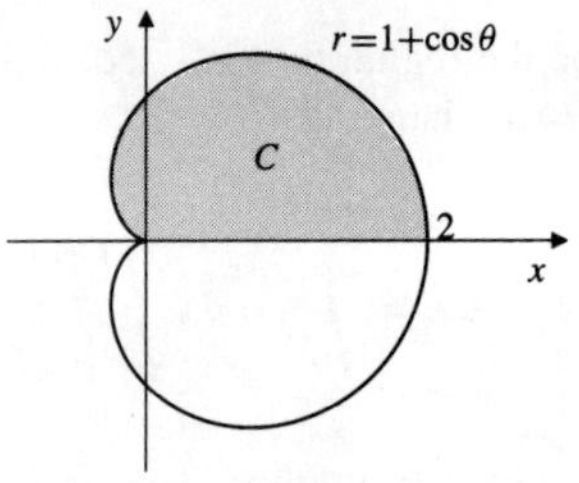

Fig. 6.4.20

**22.**

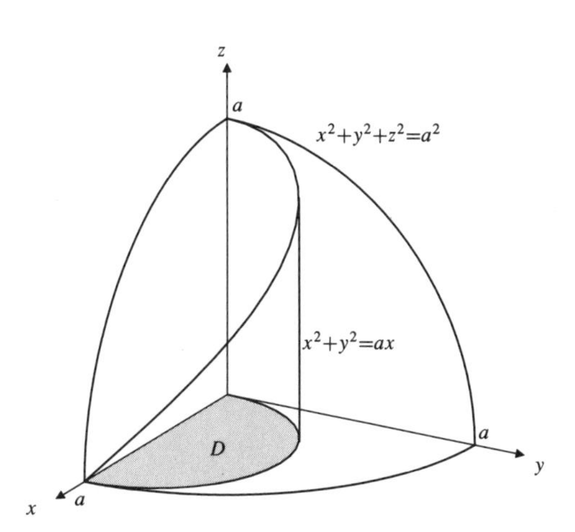


Fig. 6.4.22

One quarter of the required volume lies in the first octant. (See the figure.) In polar coordinates the cylinder $x^2 + y^2 = ax$ becomes $r = a\cos\theta$. Thus, the required volume is

$$\begin{aligned}
V &= 4\iint_D \sqrt{a^2 - x^2 - y^2}\, dA \\
&= 4\int_0^{\pi/2} d\theta \int_0^{a\cos\theta} \sqrt{a^2 - r^2}\, r\, dr \quad \text{Let } u = a^2 - r^2 \\
&\qquad\qquad\qquad\qquad\qquad\qquad\qquad du = -2r\, dr \\
&= 2\int_0^{\pi/2} d\theta \int_{a^2\sin^2\theta}^{a^2} u^{1/2}\, du \\
&= \frac{4}{3}\int_0^{\pi/2} d\theta \left( u^{3/2}\Big|_{a^2\sin^2\theta}^{a^2} \right) \\
&= \frac{4}{3}a^3 \int_0^{\pi/2} (1 - \sin^3\theta)\, d\theta \\
&= \frac{4}{3}a^3 \left( \frac{\pi}{2} - \int_0^{\pi/2} \sin\theta(1 - \cos^2\theta)\, d\theta \right) \\
&\qquad \text{Let } v = \cos\theta \\
&\qquad dv = -\sin\theta\, d\theta \\
&= \frac{2\pi a^3}{3} - \frac{4a^3}{3}\int_0^1 (1 - v^2)\, dv \\
&= \frac{2\pi a^3}{3} - \frac{4a^3}{3}\left( v - \frac{v^3}{3} \right)\Big|_0^1 \\
&= \frac{2\pi a^3}{3} - \frac{8a^3}{9} = \frac{2}{9}a^3(3\pi - 4) \text{ cu. units.}
\end{aligned}$$

**24.**

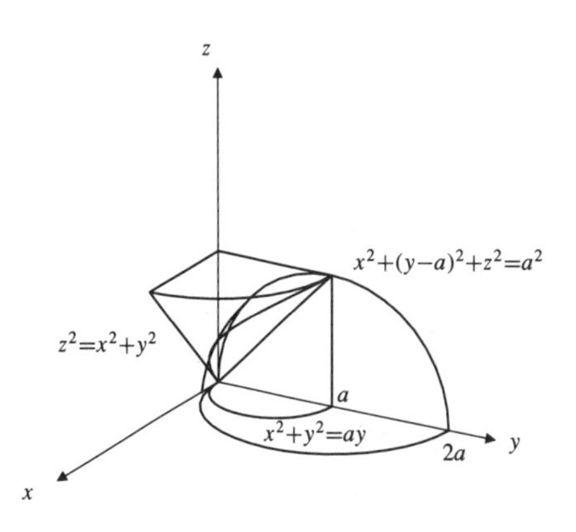


Fig. 6.4.24

The cone $z^2 = x^2 + y^2$ and the sphere $x^2 + (y - a)^2 + z^2 = a^2$ intersect where

$$x^2 + y^2 = a^2 - x^2 - y^2 + 2ay - a^2,$$

that is, on the cylinder $x^2 + y^2 = ay$, or, in polar coordinates, $r = a\sin\theta$. The volume $V$ lying outside the cone and inside the sphere lies on four octants; one quarter of it is in the first octant. To calculate $V$ we first calculate the following first octant volumes:

a) the volume $V_1$ under the cone and inside the cylinder,

b) the volume $V_2$ inside the sphere and inside the cylinder,

c) the volume $V_3$ inside the sphere and outside the cylinder.

We have

$$\begin{aligned}
V_1 &= \int_0^{\pi/2} d\theta \int_0^{a\sin\theta} r^2\, dr \\
&= \frac{a^3}{3}\int_0^{\pi/2} \sin^3\theta\, d\theta = \frac{2a^3}{9}.
\end{aligned}$$

Now $4V_2$ is the volume inside a sphere of radius $a$ and a cylinder of radius $a/2$ where the centre of the sphere lies on the cylinder. Such a volume has already been calculated in Example 5, except that the radii of the sphere and cylinder were twice as large in that example. Hence $4V_2$ is one-eighth of the volume calculated there.

$$4V_2 = \frac{2\pi a^3}{3} - \frac{8a^3}{9}.$$

Since the volume of the sphere is $4\pi a^3/3$, we have

$$4V_3 = \frac{4\pi a^3}{3} - 4V_2.$$

Therefore, the required volume $V$ is

$$\begin{aligned} V &= 4(V_1+V_3) = 4V_1 + \frac{4\pi a^3}{3} - 4V_2 \\ &= \frac{8a^3}{9} + \frac{4\pi a^3}{3} - \frac{2\pi a^3}{9} + \frac{8a^3}{9} \\ &= \frac{16a^3}{9} + \frac{2\pi a^3}{3} \text{ cu. units.} \end{aligned}$$

**26.** 
$$\begin{aligned} \text{Volume} &= \int_0^{2\pi} d\theta \int_0^2 (r\cos\theta + r\sin\theta + 4)r\,dr \\ &= \int_0^{2\pi} (\cos\theta+\sin\theta)\,d\theta \int_0^2 r^2\,dr + 8\pi \int_0^2 r\,dr \\ &= 0 + 4\pi(2^2) = 16\pi \text{ cu. units.} \end{aligned}$$

**28.** One quarter of the required volume $V$ is shown in the figure. We have

$$\begin{aligned} V &= 4\iint_D \sqrt{y}\,dA \\ &= 4\int_0^{\pi/2} d\theta \int_0^{2\sin\theta} \sqrt{r\sin\theta}\,r\,dr \\ &= 4\int_0^{\pi/2} \sqrt{\sin\theta}\,d\theta \left(\frac{2}{5}r^{5/2}\Big|_0^{2\sin\theta}\right) \\ &= \frac{32\sqrt{2}}{5}\int_0^{\pi/2} \sin^3 d\theta = \frac{64\sqrt{2}}{15} \text{ cu. units.} \end{aligned}$$

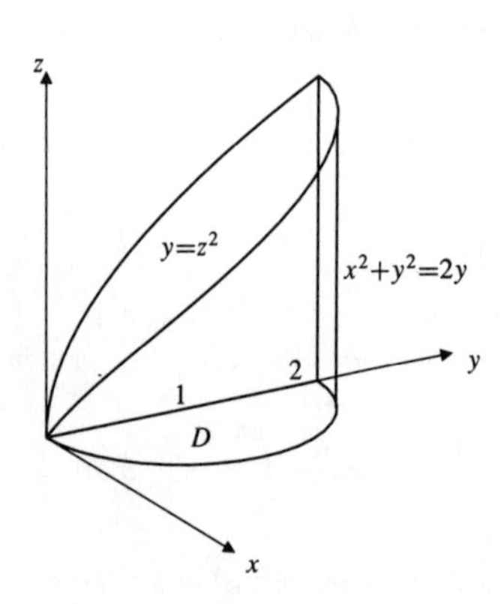


Fig. 6.4.28

**30.** The area of $S$ is $(4\pi - 3\sqrt{3})/3$ sq. units. Thus

$$\begin{aligned} \overline{x} &= \frac{3}{4\pi - 3\sqrt{3}} \iint_S x\,dA \\ &= \frac{6}{4\pi - 3\sqrt{3}} \int_0^{\pi/3} d\theta \int_{\sec\theta}^2 r\cos\theta\,r\,dr \\ &= \frac{2}{4\pi - 3\sqrt{3}} \int_0^{\pi/3} \cos\theta(8 - \sec^3\theta)\,d\theta \\ &= \frac{2}{4\pi - 3\sqrt{3}} \left(4\sqrt{3} - \tan\theta\Big|_0^{\pi/3}\right) = \frac{6\sqrt{3}}{4\pi - 3\sqrt{3}}. \end{aligned}$$

The segment has centroid $\left(\dfrac{6\sqrt{3}}{4\pi - 3\sqrt{3}}, 0\right)$.

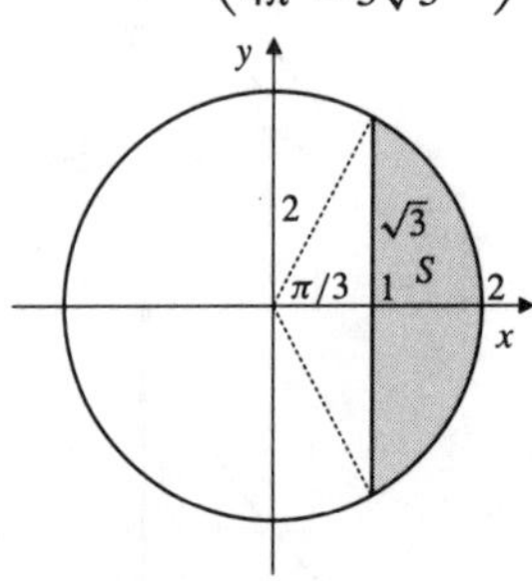


Fig. 6.4.30

**32.** Using the transformation $x = au$, $y = bv$, the required volume is

$$\begin{aligned} V &= \iint_{\frac{x^2}{a^2}+\frac{y^2}{b^2}\le 1} \left(1 - \frac{x^2}{a^2} - \frac{y^2}{b^2}\right) dx\,dy \\ &= ab\iint_{u^2+v^2\le 1} (1 - u^2 - v^2)\,du\,dv. \end{aligned}$$

Now transform to polar coordinates in the $uv$-plane: $u = r\cos\theta$, $v = r\sin\theta$.

$$\begin{aligned} V &= ab\int_0^{\pi/2} d\theta \int_0^1 (1-r^2)r\,dr \\ &= \frac{\pi ab}{2}\left(\frac{r^2}{2} - \frac{r^4}{4}\right)\Bigg|_0^1 = \frac{\pi ab}{8} \text{ cu. units.} \end{aligned}$$

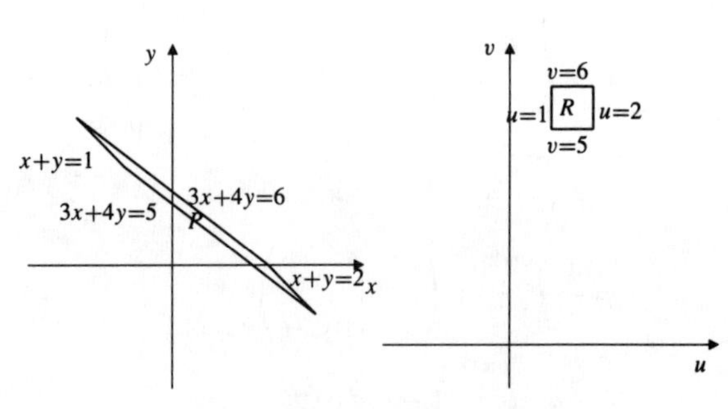


Fig. 6.4.34(a) Fig. 6.4.34(b)

**34.** The parallelogram $P$ bounded by $x+y=1$, $x+y=2$, $3x+4y=5$, and $3x+4y=6$ corresponds to the square $S$ bounded by $u=1$, $u=2$, $v=5$, and $v=6$ under the transformation

$$u = x + y, \qquad v = 3x + 4y,$$

or, equivalently,

$$x = 4u - v, \qquad y = v - 3u.$$

We have
$$\frac{\partial(x,y)}{\partial(u,v)} = \begin{vmatrix} 4 & -1 \\ -3 & 1 \end{vmatrix} = 1,$$
so $dx\,dy = du\,dv$. Also
$$x^2 + y^2 = (4u - v)^2 + (v - 3u)^2 = 25u^2 - 14uv + 2v^2.$$

Thus we have
$$\begin{aligned}\iint_P (x^2 + y^2)\,dx\,dy &= \iint_S (25u^2 - 14uv + 2v^2)\,du\,dv \\ &= \int_1^2 du \int_5^6 (25u^2 - 14uv + 2v^2)\,dv = \frac{7}{2}.\end{aligned}$$

**36.** Under the transformation $u = x^2 - y^2$, $v = xy$, the region $R$ in the first quadrant of the $xy$-plane bounded by $y = 0$, $y = x$, $xy = 1$, and $x^2 - y^2 = 1$ corresponds to the square $S$ in the $uv$-plane bounded by $u = 0$, $u = 1$, $v = 0$, and $v = 1$. Since
$$\frac{\partial(u,v)}{\partial(x,y)} = \begin{vmatrix} 2x & -2y \\ y & x \end{vmatrix} = 2(x^2 + y^2),$$
we therefore have
$$(x^2 + y^2)\,dx\,dy = \frac{1}{2}\,du\,dv.$$

Hence,
$$\iint_R (x^2 + y^2)\,dx\,dy = \iint_S \frac{1}{2}\,du\,dv = \frac{1}{2}.$$

**38.**

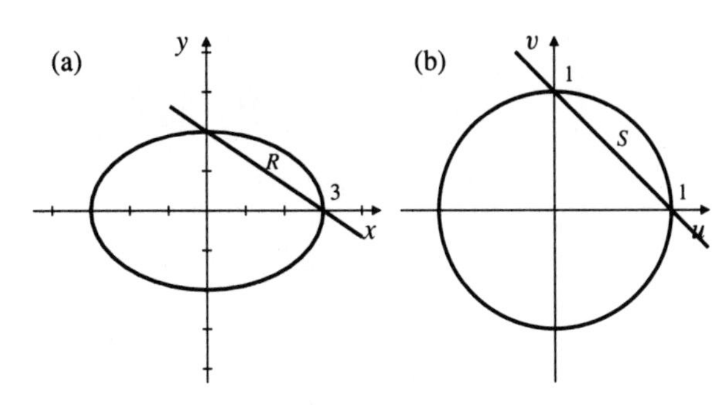


Fig. 6.4.38

The region $R$ whose area we must find is shown in part (a) of the figure. The change of variables $x = 3u$, $y = 2v$ maps the ellipse $4x^2 + 9y^2 = 36$ to the circle $u^2 + v^2 = 1$, and the line $2x + 3y = 1$ to the line $u + v = 1$. Thus it maps $R$ to the region $S$ in part (b) of the figure. Since
$$dx\,dy = \left|\begin{vmatrix} 3 & 0 \\ 0 & 2 \end{vmatrix}\right| du\,dv = 6\,du\,dv,$$
the area of $R$ is
$$A = \iint_R dx\,dy = 6 \iint_S du\,dv.$$

But the area of $S$ is $(\pi/4) - (1/2)$, so $A = (3\pi/2) - 3$ square units.

**40.** a) $\Gamma(x) = \displaystyle\int_0^\infty t^{x-1} e^{-t}\,dt$ Let $t = s^2$, $dt = 2s\,ds$
$$= 2\int_0^\infty s^{2x-1} e^{-s^2}\,ds.$$

b) $\Gamma\left(\tfrac{1}{2}\right) = 2\displaystyle\int_0^\infty e^{-s^2}\,ds = 2\frac{\sqrt{\pi}}{2} = \sqrt{\pi}$
$$\Gamma\left(\tfrac{3}{2}\right) = \frac{1}{2}\Gamma\left(\tfrac{1}{2}\right) = \frac{1}{2}\sqrt{\pi}.$$

c) $B(x,y) = \displaystyle\int_0^1 t^{x-1}(1-t)^{y-1}\,dt \qquad (x > 0,\ y > 0)$

let $t = \cos^2\theta$, $dt = -2\sin\theta\cos\theta\,d\theta$
$$= 2\int_0^{\pi/2} \cos^{2x-1}\theta \sin^{2y-1}\theta\,d\theta.$$

d) If $Q$ is the first quadrant of the $st$-plane,
$$\begin{aligned}\Gamma(x)\Gamma(y) &= \left(2\int_0^\infty s^{2x-1} e^{-s^2}\,ds\right)\left(2\int_0^\infty t^{2y-1} e^{-t^2}\,dt\right) \\ &= 4\iint_Q s^{2x-1} t^{2y-1} e^{-(s^2+t^2)}\,ds\,dt \\ &\quad \text{(change to polar coordinates)} \\ &= 4\int_0^{\pi/2} d\theta \int_0^\infty r^{2x-1}\cos^{2x-1}\theta\, r^{2y-1}\sin^{2y-1}\theta\, e^{-r^2}\, r\,dr \\ &= \left(2\int_0^{\pi/2} \cos^{2x-1}\theta \sin^{2y-1}\theta\,d\theta\right) \\ &\quad \times \left(2\int_0^\infty r^{2(x+y)-1} e^{-r^2}\,dr\right) \\ &= B(x,y)\Gamma(x+y) \qquad \text{by (a) and (c).}\end{aligned}$$

Thus $B(x,y) = \dfrac{\Gamma(x)\Gamma(y)}{\Gamma(x+y)}$.

## Section 6.5 Triple Integrals (page 300)

**2.**
$$\begin{aligned}\iiint_B xyz\,dV &= \int_0^1 x\,dx \int_{-2}^0 y\,dy \int_1^4 z\,dz \\ &= \frac{1}{2}\left(-\frac{4}{2}\right)\left(\frac{16-1}{2}\right) = -\frac{15}{2}.\end{aligned}$$

**4.** 
$$\begin{aligned}
\iiint_R x\,dV &= \int_0^a x\,dx \int_0^{b\left(1-\frac{x}{a}\right)} dy \int_0^{c\left(1-\frac{x}{a}-\frac{y}{b}\right)} dz \\
&= c\int_0^a x\,dx \int_0^{b\left(1-\frac{x}{a}\right)} \left(1-\frac{x}{a}-\frac{y}{b}\right) dy \\
&= c\int_0^a x\left[b\left(1-\frac{x}{a}\right)^2 - \frac{b^2}{2b}\left(1-\frac{x}{a}\right)^2\right] dx \\
&= \frac{bc}{2}\int_0^a \left(1-\frac{x}{a}\right)^2 x\,dx \quad \text{Let } u = 1-(x/a) \\
&\qquad\qquad\qquad\qquad\quad du = -(1/a)\,dx \\
&= \frac{a^2bc}{2}\int_0^1 u^2(1-u)\,du = \frac{a^2bc}{24}.
\end{aligned}$$

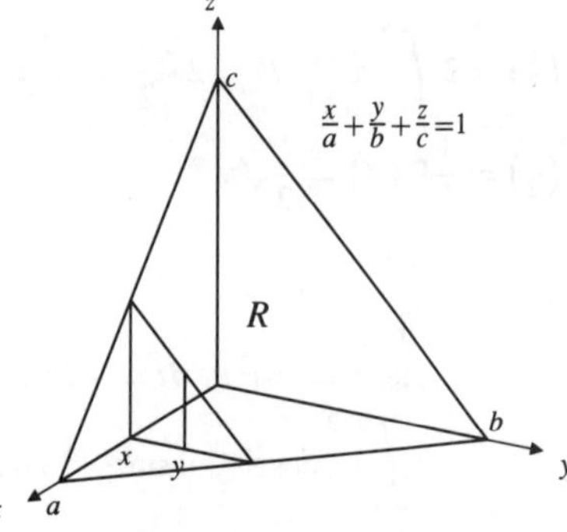


Fig. 6.5.4

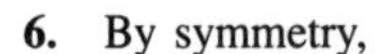

**6.** By symmetry,

$$\iiint_R (x^2+y^2+z^2)\,dV = 3\iiint_R x^2\,dV = \frac{3}{3} = 1.$$

**8.** $R$ is the cube $0 \le x, y, z \le 1$. We have

$$\begin{aligned}
&\iiint_R yz^2 e^{-xyz}\,dV \\
&= \int_0^1 z\,dz \int_0^1 dy \left(-e^{-xyz}\right)\Big|_{x=0}^{x=1} \\
&= \int_0^1 z\,dz \int_0^1 (1-e^{-yz})\,dy \\
&= \int_0^1 z\left(1+\frac{1}{z}e^{-yz}\Big|_{y=0}^{y=1}\right) dz \\
&= \frac{1}{2} + \int_0^1 (e^{-z}-1)\,dz \\
&= \frac{1}{2} - 1 - e^{-z}\Big|_0^1 = \frac{1}{2} - \frac{1}{e}.
\end{aligned}$$

**10.** 

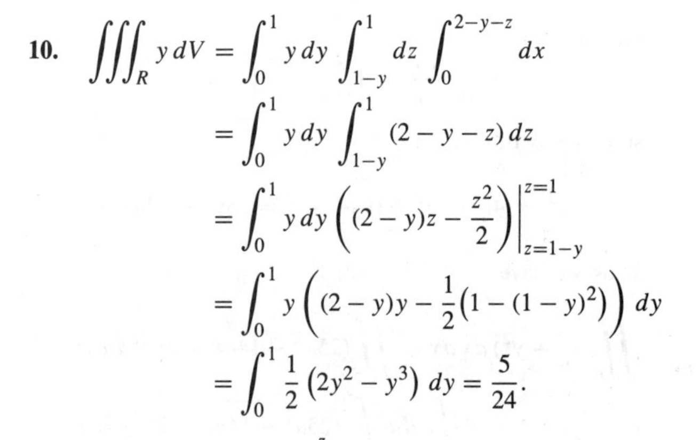

$$\begin{aligned}
\iiint_R y\,dV &= \int_0^1 y\,dy \int_{1-y}^1 dz \int_0^{2-y-z} dx \\
&= \int_0^1 y\,dy \int_{1-y}^1 (2-y-z)\,dz \\
&= \int_0^1 y\,dy \left((2-y)z - \frac{z^2}{2}\right)\Big|_{z=1-y}^{z=1} \\
&= \int_0^1 y\left((2-y)y - \frac{1}{2}\left(1-(1-y)^2\right)\right) dy \\
&= \int_0^1 \frac{1}{2}\left(2y^2-y^3\right) dy = \frac{5}{24}.
\end{aligned}$$

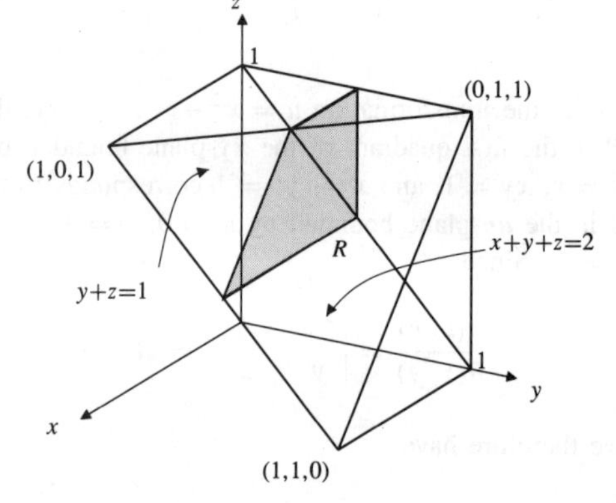


Fig. 6.5.10

**12.** We have

$$\begin{aligned}
&\iiint_R \cos x\cos y\cos z\,dV \\
&= \int_0^\pi \cos x\,dx \int_0^{\pi-x} \cos y\,dy \int_0^{\pi-x-y} \cos z\,dz \\
&= \int_0^\pi \cos x\,dx \int_0^{\pi-x} \cos y\,dy\,(\sin z)\Big|_{z=0}^{z=\pi-x-y} \\
&= \int_0^\pi \cos x\,dx \int_0^{\pi-x} \cos y\sin(x+y)\,dy \\
&\quad \text{recall that } \sin a\cos b = \frac{1}{2}\big(\sin(a+b)+\sin(a-b)\big) \\
&= \int_0^\pi \cos x\,dx \int_0^{\pi-x} \frac{1}{2}\big[\sin(x+2y)+\sin x\big]\,dy \\
&= \frac{1}{2}\int_0^\pi \cos x\,dx \left[-\frac{\cos(x+2y)}{2} + y\sin x\right]\Bigg|_{y=0}^{y=\pi-x} \\
&= \frac{1}{2}\int_0^\pi \Bigg(-\frac{\cos x\cos(2\pi-x)}{2} + \frac{\cos^2 x}{2} \\
&\qquad\qquad + (\pi-x)\cos x\sin x\Bigg)\,dx \\
&= \frac{1}{2}\int_0^\pi \frac{\pi-x}{2}\sin 2x\,dx
\end{aligned}$$

$$U = \pi - x \quad dV = \sin 2x\, dx$$
$$dU = -dx \quad V = -\frac{\cos 2x}{2}$$
$$= \frac{1}{4}\left[-\frac{\pi - x}{2}\cos 2x\Big|_0^{\pi} - \frac{1}{2}\int_0^{\pi}\cos 2x\, dx\right]$$
$$= \frac{1}{8}\left[\pi - \frac{\sin 2x}{2}\Big|_0^{\pi}\right] = \frac{\pi}{8}.$$

**14.** Let $E$ be the elliptic disk bounded by $x^2 + 4y^2 = 4$. Then $E$ has area $\pi(2)(1) = 2\pi$ square units. The volume of the region of 3-space lying above $E$ and beneath the plane $z = 2 + x$ is

$$V = \iint_E (2 + x)\, dA = 2\iint_E dA = 4\pi \text{ cu. units,}$$

since $\iint_E x\, dA = 0$ by symmetry.

**16.**

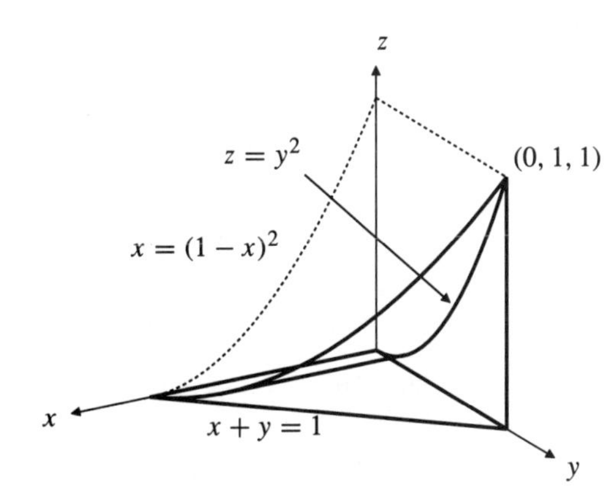

Fig. 6.5.16

$$\iiint_R f(x,y,z)\, dV = \int_0^1 dx \int_0^{1-x} dy \int_0^{y^2} f(x,y,z)\, dz$$
$$= \int_0^1 dy \int_0^{1-y} dx \int_0^{y^2} f(x,y,z)\, dz$$
$$= \int_0^1 dy \int_0^{y^2} dz \int_0^{1-y} f(x,y,z)\, dx$$
$$= \int_0^1 dz \int_{\sqrt{z}}^{1} dy \int_0^{1-y} f(x,y,z)\, dx$$
$$= \int_0^1 dx \int_0^{(1-x)^2} dz \int_{\sqrt{z}}^{1-x} f(x,y,z)\, dy$$
$$= \int_0^1 dz \int_0^{1-\sqrt{z}} dx \int_{\sqrt{z}}^{1-x} f(x,y,z)\, dy.$$

**18.** $$\int_0^1 dz \int_z^1 dy \int_0^y f(x,y,z)\, dx$$
$$= \iiint_R f(x,y,z)\, dV \qquad (R \text{ is the pyramid in the figure})$$
$$= \int_0^1 dx \int_x^1 dy \int_0^y f(x,y,z)\, dz.$$

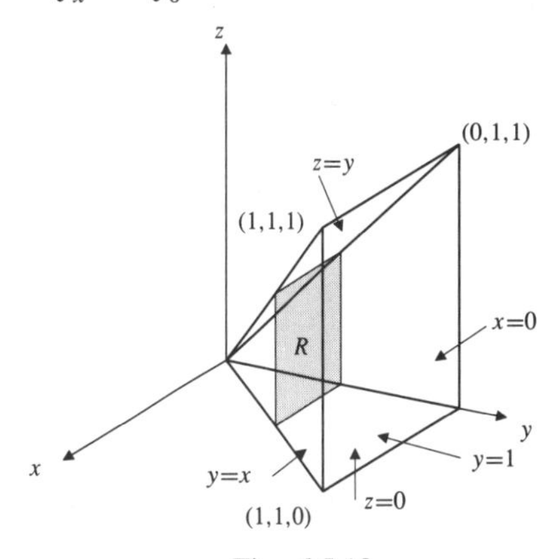

Fig. 6.5.18

**20.** $$\int_0^1 dy \int_0^{\sqrt{1-y^2}} dz \int_{y^2+z^2}^{1} f(x,y,z)\, dx$$
$$= \iiint_R f(x,y,z)\, dV \quad (R \text{ is the paraboloid in the figure})$$
$$= \int_0^1 dx \int_0^{\sqrt{x}} dy \int_0^{\sqrt{x-y^2}} f(x,y,z)\, dz.$$

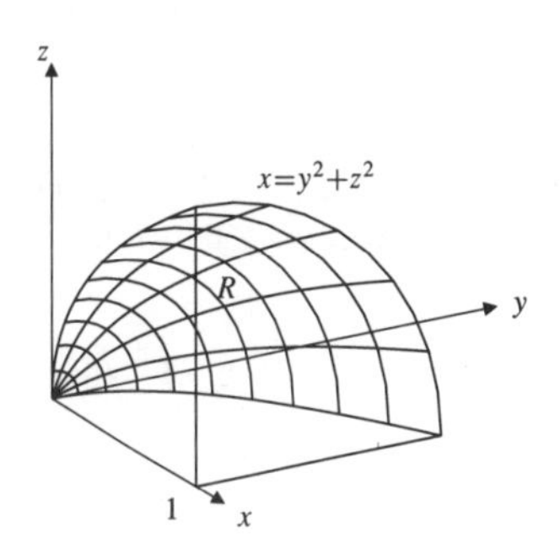

Fig. 6.5.20

**22.** $I = \int_0^1 dz \int_z^1 dy \int_0^y f(x,y,z)\, dx.$
The given iteration corresponds to

$$0 \le z \le 1, \quad z \le y \le 1, \quad 0 \le x \le y.$$

Thus $0 \le x \le 1$, $x \le y \le 1$, $0 \le z \le y$, and

$$I = \int_0^1 dx \int_x^1 dy \int_0^y f(x,y,z)\, dz.$$

**24.** $I = \int_0^1 dy \int_0^{\sqrt{1-y^2}} dz \int_{y^2+z^2}^1 f(x, y, z)\, dx.$
The given iteration corresponds to

$$0 \le y \le 1, \quad 0 \le z \le \sqrt{1 - y^2}, \quad y^2 + z^2 \le x \le 1.$$

Thus $0 \le x \le 1$, $0 \le y \le \sqrt{x}$, $0 \le z \le \sqrt{x - y^2}$, and

$$I = \int_0^1 dx \int_0^{\sqrt{x}} dy \int_0^{\sqrt{x-y^2}} f(x, y, z)\, dz.$$

**26.**

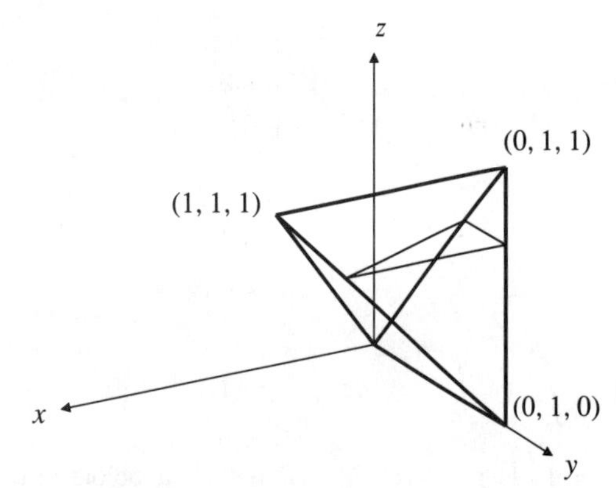


Fig. 6.5.26

$$I = \int_0^1 dx \int_x^1 dy \int_x^y f(x, y, z)\, dz = \iiint_P f(x, y, z)\, dV,$$

where $P$ is the triangular pyramid (see the figure) with vertices at $(0, 0, 0)$, $(0, 1, 0)$, $(0, 1, 1)$, and $(1, 1, 1)$. If we we reiterate $I$ to correspond to the horizontal slice shown then

$$\int_0^1 dz \int_z^1 dy \int_0^z f(x, y, z)\, dx.$$

**28.**

$$\int_0^1 dx \int_0^{1-x} dy \int_y^1 \frac{\sin(\pi z)}{z(2-z)}\, dz$$
$$= \iiint_R \frac{\sin(\pi z)}{z(2-z)}\, dV \qquad (R \text{ is the pyramid in the figure})$$
$$= \int_0^1 \frac{\sin(\pi z)}{z(2-z)}\, dz \int_0^z dy \int_0^{1-y} dx$$
$$= \int_0^1 \frac{\sin(\pi z)}{z(2-z)}\, dz \int_0^z (1 - y)\, dy$$
$$= \int_0^1 \frac{\sin(\pi z)}{z(2-z)} \left(z - \frac{z^2}{2}\right) dz$$
$$= \frac{1}{2} \int_0^1 \sin(\pi z)\, dz = \frac{1}{\pi}.$$

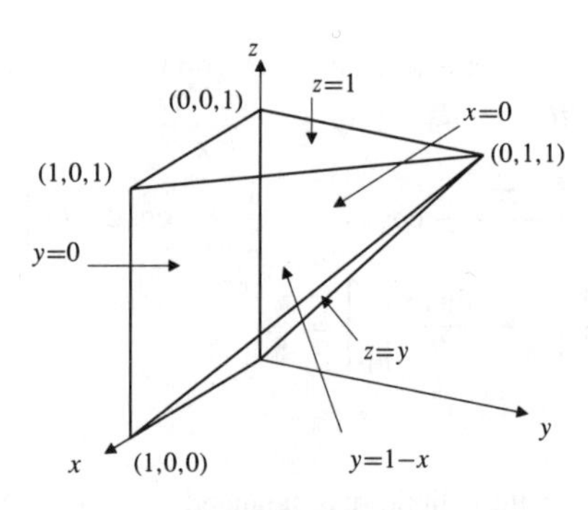


Fig. 6.5.28

**30.** If the function $f(x, y, z)$ is continuous on a closed, bounded, connected set $D$ in 3-space, then there exists a point $(x_0, y_0, z_0)$ in $D$ such that

$$\iiint_D f(x, y, z)\, dV = f(x_0, y_0, z_0) \times (\text{volume of } D).$$

Apply this with $D = B_\epsilon(a, b, c)$, which has volume $\frac{4}{3}\pi\epsilon^3$, to get

$$\iiint_{B_\epsilon(a,b,c)} f(x, y, z)\, dV = f(x_0, y_0, z_0)\frac{4}{3}\pi\epsilon^3$$

for some $(x_0, y_0, z_0)$ in $B_\epsilon(a, b, c)$. Thus

$$\lim_{\epsilon\to 0} \frac{3}{4\pi\epsilon^3} \iiint_{B_\epsilon(a,b,c)} f(x, y, z)\, dV$$
$$= \lim_{\epsilon\to 0} f(x_0, y_0, z_0) = f(a, b, c)$$

since $f$ is continuous at $(a, b, c)$.

## Section 6.6 Change of Variables in Triple Integrals (page 308)

**2.** Cartesian: $(2, -2, 1)$;
Cylindrical: $[2\sqrt{2}, -\pi/4, 1]$;
Spherical: $[3, \cos^{-1}(1/3), -\pi/4]$.

**4.** Spherical: $[1, \phi, \theta]$; Cylindrical: $[r, \pi/4, r]$.

$$x = \sin\phi\cos\theta = r\cos\pi/4 = r/\sqrt{2}$$
$$y = \sin\phi\sin\theta = r\sin\pi/4 = r/\sqrt{2}$$
$$z = \cos\phi = r.$$

Thus $x = y$, $\theta = \pi/4$, and $r = \sin\phi = \cos\phi$. Hence $\phi = \pi/4$, so $r = 1/\sqrt{2}$. Finally: $x = y = 1/2$, $z = 1/\sqrt{2}$. Cartesian: $(1/2, 1/2, 1/\sqrt{2})$.

**6.** $\phi = 2\pi/3$ represents the lower half of the right-circular cone with vertex at the origin, axis along the $z$-axis, and semi-vertical angle $\pi/3$. Its Cartesian equation is $z = -\sqrt{(x^2 + y^2)/3}$.

**8.** $\rho = 4$ represents the sphere of radius 4 centred at the origin.

**10.** $\rho = z$ represents the positive half of the $z$-axis.

**12.** $\rho = 2x$ represents the half-cone with vertex at the origin, axis along the positive $x$-axis, and semi-vertical angle $\pi/3$. Its Cartesian equation is $x = \sqrt{(y^2 + z^2)/3}$.

**14.** $r = 2\cos\theta \Rightarrow x^2 + y^2 = r^2 = 2r\cos\theta = 2x$, or $(x-1)^2 + y^2 = 1$. Thus the given equation represents the circular cylinder of radius 1 with axis along the vertical line $x = 1$, $y = 0$.

**16.** The surface $z = \sqrt{r}$ intersects the sphere $r^2 + z^2 = 2$ where $r^2 + r - 2 = 0$. This equation has positive root $r = 1$. The required volume is

$$\begin{aligned}
V &= \int_0^{2\pi} d\theta \int_0^1 r\,dr \int_{\sqrt{r}}^{\sqrt{2-r^2}} dz \\
&= \int_0^{2\pi} d\theta \int_0^1 \left(\sqrt{2-r^2} - \sqrt{r}\right) r\,dr \\
&= 2\pi\left(\int_0^1 r\sqrt{2-r^2}\,dr - \frac{2}{5}\right) \quad \text{Let } u = 2 - r^2,\ du = -2r\,dr \\
&= \pi\int_1^2 u^{1/2}\,du - \frac{4\pi}{5} \\
&= \frac{2\pi}{3}\left(2\sqrt{2} - 1\right) - \frac{4\pi}{5} = \frac{4\sqrt{2}\pi}{3} - \frac{22\pi}{15} \text{ cu. units.}
\end{aligned}$$

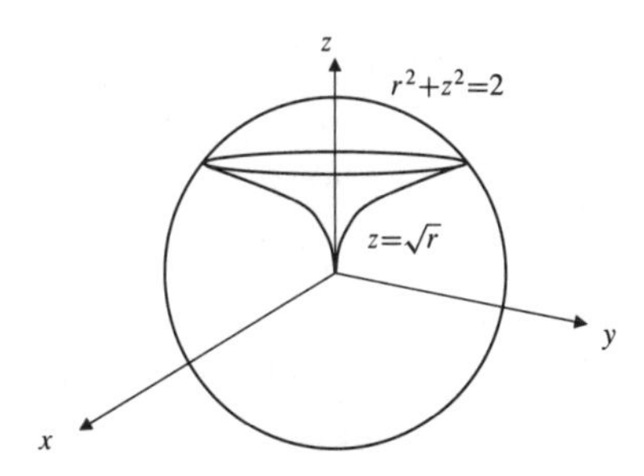


Fig. 6.6.16

**18.** The paraboloid $z = r^2$ intersects the sphere $r^2 + z^2 = 12$ where $r^4 + r^2 - 12 = 0$, that is, where $r = \sqrt{3}$. The required volume is

$$\begin{aligned}
V &= \int_0^{2\pi} d\theta \int_0^{\sqrt{3}} \left(\sqrt{12-r^2} - r^2\right) r\,dr \\
&= 2\pi\int_0^{\sqrt{3}} r\sqrt{12-r^2}\,dr - \frac{9\pi}{2} \quad \text{Let } u = 12 - r^2,\ du = -2r\,dr \\
&= \pi\int_9^{12} u^{1/2}\,du - \frac{9\pi}{2} \\
&= \frac{2\pi}{3}\left(12^{3/2} - 27\right) - \frac{9\pi}{2} = 16\sqrt{3}\pi - \frac{45\pi}{2} \text{ cu. units.}
\end{aligned}$$

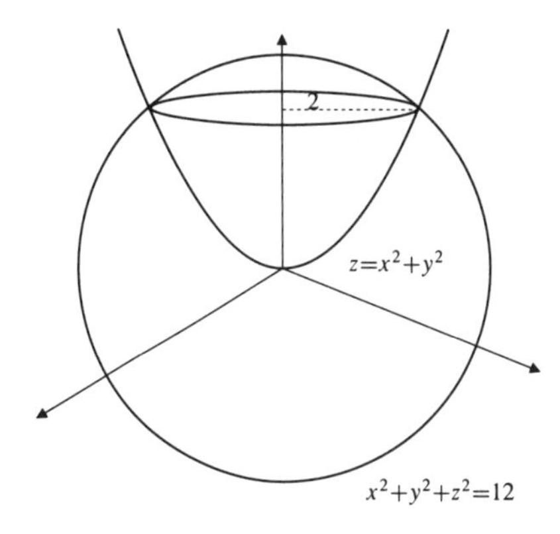


Fig. 6.6.18

**20.** The required volume $V$ lies above $z = 0$, below $z = 1 - r^2$, and between $\theta = -\pi/4$ and $\theta = \pi/3$. Thus

$$\begin{aligned}
V &= \int_{-\pi/4}^{\pi/3} d\theta \int_0^1 (1 - r^2) r\,dr \\
&= \frac{7\pi}{12}\left(\frac{1}{2} - \frac{1}{4}\right) = \frac{7\pi}{48} \text{ cu. units.}
\end{aligned}$$

**22.** One eighth of the required volume $V$ lies in the first octant. Call this region $R$. Under the transformation

$$x = au, \qquad y = bv, \qquad z = cw,$$

$R$ corresponds to the region $S$ in the first octant of $uvw$-space bounded by $w = 0$, $w = 1$, and $u^2 + v^2 - w^2 = 1$. Thus

$$V = 8abc \times (\text{volume of } S).$$

The volume of $S$ can be determined by using horizontal slices:

$$V = 8abc\int_0^1 \frac{\pi}{4}(1 + w^2)\,dw = \frac{8}{3}\pi abc \text{ cu. units.}$$

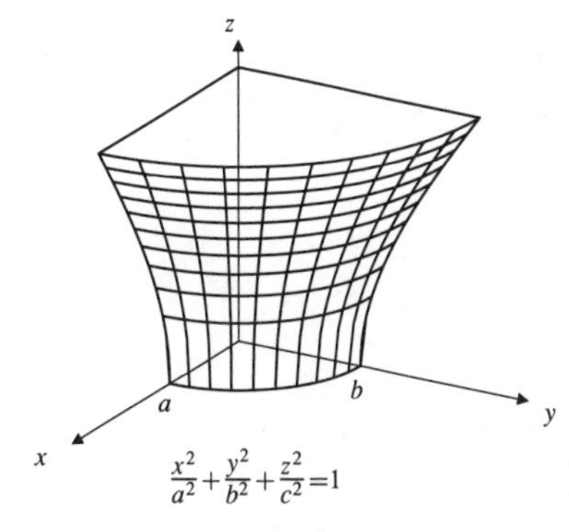


Fig. 6.6.22

**24.** Let $E$ be the ellipse in which the plane $z = 1-x$ intersects the cone $z = 2\sqrt{x^2+y^2}$. Then $E$ lies on the vertical cylinder

$$\begin{aligned}4x^2+4y^2 &= (1-x)^2\\ 3x^2+2x+4y^2 &= 1\\ 3\left(x+\tfrac{1}{3}\right)^2+4y^2 &= \frac{4}{3}\\ \frac{\left(x+\frac{1}{3}\right)^2}{\frac{4}{9}}+\frac{y^2}{\frac{1}{3}} &= 1.\end{aligned}$$

The horizontal cross-section of this vertical elliptic cylinder has area

$$\pi\left(\frac{2}{3}\right)\left(\frac{1}{\sqrt{3}}\right) = \frac{2\pi}{3\sqrt{3}} \text{ sq. units.}$$

Since $E$ lies in a plane inclined at a 45° angle to the horizontal, the area of $E$ is $2\sqrt{2}\pi/(3\sqrt{3})$ square units. The cone whose volume we are trying to find has base $E$ and vertex at the origin. The height of the cone is the distance from the origin to the plane $x+z=1$, that is, $1/\sqrt{2}$ units (by Example 7 of Section 3.4). Therefore, the volume of the cone is

$$V = \frac{1}{3}\left(\frac{2\sqrt{2}\pi}{3\sqrt{3}}\right)\left(\frac{1}{\sqrt{2}}\right) = \frac{2\pi}{9\sqrt{3}} \text{ cu. units.}$$

**26.**
$$\begin{aligned}&\iiint_R (x^2+y^2+z^2)\,dV\\ &= \int_0^{2\pi} d\theta \int_0^a r\,dr \int_0^h (r^2+z^2)\,dz\\ &= 2\pi\int_0^a \left(r^3h+\frac{1}{3}rh^3\right)dr\\ &= 2\pi\left(\frac{a^4h}{4}+\frac{a^2h^3}{6}\right) = \frac{\pi a^4 h}{2}+\frac{\pi a^2h^3}{3}.\end{aligned}$$

**28.**
$$\begin{aligned}&\iiint_B (x^2+y^2+z^2)\,dV\\ &= \int_0^{2\pi} d\theta \int_0^\pi \sin\phi\,d\phi \int_0^a R^4\,dR = \frac{4\pi a^5}{5}.\end{aligned}$$

**30.**
$$\begin{aligned}&\iiint_R (x^2+y^2)\,dV\\ &= \int_0^{2\pi} d\theta \int_0^{\tan^{-1}(1/c)} \sin^3\phi\,d\phi \int_0^a R^4\,dR\\ &= \frac{2\pi a^5}{5}\int_0^{\tan^{-1}(1/c)} \sin\phi(1-\cos^2\phi)\,d\phi \qquad \text{Let } u=\cos\phi,\ du=-\sin\phi\,d\phi\\ &= \frac{2\pi a^5}{5}\int_{c/\sqrt{c^2+1}}^1 (1-u^2)\,du\\ &= \frac{2\pi a^5}{5}\left(u-\frac{u^3}{3}\right)\Bigg|_{c/\sqrt{c^2+1}}^1\\ &= \frac{2\pi a^5}{5}\left(\frac{2}{3}-\frac{c}{\sqrt{c^2+1}}+\frac{c^3}{3(c^2+1)^{3/2}}\right).\end{aligned}$$

**32.** By symmetry, both integrals have the same value:

$$\begin{aligned}\iiint_R x\,dV = \iiint_R z\,dV &= \int_0^{\pi/2} d\theta \int_0^{\pi/2} \cos\phi\sin\phi\,d\phi \int_0^a R^3\,dR\\ &= \frac{\pi}{2}\left(\frac{1}{2}\right)\frac{a^4}{4} = \frac{\pi a^4}{16}.\end{aligned}$$

**34.**
$$\begin{aligned}&\iiint_{\mathbb{R}^3} \frac{dV}{(x^2+y^2)^\lambda(1+x^2+y^2+z^2)^\mu}\\ &= \int_0^\pi d\theta \int_0^\pi \sin\phi\,d\phi \int_0^\infty \frac{R^2\,dR}{(R\sin\phi)^{2\lambda}(1+R^2)^\mu}\\ &= 4\pi\int_0^{\pi/2} \sin^{1-2\lambda}\phi\,d\phi \int_0^\infty \frac{R^{2(1-\lambda)}\,dR}{(1+R^2)^\mu}.\end{aligned}$$

The $\phi$ integral may be improper at $\phi = 0$. Since $\sin\phi \approx \phi$ for $\phi$ near 0, this integral will converge if $1-2\lambda > -1$, that is, if $\lambda < 1$.

For such $\lambda$, the $R$ integral is improper only at infinity. It will converge if $2\mu+2\lambda-2 > 1$, that is, if $\mu+\lambda > 3/2$.

Both conditions are required for the given triple integral to converge.

**36.** If

$$x = au, \qquad y = bv, \qquad z = cw,$$

then the volume of a region $R$ in $xyz$-space is $abc$ times the volume of the corresponding region $S$ in $uvw$-space.

If $R$ is the region inside the ellipsoid

$$\frac{x^2}{a^2}+\frac{y^2}{b^2}+\frac{z^2}{c^2} = 1$$

and above the plane $y+z=b$, then the corresponding region $S$ lies inside the sphere

$$u^2+v^2+w^2 = 1$$

and above the plane $bv + cw = b$. The distance from the origin to this plane is

$$D = \frac{b}{\sqrt{b^2+c^2}} \qquad \text{(assuming } b > 0\text{)}$$

by Example 7 of Section 3.4. By symmetry, the volume of $S$ is equal to the volume lying inside the sphere $u^2 + v^2 + w^2 = 1$ and above the plane $w = D$. We calculate this latter volume by slicing; it is

$$\pi \int_D^1 (1 - w^2)\,dw = \pi \left( w - \frac{w^3}{3} \right) \Bigg|_D^1 = \pi \left( \frac{2}{3} - D + \frac{D^3}{3} \right).$$

Hence, the volume of $R$ is

$$\pi abc \left( \frac{2}{3} - \frac{b}{\sqrt{b^2+c^2}} + \frac{b^3}{3(b^2+c^2)^{3/2}} \right) \text{ cu. units.}$$

**38.** Cylindrical and spherical coordinates are related by

$$z = \rho\cos\phi, \qquad r = \rho\sin\phi.$$

(The $\theta$ coordinates are identical in the two systems.) Observe that $z$, $r$, $\rho$, and $\phi$ play, respectively, the same roles that $x$, $y$, $r$, and $\theta$ play in the transformation from Cartesian to polar coordinates in the plane. We can exploit this correspondence to avoid repeating the calculations of partial derivatives of a function $u$, since the results correspond to calculations made (for a function $z$) in Example 10 of Section 4.5. Comparing with the calculations in that Example, we have

$$\frac{\partial u}{\partial \rho} = \cos\phi \frac{\partial u}{\partial z} + \sin\phi \frac{\partial u}{\partial r}$$
$$\frac{\partial u}{\partial \phi} = -\rho\sin\phi \frac{\partial u}{\partial z} + \rho\cos\phi \frac{\partial u}{\partial r}$$
$$\frac{\partial^2 u}{\partial \rho^2} = \cos^2\phi \frac{\partial^2 u}{\partial z^2} + 2\cos\phi\sin\phi \frac{\partial^2 u}{\partial z \partial r} + \sin^2\phi \frac{\partial^2 u}{\partial r^2}$$
$$\frac{\partial^2 u}{\partial \phi^2} = -\rho \frac{\partial u}{\partial \rho} + \rho^2 \left( \sin^2\phi \frac{\partial^2 u}{\partial z^2} - 2\cos\phi\sin\phi \frac{\partial^2 u}{\partial z \partial r} + \cos^2\phi \frac{\partial^2 u}{\partial r^2} \right).$$

Substituting these expressions into the expression for $\Delta u$ given in the statement of this exercise in terms of spherical coordinates, we obtain the expression in terms of cylindrical coordinates established in the previous exercise:

$$\frac{\partial^2 u}{\partial \rho^2} + \frac{2}{\rho}\frac{\partial u}{\partial \rho} + \frac{\cot\phi}{\rho^2}\frac{\partial u}{\partial \phi} + \frac{1}{\rho^2}\frac{\partial^2 u}{\partial \phi^2} + \frac{1}{\rho^2\sin^2\phi}\frac{\partial^2 u}{\partial \theta^2}$$
$$= \frac{\partial^2 u}{\partial r^2} + \frac{1}{r}\frac{\partial u}{\partial r} + \frac{1}{r^2}\frac{\partial^2 u}{\partial \theta^2} + \frac{\partial^2 u}{\partial z^2}$$
$$= \frac{\partial^2 u}{\partial x^2} + \frac{\partial^2 u}{\partial y^2} = \Delta u$$

by Exercise 37.

## Section 6.7 Applications of Multiple Integrals (page 318)

**2.** $z = (3x - 4y)/5$, $\quad \dfrac{\partial z}{\partial x} = \dfrac{3}{5}$, $\quad \dfrac{\partial z}{\partial y} = \dfrac{4}{5}$

$$dS = \sqrt{1 + \frac{3^2+4^2}{5^2}}\,dA = \sqrt{2}\,dA$$
$$S = \iint_{(x/2)^2+y^2\le 1} \sqrt{2}\,dA = \sqrt{2}\pi(2)(1) = 2\sqrt{2}\pi \text{ sq. units.}$$

**4.** $z = 2\sqrt{1-x^2-y^2}$

$$\frac{\partial z}{\partial x} = -\frac{2x}{\sqrt{1-x^2-y^2}}, \quad \frac{\partial z}{\partial y} = -\frac{2y}{\sqrt{1-x^2-y^2}}$$
$$dS = \sqrt{1 + \frac{4(x^2+y^2)}{1-x^2-y^2}}\,dA = \sqrt{\frac{1+3(x^2+y^2)}{1-x^2-y^2}}\,dA$$
$$S = \iint_{x^2+y^2\le 1} dS$$
$$= \int_0^{2\pi} d\theta \int_0^1 \sqrt{\frac{1+3r^2}{1-r^2}}\, r\,dr \quad \text{Let } u^2 = 1 - r^2,\ u\,du = -r\,dr$$
$$= 2\pi \int_0^1 \sqrt{4-3u^2}\,du \quad \text{Let } \sqrt{3}u = 2\sin v,\ \sqrt{3}\,du = 2\cos v\,dv$$
$$= 2\pi \int_0^{\pi/3} (2\cos^2 v)\frac{2\,dv}{\sqrt{3}}$$
$$= \frac{4\pi}{\sqrt{3}} \int_0^{\pi/3} (1+\cos 2v)\,dv$$
$$= \frac{4\pi}{\sqrt{3}} \left( v + \frac{\sin 2v}{2} \right) \Bigg|_0^{\pi/3} = \frac{4\pi^2}{3\sqrt{3}} + \pi \text{ sq. units.}$$

**6.** $z = 1 - x^2 - y^2$, $\quad \dfrac{\partial z}{\partial x} = -2x$, $\quad \dfrac{\partial z}{\partial y} = -2y$

$$dS = \sqrt{1+4x^2+4y^2}\,dA$$
$$S = \iint_{x^2+y^2\le 1,\ x\ge 0,\ y\ge 0} \sqrt{1+4(x^2+y^2)}\,dA$$
$$= \int_0^{\pi/2} d\theta \int_0^1 \sqrt{1+4r^2}\, r\,dr \quad \text{Let } u = 1+4r^2,\ du = 8r\,dr$$
$$= \frac{\pi}{16} \int_1^5 u^{1/2}\,du$$
$$= \frac{\pi}{16} \left( \frac{2}{3} u^{3/2} \right) \Bigg|_1^5 = \frac{\pi(5\sqrt{5}-1)}{24} \text{ sq. units.}$$

**8.** $z = \sqrt{x}, \quad \frac{\partial z}{\partial x} = \frac{1}{2\sqrt{x}}, \quad dS = \sqrt{1 + \frac{1}{4x}}\, dA$

$$\begin{aligned} S &= \int_0^1 dx \int_0^{\sqrt{x}} \sqrt{1+\frac{1}{4x}}\, dy = \int_0^1 \sqrt{\frac{4x+1}{4x}}\, \sqrt{x}\, dx \\ &= \frac{1}{2}\int_0^1 \sqrt{4x+1}\, dx \quad \text{Let } u = 4x+1 \\ &\qquad\qquad\qquad\qquad\quad du = 4\,dx \\ &= \frac{1}{8}\int_1^5 u^{1/2}\, du = \frac{1}{8}\left(\frac{2}{3}u^{3/2}\right)\bigg|_1^5 = \frac{5\sqrt{5}-1}{12} \text{ sq. units.} \end{aligned}$$

**10.** The area elements on $z = 2xy$ and $z = x^2 + y^2$, respectively, are

$$\begin{aligned} dS_1 &= \sqrt{1+(2y)^2+(2x)^2}\, dA = \sqrt{1+4x^2+4y^2}\, dx\, dy, \\ dS_2 &= \sqrt{1+(2x)^2+(2y)^2}\, dA = \sqrt{1+4x^2+4y^2}\, dx\, dy. \end{aligned}$$

Since these elements are equal, the area of the parts of both surfaces defined over any region of the $xy$-plane will be equal.

**12.**

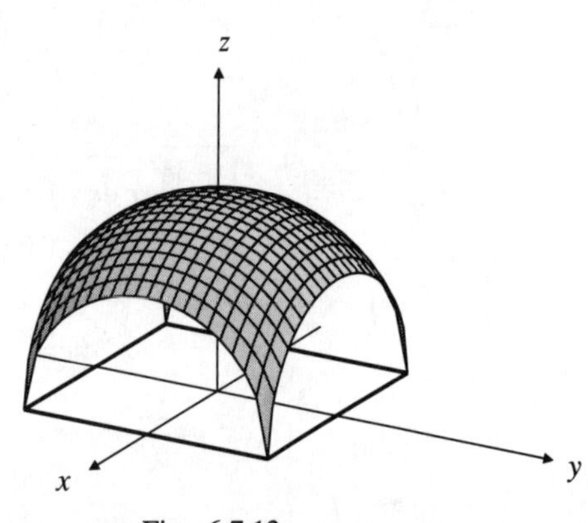


Fig. 6.7.12

As the figure suggests, the area of the canopy is the area of a hemisphere of radius $\sqrt{2}$ minus four times the area of half of a spherical cap cut off from the sphere $x^2 + y^2 + z^2 = 2$ by a plane at distance 1 from the origin, say the plane $z = 1$. Such a spherical cap, $z = \sqrt{2 - x^2 - y^2}$, lies above the disk $x^2+y^2 \le 2-1 = 1$. Since $\frac{\partial z}{\partial x} = -x/z$ and $\frac{\partial z}{\partial y} = -y/z$ on it, the area of the spherical cap is

$$\begin{aligned} &\iint_{x^2+y^2\le 1} \sqrt{1+\frac{x^2+y^2}{z^2}}\, dA \\ &= 2\sqrt{2}\pi \int_0^1 \frac{r\, dr}{\sqrt{2-r^2}} \quad \text{Let } u = 2-r^2 \\ &\qquad\qquad\qquad\qquad\qquad du = -2r\, dr \\ &= \sqrt{2}\pi \int_1^2 u^{-1/2}\, du = 2\sqrt{2}(\sqrt{2}-1) = 4 - 2\sqrt{2}. \end{aligned}$$

Thus the area of the canopy is

$$S = 2\pi(\sqrt{2})^2 - 4\times\frac{1}{2}\times(4-2\sqrt{2}) = 4(\pi+\sqrt{2}) - 8 \text{ sq. units.}$$

**14.** A slice of the ball at height $z$, having thickness $dz$, is a circular disk of radius $\sqrt{a^2 - z^2}$ and areal density $\delta\, dz$. As calculated in the text, this disk attracts mass $m$ at $(0, 0, b)$ with vertical force

$$dF = 2\pi k m \delta dz \left(1 - \frac{b-z}{\sqrt{a^2 - z^2 + (b-z)^2}}\right).$$

Thus the ball attracts $m$ with vertical force

$$\begin{aligned} F &= 2\pi k m \delta \int_{-a}^{a} \left(1 - \frac{b-z}{\sqrt{a^2+b^2-2bz}}\right) dz \\ &\qquad \text{let } v = a^2+b^2-2bz, \quad dv = -2b\, dz \\ &\qquad \text{then } b - z = b - \frac{a^2+b^2-v}{2b} = \frac{b^2-a^2+v}{2b} \\ &= 2\pi k m \delta \left[2a - \frac{1}{4b^2}\int_{(b-a)^2}^{(b+a)^2} \frac{b^2-a^2+v}{\sqrt{v}}\, dv\right] \\ &= 2\pi k m \delta \Big[2a - \frac{b^2-a^2}{2b^2}\big(b+a-(b-a)\big) \\ &\qquad - \frac{1}{6b^2}\big((b+a)^3 - (b-a)^3\big)\Big] \\ &= \frac{4\pi k m \delta a^3}{3b^2} = \frac{kmM}{b^2}, \end{aligned}$$

where $M = (4/3)\pi a^3 \delta$ is the mass of the ball. Thus the ball attracts the external mass $m$ as though the ball were a point mass $M$ located at its centre.

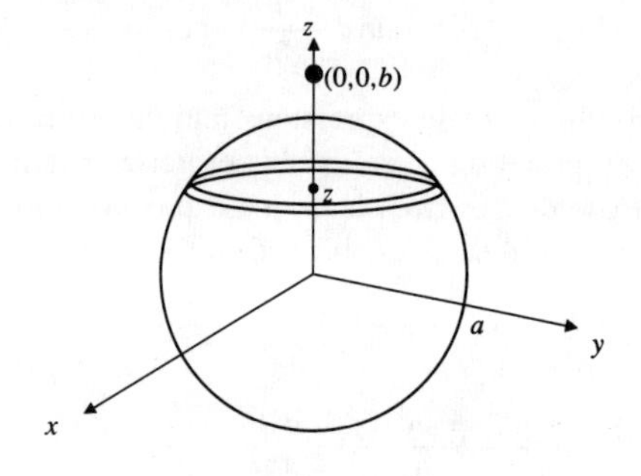


Fig. 6.7.14

**16.** The force is

$$\begin{aligned} F &= 2\pi km\delta \int_0^b \left(1 - \frac{b-z}{\sqrt{a^2(b-z)^2 + (b-z)^2}}\right) dz \\ &= 2\pi km\delta \int_0^b \left(1 - \frac{1}{\sqrt{a^2+1}}\right) dz \\ &= 2\pi km\delta b \left(1 - \frac{1}{\sqrt{a^2+1}}\right). \end{aligned}$$

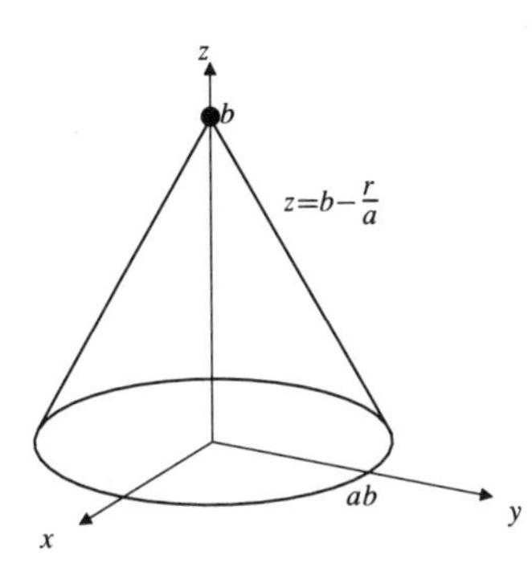


Fig. 6.7.16

**18.**

$$\begin{aligned} m &= \int_0^a dx \int_0^a dy \int_0^a (x^2+y^2+z^2)\,dz \\ &= 3\int_0^a x^2\,dx \int_0^a dy \int_0^a dz = a^5 \\ M_{x=0} &= \int_0^a x\,dx \int_0^a dy \int_0^a (x^2+y^2+z^2)\,dz \\ &= \int_0^a x\,dx \int_0^a \left(a(x^2+y^2) + \frac{a^3}{3}\right) dy \\ &= \int_0^a \left(\frac{2a^4}{3} + a^2x^2\right) x\,dx = \frac{7a^6}{12}. \end{aligned}$$

Thus $\bar{x} = M_{x=0}/m = \dfrac{7a}{12}$.

By symmetry, the centre of mass is $\left(\dfrac{7a}{12}, \dfrac{7a}{12}, \dfrac{7a}{12}\right)$.

**20.** Volume of region $= \displaystyle\int_0^{2\pi} d\theta \int_0^\infty e^{-r^2} r\,dr = \pi$. By symmetry, the moments about $x = 0$ and $y = 0$ are both zero. We have

$$\begin{aligned} M_{z=0} &= \int_0^{2\pi} d\theta \int_0^\infty r\,dr \int_0^{e^{-r^2}} z\,dz \\ &= \pi \int_0^\infty re^{-2r^2}\,dr = \frac{\pi}{4}. \end{aligned}$$

The centroid is $(0, 0, 1/4)$.

**22.** The region is half of a circular cone of base radius $a^2$ and height 1. Its volume is $V = \pi a^4/6$. By symmetry, $\bar{x} = 0$. We have

$$\begin{aligned} M_{y=0} &= \int_0^\pi d\theta \int_0^{a^2} r\,dr \int_{r/a^2}^1 r\sin\theta\,dz \\ &= 2\int_0^{a^2} \left(1 - \frac{r}{a^2}\right) r^2\,dr \\ &= 2\left(\frac{a^6}{3} - \frac{a^6}{4}\right) = \frac{a^6}{6} \\ M_{z=0} &= \int_0^\pi d\theta \int_0^{a^2} r\,dr \int_{r/a^2}^1 z\,dz \\ &= \frac{\pi}{2}\int_0^{a^2} \left(1 - \frac{r^2}{a^4}\right) r\,dr \\ &= \frac{\pi}{2}\left(\frac{a^4}{2} - \frac{a^4}{4}\right) = \frac{\pi a^4}{8}. \end{aligned}$$

Thus $\bar{y} = a^2/\pi$ and $\bar{z} = 3/4$.

The centroid is $\left(0, \dfrac{a^2}{\pi}, \dfrac{3}{4}\right)$.

**24.** The cube has centroid $(1/2, 1/2, 1/2)$. The tetrahedron lying above the plane $x + y + x = 2$ has centroid $(3/4, 3/4, 3/4)$ and volume $1/6$. Therefore the part of the cube lying below the plane has centroid $(c, c, c)$ and volume $5/6$, where

$$\frac{5}{6}c + \frac{3}{4} \times \frac{1}{6} = \frac{1}{2} \times 1.$$

Thus $c = 9/20$; the centroid is $\left(\dfrac{9}{20}, \dfrac{9}{20}, \dfrac{9}{20}\right)$.

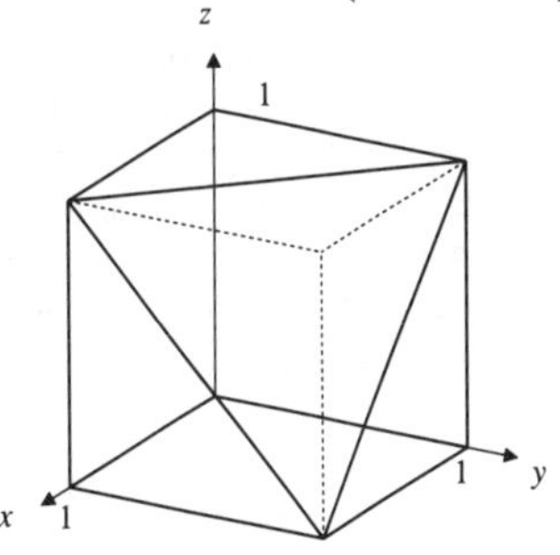


Fig. 6.7.24

**26.**

$$\begin{aligned} I &= \delta \int_0^{2\pi} d\theta \int_0^a r^3\,dr \int_0^h dz \\ &= 2\pi\delta h \left(\frac{a^4}{4}\right) = \frac{\pi\delta h a^4}{2}. \end{aligned}$$

$$m = \pi\delta a^2 h, \qquad \bar{D} = \sqrt{I/m} = \frac{a}{\sqrt{2}}.$$

**28.** $$I = \delta \int_0^{2\pi} d\theta \int_0^a r^3\,dr \int_0^{h(1-(r/a))} dz$$
$$= 2\pi\delta h \int_0^a r^3\left(1 - \frac{r}{a}\right) dr = \frac{\pi\delta a^4 h}{10},$$
$$m = \frac{\pi\delta a^2 h}{3}, \qquad \overline{D} = \sqrt{I/m} = \sqrt{\frac{3}{10}}a.$$

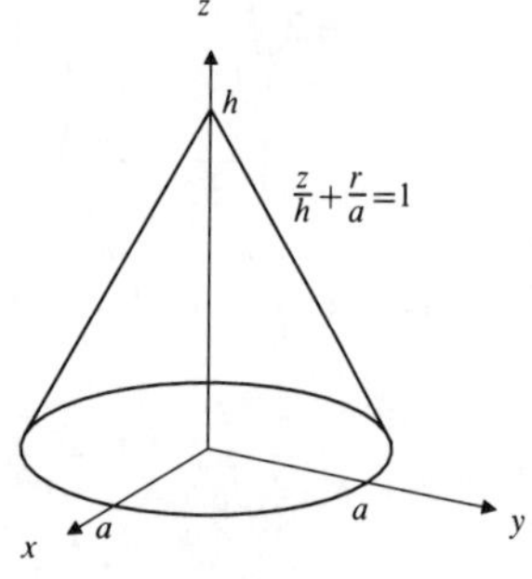


Fig. 6.7.28

**30.** $$I = \delta \iiint_Q (x^2 + y^2)\,dV$$
$$= 2\delta \int_0^a x^2\,dx \int_0^a dy \int_0^a dz = \frac{2\delta a^5}{3},$$
$$m = \delta a^3, \qquad \overline{D} = \sqrt{I/m} = \sqrt{\frac{2}{3}}a.$$

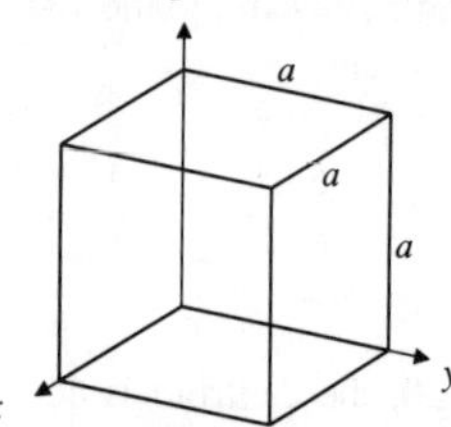


Fig. 6.7.30

**32.** The line $L$ through the origin parallel to the vector $\mathbf{v} = \mathbf{i} + \mathbf{j} + \mathbf{k}$ is a diagonal of the cube $Q$. By Example 8 of Section 3.4, the distance from the point with position vector $\mathbf{r} = x\mathbf{i} + y\mathbf{j} + z\mathbf{k}$ to $L$ is $s = |\mathbf{v} \times \mathbf{r}|/|\mathbf{v}|$. Thus, the square of the distance from $(x, y, z)$ to $L$ is

$$s^2 = \frac{(x-y)^2 + (y-z)^2 + (z-x)^2}{3}$$
$$= \frac{2}{3}\left(x^2 + y^2 + z^2 - xy - xz - yz\right).$$

We have

$$\iiint_Q x^2\,dV = \iiint_Q y^2\,dV = \iiint_Q z^2\,dV = \frac{a^5}{3}$$
$$\iiint_Q xy\,dV = \iiint_Q yz\,dV = \iiint_Q xz\,dV = \frac{a^5}{4}.$$

Therefore, the moment of inertia of $Q$ about $L$ is

$$I = \frac{2\delta}{3}\left(3 \times \frac{a^5}{3} - 3 \times \frac{a^5}{4}\right) = \frac{\delta a^5}{6}.$$

The mass of $Q$ is $m = \delta a^3$, so the radius of gyration is

$$\overline{D} = \sqrt{I/m} = \frac{a}{\sqrt{6}}.$$

**34.** $$I = \delta \int_0^{2\pi} d\theta \int_0^c dz \int_a^b r^3\,dr = \frac{\pi\delta c(b^4 - a^4)}{2},$$
$$m = \pi\delta c(b^2 - a^2), \qquad \overline{D} = \sqrt{\frac{b^2 + a^2}{2}}.$$

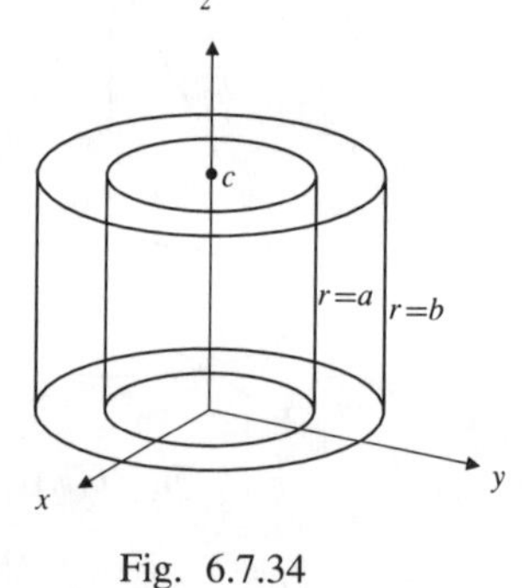


Fig. 6.7.34

**36.** By Exercise 26, the cylinder has moment of inertia

$$I = \frac{\pi\delta a^4 h}{2} = \frac{ma^2}{2},$$

where $m$ is its mass. Following the method of Example 4(b), the kinetic energy of the cylinder rolling down the inclined plane with speed $v$ is

$$KE = \frac{1}{2}mv^2 + \frac{1}{2}I\Omega^2$$
$$= \frac{1}{2}mv^2 + \frac{1}{4}ma^2\frac{v^2}{a^2} = \frac{3}{4}mv^2.$$

The potential energy of the cylinder when it is at height $h$ is $mgh$, so, by conservation of energy,

$$\frac{3}{4}mv^2 + mgh = \text{constant}.$$

Differentiating this equation with respect to time $t$, we obtain

$$0 = \frac{3}{2}mv\frac{dv}{dt} + mg\frac{dh}{dt}$$
$$= \frac{3}{2}mv\frac{dv}{dt} + mgv\sin\alpha.$$

Thus the cylinder rolls down the plane with acceleration

$$-\frac{dv}{dt} = \frac{2}{3}g\sin\alpha.$$

**38.** The kinetic energy of the oscillating pendulum is

$$KE = \frac{1}{2}I\left(\frac{d\theta}{dt}\right)^2.$$

The potential energy is $mgh$, where $h$ is the distance of $C$ above $A$. In this case, $h = -a\cos\theta$. By conservation of energy,

$$\frac{1}{2}I\left(\frac{d\theta}{dt}\right)^2 - mga\cos\theta = \text{constant}.$$

Differentiating with respect to time $t$, we obtain

$$I\left(\frac{d\theta}{dt}\right)\frac{d^2\theta}{dt^2} + mga\sin\theta\left(\frac{d\theta}{dt}\right) = 0,$$

or

$$\frac{d^2\theta}{dt^2} + \frac{mga}{I}\sin\theta = 0.$$

For small oscillations we have $\sin\theta \approx \theta$, and the above equation is approximated by

$$\frac{d^2\theta}{dt^2} + \omega^2\theta = 0,$$

where $\omega^2 = mga/I$. The period of oscillation is

$$T = \frac{2\pi}{\omega} = 2\pi\sqrt{\frac{I}{mga}}.$$

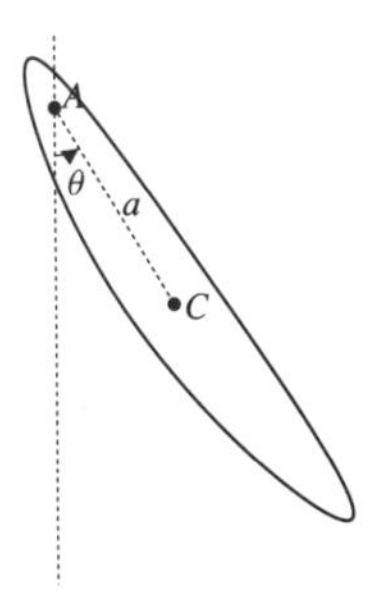


Fig. 6.7.38

**40.** The moment of inertia of the ball about the point where it contacts the plane is, by Example 4(b) and Exercise 39,

$$I = \frac{8}{15}\pi\delta a^5 + \left(\frac{4}{3}\pi\delta a^3\right)a^2$$
$$= \left(\frac{2}{5}+1\right)ma^2 = \frac{7}{5}ma^2.$$

The kinetic energy of the ball, regarded as rotating about the point of contact with the plane, is therefore

$$KE = \frac{1}{2}I\Omega^2 = \frac{7}{10}ma^2\frac{v^2}{a^2} = \frac{7}{10}mv^2.$$

## Review Exercises 6 (page 319)

**2.**
$$\iint_P (x^2+y^2)\,dA = \int_0^1 dy\int_y^{2+y}(x^2+y^2)\,dx$$
$$= \int_0^1 \left(\frac{x^3}{3}+xy^2\right)\Bigg|_{x=y}^{x=2+y} dy$$
$$= \int_0^1 \left(\frac{(2+y)^3}{3} + y^2(2+y) - \frac{y^3}{3} - y^3\right) dy$$
$$= \int_0^1 \left(\frac{8}{3}+4y+4y^2\right) dy = \frac{8}{3}+2+\frac{4}{3} = 6$$

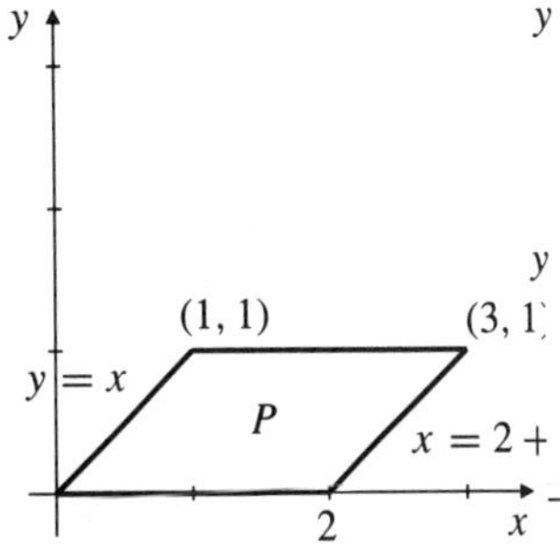

Fig. R-6.2 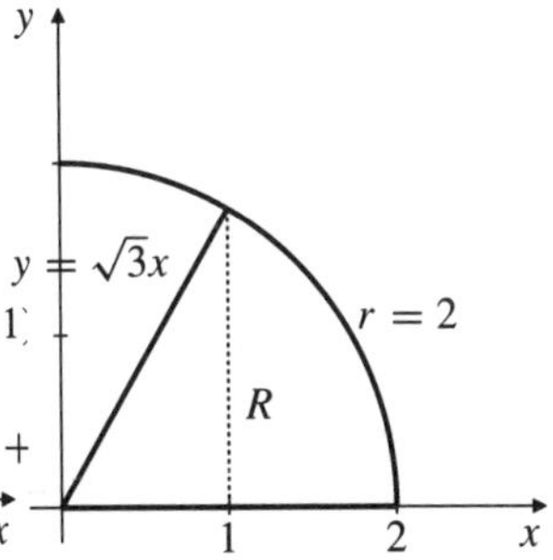 Fig. R-6.4

**4.** a)
$$I = \int_0^{\sqrt{3}} dy \int_{y/\sqrt{3}}^{\sqrt{4-y^2}} e^{-x^2-y^2}\,dx$$
$$= \iint_R e^{-x^2-y^2}\,dA$$
where $R$ is as shown in the figure.

b)
$$I = \int_0^1 dx\int_0^{\sqrt{3}x} e^{-x^2-y^2}\,dy + \int_1^2 dx\int_0^{\sqrt{4-x^2}} e^{-x^2-y^2}\,dy$$

c)
$$I = \int_0^{\pi/3} d\theta\int_0^2 e^{-r^2}r\,dr$$

d)
$$I = \frac{\pi}{3}\left(-\frac{e^{-r^2}}{2}\right)\Bigg|_0^2 = \frac{\pi(1-e^{-4})}{6}$$

**6.** $I = \int_0^2 dy \int_0^y f(x, y)\,dx + \int_2^6 dy \int_0^{\sqrt{6-y}} f(x, y)\,dx$
$= \iint_R f(x, y)\,dA,$
where $R$ is as shown in the figure. Thus

$$I = \int_0^2 dx \int_x^{6-x^2} f(x, y)\,dy.$$

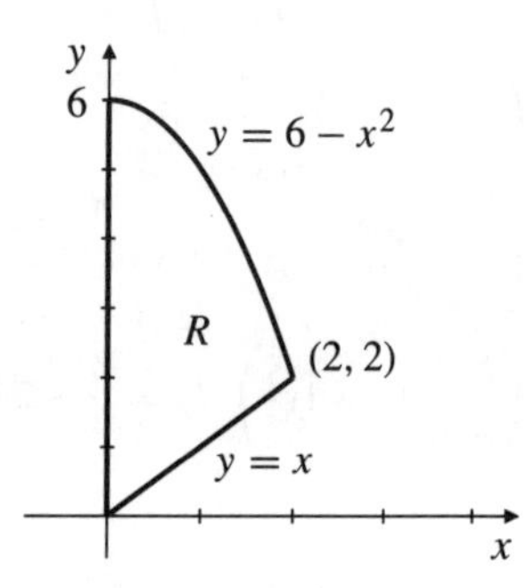


Fig. R-6.6

**8.** A horizontal slice of the object at height $z$ above the base, and having thickness $dz$, is a disk of radius $r = \frac{1}{2}(10 - z)$ m. Its volume is

$$dV = \pi \frac{(10 - z)^2}{4}\,dz \text{ m}^3.$$

The density of the slice is $\delta = kz^2$ kg/m$^3$. Since $\delta = 3,000$ when $z = 10$, we have $k = 30$.

a) The mass of the object is

$$\begin{aligned} m &= \int_0^{10} 30z^2 \frac{\pi}{4}(10 - z)^2\,dz \\ &= \frac{15\pi}{2} \int_0^{10} (100z^2 - 20z^3 + z^4)\,dz \\ &= \frac{15\pi}{2} \left( \frac{100,000}{3} - 50,000 + 20,000 \right) \approx 78,540 \text{ kg}. \end{aligned}$$

b) The moment of inertia (about its central axis) of the disk-shaped slice at height $z$ is

$$dI = 30z^2\,dz \int_0^{2\pi} d\theta \int_0^{(10-z)/2} r^3\,dr.$$

Thus the moment of inertia about the whole solid cone is

$$I = \int_0^{10} 30z^2\,dz \int_0^{2\pi} d\theta \int_0^{(10-z)/2} r^3\,dr.$$

**10.** If $f(x, y) = \lfloor x + y \rfloor$, then $f = 0$, 1, or 2, in parts of the quarter disk $Q$, as shown in the figure.

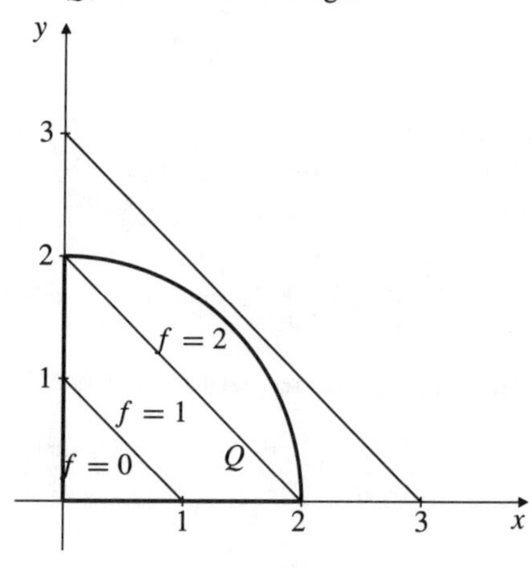


Fig. R-6.10

Thus

$$\iint_Q f(x, y)\,dA = 0\left(\frac{1}{2}\right) + 1\left(\frac{3}{2}\right) + 2(\pi - 2) = 2\pi - \frac{5}{2},$$

and $\bar{f} = \frac{1}{\pi}\left(2\pi - \frac{5}{2}\right) = 2 - \frac{5}{2\pi}$.

**12.** The solid $S$ lies above the region in the $xy$-plane bounded by the circle $x^2 + y^2 = 2ay$, which has polar equation $r = 2a\sin\theta$, $(0 \le \theta \le \pi)$. It lies below the cone $z = \sqrt{x^2 + y^2} = r$. The moment of inertia of $S$ about the $z$-axis is

$$\begin{aligned} I &= \iiint_S (x^2 + y^2)\,dV = \int_0^\pi d\theta \int_0^{2a\sin\theta} r^3\,dr \int_0^r dz \\ &= \int_0^\pi d\theta \int_0^{2a\sin\theta} r^4\,dr = \frac{32a^5}{5} \int_0^\pi \sin^5\theta\,d\theta \\ &= \frac{32a^5}{5} \int_0^\pi (1 - \cos^2\theta)^2 \sin\theta\,d\theta \quad \text{Let } u = \cos\theta,\ du = -\sin\theta\,d\theta \\ &= \frac{32a^5}{5} \int_{-1}^1 (1 - 2u^2 + u^4)\,du \\ &= \frac{64a^5}{5}\left(1 - \frac{2}{3} + \frac{1}{5}\right) = \frac{512a^5}{75}. \end{aligned}$$

**14.**

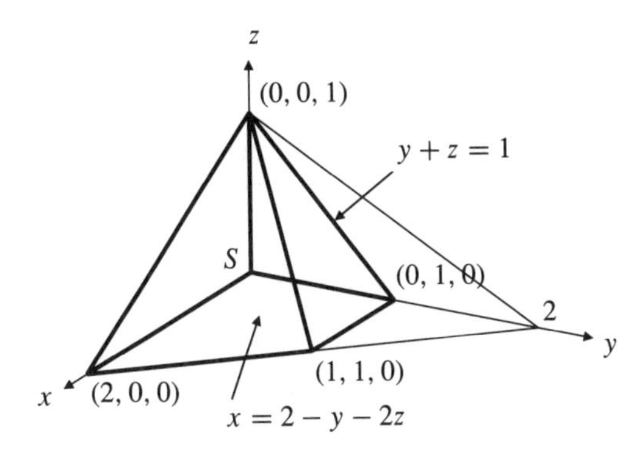


Fig. R-6.14

$$\begin{aligned}
V &= \iiint_S dV = \int_0^1 dy \int_0^{1-y} dz \int_0^{2-y-2z} dx \\
&= \int_0^1 dy \int_0^{1-y} (2-y-2z)\,dz \\
&= \int_0^1 [(2-y)(1-y)-(1-y)^2]\,dy \\
&= \int_0^1 (1-y)\,dy = \frac{1}{2}
\end{aligned}$$

$$\begin{aligned}
M_{x=0} &= \iiint_S x\,dV = \int_0^1 dy \int_0^{1-y} dz \int_0^{2-y-2z} x\,dx \\
&= \frac{1}{2}\int_0^1 dy \int_0^{1-y} [(2-y)^2 - 4(2-y)z + 4z^2]\,dz \\
&= \frac{1}{2}\int_0^1 \Big[(2-y)^2(1-y) - 2(2-y)(1-y)^2 \\
&\qquad + \frac{4}{3}(1-y)^3\Big]\,dy \quad \text{Let } u = 1-y,\ du = -dy \\
&= \frac{1}{2}\int_0^1 \left[(u+1)^2u - 2(u+1)u^2 + \frac{4}{3}u^3\right] du \\
&= \frac{1}{2}\int_0^1 \left[\frac{1}{3}u^3 + u\right] du = \frac{7}{24}
\end{aligned}$$

$$\bar{x} = \frac{7}{24}\Big/\frac{1}{2} = \frac{7}{12}$$

**16.** The plane $z = 2x$ intersects the paraboloid $z = x^2 + y^2$ on the circular cylinder $x^2 + y^2 = 2x$, (that is, $(x-1)^2 + y^2 = 1$), which has radius 1. Since $dS = \sqrt{1+2^2}\,dA = \sqrt{5}\,dA$ on the plane, the area of the part of the plane inside the paraboloid (and therefore inside the cylinder) is $\sqrt{5}$ times the area of a circle of radius 1, that is, $\sqrt{5}\pi$ square units.

**18.** The region $R$ inside the ellipsoid $\dfrac{x^2}{36} + \dfrac{y^2}{9} + \dfrac{z^2}{4} = 1$ and above the plane $x+y+z = 1$ is transformed by the change of variables

$$x = 6u, \quad y = 3v, \quad z = 2w$$

to the region $S$ inside the sphere $u^2 + v^2 + w^2 = 1$ and above the plane $6u + 3v + 2w = 1$. The distance from the origin to this plane is

$$D = \frac{1}{\sqrt{6^2+3^2+2^2}} = \frac{1}{7},$$

so, by symmetry, the volume of $S$ is equal to the volume inside the sphere and above the plane $w = 1/7$, that is,

$$\int_{1/7}^1 \pi(1-w^2)\,dw = \pi\left(w - \frac{w^3}{3}\right)\Bigg|_{1/7}^1 = \frac{180\pi}{343} \text{ units}^3.$$

Since $|\partial(x,y,z)/\partial(u,v,w)| = 6\cdot 3\cdot 2 = 18$, the volume of $R$ is $18 \times (180\pi/343) = 3240\pi/343 \approx 29.68$ cu. units.

## Challenging Problems 6 (page 320)

**2.** The plane $(x/a)+(y/b)+(z/c) = 1$ intersects the ellipsoid $(x/a)^2 + (y/b)^2 + (z/c)^2 = 1$ above the region $R$ in the $xy$-plane bounded by the ellipse

$$\frac{x^2}{a^2} + \frac{y^2}{b^2} + \left(1 - \frac{x}{a} - \frac{y}{b}\right)^2 = 1,$$

or, equivalently,

$$\frac{x^2}{a^2} + \frac{y^2}{b^2} + \frac{xy}{ab} - \frac{x}{a} - \frac{y}{b} = 0.$$

Thus the area of the part of the plane lying inside the ellipsoid is

$$\begin{aligned}
S &= \iint_R \sqrt{1 + \frac{c^2}{a^2} + \frac{c^2}{b^2}}\,dx\,dy \\
&= \frac{\sqrt{a^2b^2 + a^2c^2 + b^2c^2}}{ab}(\text{area of } R).
\end{aligned}$$

Under the transformation $x = a(u+v)$, $y = b(u-v)$, $R$ corresponds to the ellipse in the $uv$-plane bounded by

$$\begin{aligned}
&(u+v)^2 + (u-v)^2 + (u^2-v^2) - (u+v) - (u-v) = 0 \\
&3u^2 + v^2 - 2u = 0 \\
&3\left(u^2 - \frac{2}{3}u + \frac{1}{9}\right) + v^2 = \frac{1}{3} \\
&\frac{(u-1/3)^2}{1/9} + \frac{v^2}{1/3} = 1,
\end{aligned}$$

an ellipse with area $\pi(1/3)(1/\sqrt{3}) = \pi/(3\sqrt{3})$ sq. units. Since

$$dx\,dy = \left|\begin{vmatrix} a & a \\ b & -b \end{vmatrix}\right| du\,dv = 2ab\,du\,dv,$$

we have

$$S = \frac{2\pi}{3\sqrt{3}}\sqrt{a^2b^2 + a^2c^2 + b^2c^2} \text{ sq. units.}$$

**4.** a) $G(y) = \displaystyle\int_0^\infty \frac{\tan^{-1}(xy)}{x}\,dx$

$$G'(y) = \int_0^\infty \frac{1}{x}\,\frac{x}{1+x^2y^2}\,dx \quad \text{Let } u = xy, \; du = y\,dx$$

$$= \frac{1}{y}\int_0^\infty \frac{du}{1+u^2} = \frac{\pi}{2y} \quad \text{for } y > 0.$$

b) $\displaystyle\int_0^\infty \frac{\tan^{-1}(\pi x) - \tan^{-1}x}{x}\,dx$

$$= G(\pi) - G(1) = \int_1^\pi G'(y)\,dy = \frac{\pi}{2}\int_1^\pi \frac{dy}{y} = \frac{\pi \ln \pi}{2}.$$

**6.** The volume $V_0$ removed from the ball is eight times the part in the first octant, which is itself split into two equal parts by the plane $x = y$:

$$V_0 = 16\int_0^1 dx \int_0^x \sqrt{4 - x^2 - y^2}\,dy$$

$$= 16\int_0^{\pi/4} d\theta \int_0^{\sec\theta} \sqrt{4 - r^2}\,r\,dr \quad \text{Let } u = 4 - r^2, \; du = -2r\,dr$$

$$= 8\int_0^{\pi/4} d\theta \int_{4-\sec^2\theta}^4 u^{1/2}\,du$$

$$= \frac{16}{3}\int_0^{\pi/4} \left[8 - (4 - \sec^2\theta)^{3/2}\right] d\theta$$

$$= \frac{32\pi}{3} - \frac{16}{3}\int_0^{\pi/4} \frac{(4\cos^2\theta - 1)^{3/2}}{\cos^3\theta}\,d\theta.$$

Now the volume of the whole ball is $(4\pi/3)2^3 = 32\pi/3$, so the volume remaining after the hole is cut is

$$V = \frac{32\pi}{3} - V_0$$

$$= \frac{16}{3}\int_0^{\pi/4} \frac{(3 - 4\sin^2\theta)^{3/2}}{(1 - \sin^2\theta)^2}\cos\theta\,d\theta \quad \text{Let } v = \sin\theta, \; dv = \cos\theta\,d\theta$$

$$= \frac{16}{3}\int_0^{1/\sqrt{2}} \frac{(3 - 4v^2)^{3/2}}{(1 - v^2)^2}\,dv.$$

We submitted this last integral to Mathematica to obtain

$$V = \frac{4}{3}\left(32\sin^{-1}\sqrt{\frac{2}{3}} - 2^{3/2} + 11\tan^{-1}(3 - 2^{3/2}) - 11\tan^{-1}(3 + 2^{3/2})\right) \approx 18.9349.$$

**8.** One-eighth of the required volume lies in the first octant. Under the transformation $x = u^6$, $y = v^6$, $z = w^6$, the region first-octant $R$ bounded by the surface $x^{1/3} + y^{1/3} + z^{1/3} = a^{1/3}$ and the coordinate planes gets mapped to the first octant part $B$ of the ball bounded by $u^2 + v^2 + w^2 \le a^{1/3}$. Assume that $a > 0$. Since

$$\frac{\partial(x, y, z)}{\partial(u, v, w)} = 6^3 u^5 v^5 w^5,$$

the required volume is

$$V = 8(6^3)\iiint_B u^5 v^5 w^5\,du\,dv\,dw.$$

Now switch to polar coordinates $[\rho, \phi, \theta]$ in $uvw$-space. Since

$$uvw = (\rho\sin\phi\cos\theta)(\rho\sin\phi\sin\theta)(\rho\cos\phi),$$

we have

$$V = 1,728\int_0^{\pi/2} (\cos\theta\sin\theta)^5\,d\theta \int_0^{\pi/2} (\sin^2\phi\cos\phi)^5 \sin\phi\,d\phi \times \int_0^{a^{1/6}} \rho^{17}\,d\rho$$

$$= 96a^3\int_0^{\pi/2} \frac{\sin^5(2\theta)}{32}\,d\theta \int_0^{\pi/2} \sin^{11}\phi(1 - \sin^2\phi)^2\cos\phi\,d\phi$$

Let $s = \sin\phi$, $ds = \cos\phi\,d\phi$

$$= 3a^3\int_0^{\pi/2} (1 - \cos^2(2\theta))^2\sin(2\theta)\,d\theta \int_0^1 s^{11}(1 - s^2)^2\,ds$$

Let $t = \cos(2\theta)$, $dt = -2\sin(2\theta)\,d\theta$

$$= \frac{3a^3}{2}\int_{-1}^1 (1 - 2t^2 + t^4)\,dt \int_0^1 (s^{11} - 2s^{13} + s^{15})\,ds$$

$$= 3a^3\left(1 - \frac{2}{3} + \frac{1}{5}\right)\left(\frac{1}{12} - \frac{1}{7} + \frac{1}{16}\right) = \frac{a^3}{210} \text{ cu. units.}$$

# CHAPTER 7. CURVES IN 3-SPACE

## Section 7.1 Vector Functions of One Variable (page 328)

**2.** Position: $\mathbf{r} = t^2\mathbf{i} + \mathbf{k}$
Velocity: $\mathbf{v} = 2t\mathbf{i}$
Speed: $v = 2|t|$
Acceleration : $\mathbf{a} = 2\mathbf{i}$
Path: the line $z = 1$, $y = 0$.

**4.** Position: $\mathbf{r} = \mathbf{i} + t\mathbf{j} + t\mathbf{k}$
Velocity: $\mathbf{v} = \mathbf{j} + \mathbf{k}$
Speed: $v = \sqrt{2}$
Acceleration : $\mathbf{a} = \mathbf{0}$
Path: the straight line $x = 1$, $y = z$.

**6.** Position: $\mathbf{r} = t\mathbf{i} + t^2\mathbf{j} + t^2\mathbf{k}$
Velocity: $\mathbf{v} = \mathbf{i} + 2t\mathbf{j} + 2t\mathbf{k}$
Speed: $v = \sqrt{1 + 8t^2}$
Acceleration: $\mathbf{a} = 2\mathbf{j} + 2\mathbf{k}$
Path: the parabola $y = z = x^2$.

**8.** Position: $\mathbf{r} = a\cos\omega t\mathbf{i} + b\mathbf{j} + a\sin\omega t\mathbf{k}$
Velocity: $\mathbf{v} = -a\omega\sin\omega t\mathbf{i} + a\omega\cos\omega t\mathbf{k}$
Speed: $v = |a\omega|$
Acceleration: $\mathbf{a} = -a\omega^2\cos\omega t\mathbf{i} - a\omega^2\sin\omega t\mathbf{k}$
Path: the circle $x^2 + z^2 = a^2$, $y = b$.

**10.** Position: $\mathbf{r} = 3\cos t\mathbf{i} + 4\sin t\mathbf{j} + t\mathbf{k}$
Velocity: $\mathbf{v} = -3\sin t\mathbf{i} + 4\cos t\mathbf{j} + \mathbf{k}$
Speed: $v = \sqrt{9\sin^2 t + 16\cos^2 t + 1} = \sqrt{10 + 7\cos^2 t}$
Acceleration : $\mathbf{a} = -3\cos t\mathbf{i} - 4\sin t\mathbf{j} = t\mathbf{k} - \mathbf{r}$
Path: a helix (spiral) wound around the elliptic cylinder $(x^2/9) + (y^2/16) = 1$.

**12.** Position: $\mathbf{r} = at\cos\omega t\mathbf{i} + at\sin\omega t\mathbf{j} + b\ln t\mathbf{k}$
Velocity: $\mathbf{v} = a(\cos\omega t - \omega t\sin\omega t)\mathbf{i} + a(\sin\omega t + \omega t\cos\omega t)\mathbf{j} + (b/t)\mathbf{k}$
Speed: $v = \sqrt{a^2(1 + \omega^2t^2) + (b^2/t^2)}$
Acceleration: $\mathbf{a} = -a\omega(2\sin\omega t + \omega\cos\omega t)\mathbf{i} + a\omega(2\cos\omega t - \omega\sin\omega t)\mathbf{j} - (b/t^2)\mathbf{k}$
Path: a spiral on the surface $x^2 + y^2 = a^2e^{z/b}$.

**14.** Position:
$$\mathbf{r} = a\cos t\sin t\mathbf{i} + a\sin^2 t\mathbf{j} + a\cos t\mathbf{k} = \frac{a}{2}\sin 2t\mathbf{i} + \frac{a}{2}\big(1 - \cos 2t\big)\mathbf{j} + a\cos t\mathbf{k}$$
Velocity: $\mathbf{v} = a\cos 2t\mathbf{i} + a\sin 2t\mathbf{j} - a\sin t\mathbf{k}$
Speed: $v = a\sqrt{1 + \sin^2 t}$
Acceleration: $\mathbf{a} = -2a\sin 2t\mathbf{i} + 2a\cos 2t\mathbf{j} - a\cos t\mathbf{k}$
Path: the path lies on the sphere $x^2 + y^2 + z^2 = a^2$, on the surface defined in terms of spherical polar coordinates by $\phi = \theta$, on the circular cylinder $x^2 + y^2 = ay$, and on the parabolic cylinder $ay + z^2 = a^2$. Any two of these surfaces serve to pin down the shape of the path.

**16.** When its $x$-coordinate is $x$, the particle is at position $\mathbf{r} = x\mathbf{i} + (3/x)\mathbf{j}$, and its velocity and speed are
$$\mathbf{v} = \frac{d\mathbf{r}}{dt} = \frac{dx}{dt}\mathbf{i} - \frac{3}{x^2}\frac{dx}{dt}\mathbf{j}$$
$$v = \left|\frac{dx}{dt}\right|\sqrt{1 + \frac{9}{x^4}}.$$
We know that $dx/dt > 0$ since the particle is moving to the right. When $x = 2$, we have $10 = v = (dx/dt)\sqrt{1 + (9/16)} = (5/4)(dx/dt)$. Thus $dx/dt = 8$. The velocity at that time is $\mathbf{v} = 8\mathbf{i} - 6\mathbf{j}$.

**18.** The position of the object when its $x$-coordinate is $x$ is
$$\mathbf{r} = x\mathbf{i} + x^2\mathbf{j} + x^3\mathbf{k},$$
so its velocity is $\mathbf{v} = \dfrac{dx}{dt}\big[\mathbf{i} + 2x\mathbf{j} + 3x^2\mathbf{k}\big]$. Since $dz/dt = 3x^2\,dx/dt = 3$, when $x = 2$ we have $12\,dx/dt = 3$, so $dx/dt = 1/4$. Thus
$$\mathbf{v} = \frac{1}{4}\mathbf{i} + \mathbf{j} + 3\mathbf{k}.$$

**20.** If $\mathbf{u}(t) = u_1(t)\mathbf{i} + u_2(t)\mathbf{j} + u_3(t)\mathbf{k}$
$\mathbf{v}(t) = v_1(t)\mathbf{i} + v_2(t)\mathbf{j} + v_3(t)\mathbf{k}$
then $\mathbf{u} \bullet \mathbf{v} = u_1v_1 + u_2v_2 + u_3v_3$, so
$$\begin{aligned}\frac{d}{dt}\mathbf{u} \bullet \mathbf{v} &= \frac{du_1}{dt}v_1 + u_1\frac{dv_1}{dt} + \frac{du_2}{dt}v_2 + u_2\frac{dv_2}{dt} \\ &\quad + \frac{du_3}{dt}v_3 + u_3\frac{dv_3}{dt} \\ &= \frac{d\mathbf{u}}{dt} \bullet \mathbf{v} + \mathbf{u} \bullet \frac{d\mathbf{v}}{dt}.\end{aligned}$$

**22.** $\dfrac{d}{dt}|\mathbf{r}|^2 = \dfrac{d}{dt}\mathbf{r} \bullet \mathbf{r} = 2\mathbf{r} \bullet \mathbf{v} = 0$ implies that $|\mathbf{r}|$ is constant. Thus $\mathbf{r}(t)$ lies on a sphere centred at the origin.

**24.** If $\mathbf{r}\bullet\mathbf{v} > 0$ then $|\mathbf{r}|$ is increasing. (See Exercise 16 above.) Thus $\mathbf{r}$ is moving farther away from the origin. If $\mathbf{r}\bullet\mathbf{v} < 0$ then $\mathbf{r}$ is moving closer to the origin.

**26.**
$$\begin{aligned}&\frac{d}{dt}\big(\mathbf{u} \bullet (\mathbf{v} \times \mathbf{w})\big) \\ &= \mathbf{u}' \bullet (\mathbf{v} \times \mathbf{w}) + \mathbf{u} \bullet (\mathbf{v}' \times \mathbf{w}) + \mathbf{u} \bullet (\mathbf{v} \times \mathbf{w}').\end{aligned}$$

**28.**
$$\begin{aligned}&\frac{d}{dt}\left(\mathbf{u} \times \left(\frac{d\mathbf{u}}{dt} \times \frac{d^2\mathbf{u}}{dt^2}\right)\right) \\ &= \frac{d\mathbf{u}}{dt} \times \left(\frac{d\mathbf{u}}{dt} \times \frac{d^2\mathbf{u}}{dt^2}\right) + \mathbf{u} \times \left(\frac{d^2\mathbf{u}}{dt^2} \times \frac{d^2\mathbf{u}}{dt^2}\right) \\ &\quad + \mathbf{u} \times \left(\frac{d\mathbf{u}}{dt} \times \frac{d^3\mathbf{u}}{dt^3}\right) \\ &= \frac{d\mathbf{u}}{dt} \times \left(\frac{d\mathbf{u}}{dt} \times \frac{d^2\mathbf{u}}{dt^2}\right) + \mathbf{u} \times \left(\frac{d\mathbf{u}}{dt} \times \frac{d^3\mathbf{u}}{dt^3}\right).\end{aligned}$$

**30.** $$\frac{d}{dt}\left[(\mathbf{u}\times\mathbf{u}')\bullet(\mathbf{u}'\times\mathbf{u}'')\right]$$
$$= (\mathbf{u}'\times\mathbf{u}')\bullet(\mathbf{u}'\times\mathbf{u}'') + (\mathbf{u}\times\mathbf{u}'')\bullet(\mathbf{u}'\times\mathbf{u}'')$$
$$+ (\mathbf{u}\times\mathbf{u}')\bullet(\mathbf{u}''\times\mathbf{u}'') + (\mathbf{u}\times\mathbf{u}')\bullet(\mathbf{u}'\times\mathbf{u}''')$$
$$= (\mathbf{u}\times\mathbf{u}'')\bullet(\mathbf{u}'\times\mathbf{u}'') + (\mathbf{u}\times\mathbf{u}')\bullet(\mathbf{u}'\times\mathbf{u}''').$$

**32.** $$\mathbf{r} = \mathbf{r}_0\cos\omega t + \left(\frac{\mathbf{v}_0}{\omega}\right)\sin\omega t$$
$$\frac{d\mathbf{r}}{dt} = -\omega\mathbf{r}_0\sin\omega t + \mathbf{v}_0\cos\omega t$$
$$\frac{d^2\mathbf{r}}{dt^2} = -\omega^2\mathbf{r}_0\cos\omega t - \omega\mathbf{v}_0\sin\omega t = -\omega^2\mathbf{r}$$
$$\mathbf{r}(0) = \mathbf{r}_0, \qquad \left.\frac{d\mathbf{r}}{dt}\right|_{t=0} = \mathbf{v}_0.$$

Observe that $\mathbf{r}\bullet(\mathbf{r}_0\times\mathbf{v}_0) = 0$ for all $t$. Therefore the path lies in a plane through the origin having normal $\mathbf{N} = \mathbf{r}_0\times\mathbf{v}_0$.
Let us choose our coordinate system so that $\mathbf{r}_0 = a\mathbf{i}$ ($a > 0$) and $\mathbf{v}_0 = \omega b\mathbf{i} + \omega c\mathbf{j}$ ($c > 0$). Therefore, $\mathbf{N}$ is in the direction of $\mathbf{k}$. The path has parametric equations

$$x = a\cos\omega t + b\sin\omega t$$
$$y = c\sin\omega t.$$

The curve is a conic section since it has a quadratic equation:
$$\frac{1}{a^2}\left(x - \frac{by}{c}\right)^2 + \frac{y^2}{c^2} = 1.$$

Since the path is bounded ($|\mathbf{r}(t)| \le |\mathbf{r}_0| + (|\mathbf{v}_0|/\omega)$), it must be an ellipse.

If $\mathbf{r}_0$ is perpendicular to $\mathbf{v}_0$, then $b = 0$ and the path is the ellipse $(x/a)^2 + (y/c)^2 = 1$ having semi-axes $a = |\mathbf{r}_0|$ and $c = |\mathbf{v}_0|/\omega$.

## Section 7.2 Some Applications of Vector Differentiation (page 334)

**2.** Let $v(t)$ be the speed of the tank car at time $t$ seconds. The mass of the car at time $t$ is $m(t) = M - kt$ kg. At full power, the force applied to the car is $F = Ma$ (since the motor can accelerate the full car at $a$ m/s$^2$). By Newton's Law, this force is the rate of change of the momentum of the car. Thus

$$\frac{d}{dt}\left[(M-kt)v\right] = Ma$$
$$(M-kt)\frac{dv}{dt} - kv = Ma$$
$$\frac{dv}{Ma+kv} = \frac{dt}{M-kt}$$
$$\frac{1}{k}\ln(Ma+kv) = -\frac{1}{k}\ln(M-kt) + \frac{1}{k}\ln C$$
$$Ma + kv = \frac{C}{M-kt}.$$

At $t = 0$ we have $v = 0$, so $Ma = C/M$. Thus $C = M^2a$ and
$$kv = \frac{M^2a}{M-kt} - Ma = \frac{Makt}{M-kt}.$$
The speed of the tank car at time $t$ (before it is empty) is
$$v(t) = \frac{Mat}{M-kt}\ \text{m/s}.$$

**4.** First observe that
$$\frac{d}{dt}|\mathbf{r}-\mathbf{b}|^2 = 2(\mathbf{r}-\mathbf{b})\bullet\frac{d\mathbf{r}}{dt} = 2(\mathbf{r}-\mathbf{b})\bullet\big(\mathbf{a}\times(\mathbf{r}-\mathbf{b})\big) = 0,$$
so $|\mathbf{r}-\mathbf{b}|$ is constant; for all $t$ the object lies on the sphere centred at the point with position vector $\mathbf{b}$ having radius $\mathbf{r}_0 - \mathbf{b}$.
Next, observe that
$$\frac{d}{dt}(\mathbf{r}-\mathbf{r}_0)\bullet\mathbf{a} = \big(\mathbf{a}\times(\mathbf{r}-\mathbf{b})\big)\bullet\mathbf{a} = 0,$$
so $\mathbf{r}-\mathbf{r}_0 \perp \mathbf{a}$; for all $t$ the object lies on the plane through $\mathbf{r}_0$ having normal $\mathbf{a}$. Hence the path of the object lies on the circle in which this plane intersects the sphere described above. The angle between $\mathbf{r}-\mathbf{b}$ and $\mathbf{a}$ must therefore also be constant, and so the object's speed $|d\mathbf{r}/dt|$ is constant. Hence the path must be the whole circle.

**6.** The angular velocity of the earth is $\boldsymbol{\Omega}$, pointing due north. For a particle moving with horizontal velocity $\mathbf{v}$, the tangential and normal components of the Coriolis force $\mathbf{C}$, and of $\boldsymbol{\Omega}$, are related by
$$\mathbf{C}_T = -2\boldsymbol{\Omega}_N\times\mathbf{v}, \qquad \mathbf{C}_N = -2\boldsymbol{\Omega}_T\times\mathbf{v}.$$

At the north or south pole, $\boldsymbol{\Omega}_T = \mathbf{0}$ and $\boldsymbol{\Omega}_N = \boldsymbol{\Omega}$. Thus $\mathbf{C}_N = \mathbf{0}$ and $\mathbf{C}_T = -2\boldsymbol{\Omega}\times\mathbf{v}$. The Coriolis force is horizontal. It is 90° east of $\mathbf{v}$ at the north pole and 90° west of $\mathbf{v}$ at the south pole.

At the equator, $\boldsymbol{\Omega}_N = \mathbf{0}$ and $\boldsymbol{\Omega}_T = \boldsymbol{\Omega}$. Thus $\mathbf{C}_T = \mathbf{0}$ and $\mathbf{C}_N = -2\boldsymbol{\Omega}\times\mathbf{v}$. The Coriolis force is vertical.

## Section 7.3 Curves and Parametrizations (page 341)

**2.** On the first quadrant part of the circle $x^2 + y^2 = a^2$ we have $y = \sqrt{a^2 - x^2}$, $0 \le x \le a$. The required parametrization is
$$\mathbf{r} = \mathbf{r}(x) = x\mathbf{i} + \sqrt{a^2 - x^2}\,\mathbf{j}, \quad (0 \le x \le a).$$

**4.** $$x = a\sin\frac{s}{a}, \quad y = a\cos\frac{s}{a}, \quad 0 \le \frac{s}{a} \le \frac{\pi}{2}$$
$$\mathbf{r} = a\sin\frac{s}{a}\mathbf{i} + a\cos\frac{s}{a}\mathbf{j}, \quad \left(0 \le s \le \frac{a\pi}{2}\right).$$

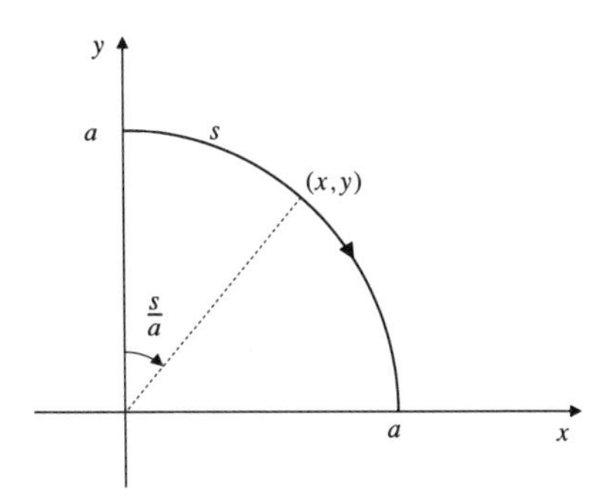


Fig. 7.3.4

**6.** $z = x^2, x + y + z = 1$. If $t = x$, then $z = t^2$ and $y = 1 - t - t^2$. The parametrization is $\mathbf{r} = t\mathbf{i} + (1 - t - t^2)\mathbf{j} + t^2\mathbf{k}$.

**8.** $x + y = 1, z = \sqrt{1 - x^2 - y^2}$. If $x = t$, then $y = 1 - t$ and
$z = \sqrt{1 - t^2 - (1 - t)^2} = \sqrt{2(t - t^2)}$. One possible parametrization is

$$\mathbf{r} = t\mathbf{i} + (1 - t)\mathbf{j} + \sqrt{2(t - t^2)}\mathbf{k}.$$

**10.** $yz + x = 1, xz - x = 1$. One possible parametrization is $x = t, z = (1 + t)/t$, and $y = (1 - t)/z = (1 - t)t/(1 + t)$, that is,

$$\mathbf{r} = t\mathbf{i} + \frac{t - t^2}{1 + t}\mathbf{j} + \frac{1 + t}{t}\mathbf{k}.$$

**12.** By symmetry, the centre of the circle $\mathcal{C}$ of intersection of the plane $x + y + z = 1$ and the sphere $x^2 + y^2 + z^2 = 1$ must lie on the plane and must have its three coordinates equal. Thus the centre has position vector

$$\mathbf{r}_0 = \frac{1}{3}(\mathbf{i} + \mathbf{j} + \mathbf{k}).$$

Since $\mathcal{C}$ passes through the point $(0, 0, 1)$, its radius is

$$\sqrt{\left(0 - \frac{1}{3}\right)^2 + \left(0 - \frac{1}{3}\right)^2 + \left(1 - \frac{1}{3}\right)^2} = \sqrt{\frac{2}{3}}.$$

Any vector $\mathbf{v}$ that satisfies $\mathbf{v} \bullet (\mathbf{i} + \mathbf{j} + \mathbf{k}) = 0$ is parallel to the plane $x + y + z = 1$ containing $\mathcal{C}$. One such vector is $\mathbf{v}_1 = \mathbf{i} - \mathbf{j}$. A second one, perpendicular to $\mathbf{v}_1$, is

$$\mathbf{v}_2 = (\mathbf{i} + \mathbf{j} + \mathbf{k}) \times (\mathbf{i} - \mathbf{j}) = \mathbf{i} + \mathbf{j} - 2\mathbf{k}.$$

Two perpendicular unit vectors that are parallel to the plane of $\mathcal{C}$ are

$$\hat{\mathbf{v}}_1 = \frac{\mathbf{i} - \mathbf{j}}{\sqrt{2}}, \quad \hat{\mathbf{v}}_2 = \frac{\mathbf{i} + \mathbf{j} - 2\mathbf{k}}{\sqrt{6}}.$$

Thus one possible parametrization of $\mathcal{C}$ is

$$\begin{aligned}\mathbf{r} &= \mathbf{r}_0 + \sqrt{\frac{2}{3}}(\cos t\,\hat{\mathbf{v}}_1 + \sin t\,\hat{\mathbf{v}}_2) \\ &= \frac{\mathbf{i} + \mathbf{j} + \mathbf{k}}{3} + \frac{\cos t}{\sqrt{3}}(\mathbf{i} - \mathbf{j}) + \frac{\sin t}{3}(\mathbf{i} + \mathbf{j} - 2\mathbf{k}).\end{aligned}$$

**14.** $\mathbf{r} = t\mathbf{i} + \lambda t^2\mathbf{j} + t^3\mathbf{k}, \quad (0 \le t \le T)$
$v = \sqrt{1 + (2\lambda t)^2 + 9t^4} = \sqrt{(1 + 3t^2)^2}$
if $4\lambda^2 = 6$, that is, if $\lambda = \pm\sqrt{3/2}$. In this case, the length of the curve is

$$s(T) = \int_0^T (1 + 3t^2)\,dt = T + T^3.$$

**16.** $x = a\cos t \sin t = \dfrac{a}{2}\sin 2t$,
$y = a\sin^2 t = \dfrac{a}{2}(1 - \cos 2t)$,
$z = bt$.
The curve is a circular helix lying on the cylinder

$$x^2 + \left(y - \frac{a}{2}\right)^2 = \frac{a^2}{4}.$$

Its length, from $t = 0$ to $t = T$, is

$$\begin{aligned}L &= \int_0^T \sqrt{a^2\cos^2 2t + a^2\sin^2 2t + b^2}\,dt \\ &= T\sqrt{a^2 + b^2} \text{ units.}\end{aligned}$$

**18.**

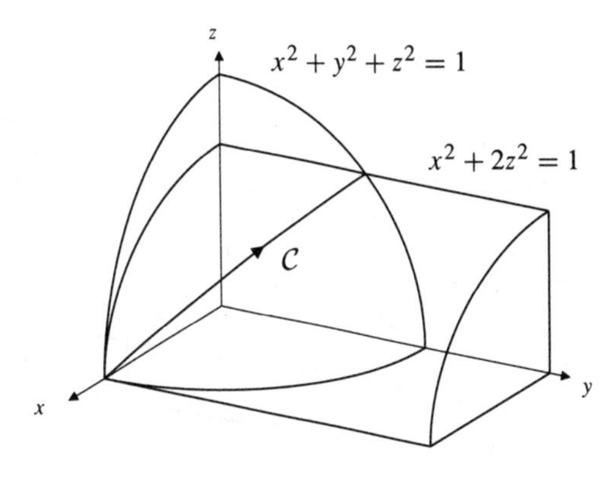


Fig. 7.3.18

One-eighth of the curve $\mathcal{C}$ lies in the first octant. That part can be parametrized

$$x = \cos t, \quad z = \frac{1}{\sqrt{2}}\sin t, \quad (0 \le t \le \pi/2)$$

$$y = \sqrt{1 - \cos^2 t - \frac{1}{2}\sin^2 t} = \frac{1}{\sqrt{2}}\sin t.$$

Since the first octant part of $\mathcal{C}$ lies in the plane $y = z$, it must be a quarter of a circle of radius 1. Thus the length of all of $\mathcal{C}$ is $8 \times (\pi/2) = 4\pi$ units.
If you wish to use an integral, the length is

$$8\int_0^{\pi/2}\sqrt{\sin^2 t + \frac{1}{2}\cos^2 t + \frac{1}{2}\cos^2 t}\,dt$$
$$= 8\int_0^{\pi/2} dt = 4\pi \text{ units.}$$

**20.** $\mathbf{r} = r\cos\theta\mathbf{i} + r\sin\theta\mathbf{j} + z\mathbf{k}$

$$\mathbf{v} = \left(\frac{dr}{dt}\cos\theta - r\sin\theta\frac{d\theta}{dt}\right)\mathbf{i} + \left(\frac{dr}{dt}\sin\theta + r\cos\theta\frac{d\theta}{dt}\right)\mathbf{j} + \frac{dz}{dt}\mathbf{k}$$

$$ds = |\mathbf{v}|\,dt = \sqrt{\left(\frac{dr}{dt}\right)^2 + r^2\left(\frac{d\theta}{dt}\right)^2 + \left(\frac{dz}{dt}\right)^2}\,dt$$

**22.** The length $L$ of the spiral $r = t$, $\theta = t$, $z = t$, $(0 \le t \le 2\pi)$ is

$$L = \int_0^{2\pi}\sqrt{\left(\frac{dr}{dt}\right)^2 + r^2\left(\frac{d\theta}{dt}\right)^2 + \left(\frac{dz}{dt}\right)^2}\,dt$$
$$= \int_0^{2\pi}\sqrt{2 + t^2}\,dt \quad \text{Let } t = \sqrt{2}\tan u, \quad dt = \sqrt{2}\sec^2 u\,du$$
$$= 2\int_0^{\tan^{-1}(\pi\sqrt{2})}\sec^3 u\,du$$
$$= \left(\sec u\tan u + \ln(\sec u + \tan u)\right)\Big|_0^{\tan^{-1}(\pi\sqrt{2})}$$
$$= \pi\sqrt{2 + 4\pi^2} + \ln\left(\frac{2\pi + \sqrt{2 + 4\pi^2}}{\sqrt{2}}\right) \text{ units.}$$

**24.** (Solution due to Roland Urbanek, a student at Okanagan College.) Suppose the spool is vertical and the cable windings make angle $\theta$ with the horizontal at each point.

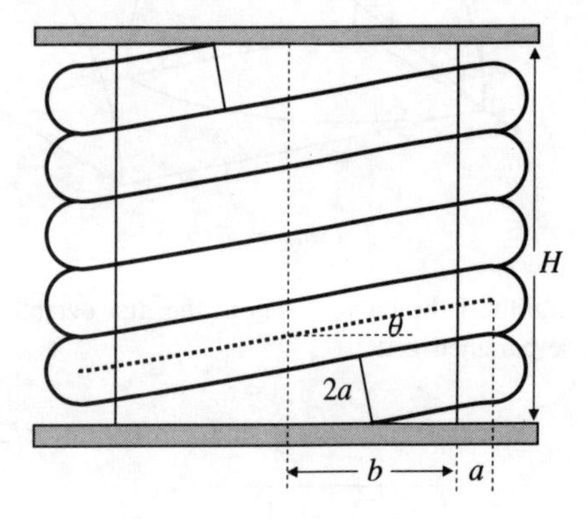


Fig. 7.3.24

The centreline of the cable is wound around a cylinder of radius $a + b$ and must rise a vertical distance $\dfrac{2a}{\cos\theta}$ in one revolution. The figure below shows the cable unwound from the spool and inclined at angle $\theta$. The total length of spool required is the total height $H$ of the cable as shown in that figure.

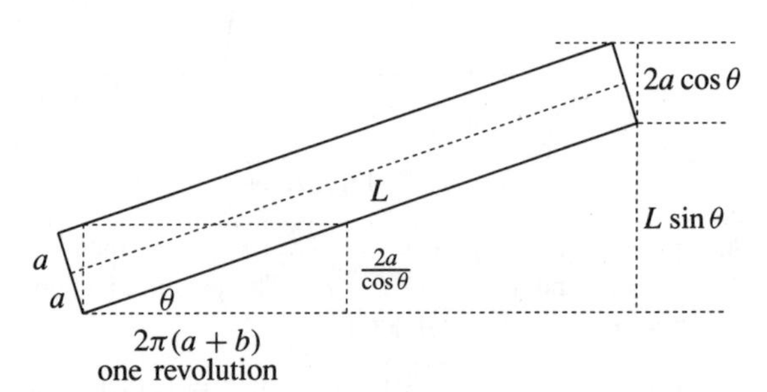


Fig. 7.3.24

Observe that $\tan\theta = \dfrac{2a}{\cos\theta} \times \dfrac{1}{2\pi(a+b)}$. Therefore

$$\sin\theta = \frac{a}{\pi(a+b)}$$
$$\cos\theta = \sqrt{1 - \frac{a^2}{\pi^2(a+b)^2}} = \frac{\sqrt{\pi^2(a+b)^2 - a^2}}{\pi(a+b)}.$$

The total length of spool required is

$$H = L\sin\theta + 2a\cos\theta$$
$$= \frac{a}{\pi(a+b)}\left(L + 2\sqrt{\pi^2(a+b)^2 - a^2}\right) \text{ units.}$$

**26.** $\mathbf{r} = e^t\mathbf{i} + \sqrt{2}t\mathbf{j} - e^{-t}\mathbf{k}$
$\mathbf{v} = e^t\mathbf{i} + \sqrt{2}\mathbf{j} + e^{-t}\mathbf{k}$
$v = |\mathbf{v}| = \sqrt{e^{2t} + 2 + e^{-2t}} = e^t + e^{-t}$.
The arc length from the point where $t = 0$ to the point corresponding to arbitrary $t$ is

$$s = s(t) = \int_0^t (e^u + e^{-u})\,du = e^t - e^{-t} = 2\sinh t.$$

Thus $t = \sinh^{-1}(s/2) = \ln\left(\dfrac{s + \sqrt{s^2+4}}{2}\right)$,

and $e^t = \dfrac{s + \sqrt{s^2+4}}{2}$. The required parametrization is

$$\mathbf{r} = \frac{s + \sqrt{s^2+4}}{2}\mathbf{i} + \sqrt{2}\ln\left(\frac{s + \sqrt{s^2+4}}{2}\right)\mathbf{j} - \frac{2\mathbf{k}}{s + \sqrt{s^2+4}}.$$

**28.** $\mathbf{r} = 3t\cos t\mathbf{i} + 3t\sin t\mathbf{j} + 2\sqrt{2}t^{3/2}\mathbf{k}, \ (t \ge 0)$

$\mathbf{v} = 3(\cos t - t\sin t)\mathbf{i} + 3(\sin t + t\cos t)\mathbf{j} + 3\sqrt{2}\sqrt{t}\mathbf{k}$

$v = |\mathbf{v}| = 3\sqrt{1+t^2+2t} = 3(1+t)$

$s = \int_0^t 3(1+u)\,du = 3\left(t + \frac{t^2}{2}\right)$

Thus $t^2 + 2t = \frac{2s}{3}$, so $t = -1 + \sqrt{1 + \frac{2s}{3}}$ since $t \ge 0$. The required parametrization is the given one with $t$ replaced by $-1 + \sqrt{1 + (2s)/3}$.

**30.** If $\mathbf{r} = \mathbf{r}(t)$ has nonvanishing velocity $\mathbf{v} = d\mathbf{r}/dt$ on $[a, b]$, then for any $t_0$ in $[a, b]$, the function

$$s = g(t) = \int_{t_0}^t |\mathbf{v}(u)|\,du,$$

which gives the (signed) arc length $s$ measured from $\mathbf{r}(t_0)$ along the curve, is an increasing function:

$$\frac{ds}{dt} = g'(t) = |\mathbf{v}(t)| > 0$$

on $[a, b]$, by the Fundamental Theorem of Calculus. Hence $g$ is invertible, and defines $t$ as a function of arc length $s$:

$$t = g^{-1}(s) \Leftrightarrow s = g(t).$$

Then

$$\mathbf{r} = \mathbf{r}_2(s) = \mathbf{r}\big(g^{-1}(s)\big)$$

is a parametrization of the curve $\mathbf{r} = \mathbf{r}(t)$ in terms of arc length.

## Section 7.4 Curvature, Torsion, and the Frenet Frame (page 350)

**2.** $\mathbf{r} = a\sin\omega t\mathbf{i} + a\cos\omega t\mathbf{k}$

$\mathbf{v} = a\omega\cos\omega t\mathbf{i} - a\omega\sin\omega t\mathbf{k}, \quad v = |a\omega|$

$\hat{\mathbf{T}} = \mathrm{sgn}(a\omega)\big[\cos\omega t\mathbf{i} - \sin\omega t\mathbf{k}\big].$

**4.** $\mathbf{r} = a\cos t\mathbf{i} + b\sin t\mathbf{j} + t\mathbf{k}$

$\mathbf{v} = -a\sin t\mathbf{i} + b\cos t\mathbf{j} + \mathbf{k}$

$v = \sqrt{a^2\sin^2 t + b^2\cos^2 t + 1}$

$\hat{\mathbf{T}} = \frac{\mathbf{v}}{v} = \frac{-a\sin t\mathbf{i} + b\cos t\mathbf{j} + \mathbf{k}}{\sqrt{a^2\sin^2 t + b^2\cos^2 t + 1}}.$

**6.** If $\tau(s) = 0$ for all $s$, then $\frac{d\hat{\mathbf{B}}}{ds} = -\tau\hat{\mathbf{N}} = 0$, so $\hat{\mathbf{B}}(s) = \hat{\mathbf{B}}(0)$ is constant. Therefore,

$$\frac{d}{ds}\big(\mathbf{r}(s) - \mathbf{r}(0)\big) \bullet \hat{\mathbf{B}}(s) = \frac{d\mathbf{r}}{ds} \bullet \hat{\mathbf{B}}(s) = \hat{\mathbf{T}}(s) \bullet \hat{\mathbf{B}}(s) = 0.$$

It follows that

$$\big(\mathbf{r}(s) - \mathbf{r}(0)\big) \bullet \hat{\mathbf{B}}(0) = \big(\mathbf{r}(s) - \mathbf{r}(0)\big) \bullet \hat{\mathbf{B}}(s) = 0$$

for all $s$. This says that $\mathbf{r}(s)$ lies in the plane through $\mathbf{r}(0)$ having normal $\hat{\mathbf{B}}(0)$.

**8.** The circular helix $C_1$ given by

$$\mathbf{r} = a\cos t\mathbf{i} + a\sin t\mathbf{j} + bt\mathbf{k}$$

has curvature and torsion given by

$$\kappa(s) = \frac{a}{a^2+b^2}, \qquad \tau(s) = \frac{b}{a^2+b^2},$$

by Example 3.
if a curve $\mathcal{C}$ has constant curvature $\kappa(s) = C > 0$, and constant torsion $\tau(s) = T \ne 0$, then we can choose $a$ and $b$ so that

$$\frac{a}{a^2+b^2} = C, \qquad \frac{b}{a^2+b^2} = T.$$

(Specifically, $a = \frac{C}{C^2+T^2}$, and $b = \frac{T}{C^2+T^2}$.) By Theorem 3, $\mathcal{C}$ is itself a circular helix, congruent to $C_1$.

## Section 7.5 Curvature and Torsion for General Parametrizations (page 356)

**2.** For $y = \cos$ we have

$$\kappa(x) = \frac{|d^2y/dx^2|}{(1 + (dy/dx)^2)^{3/2}} = \frac{|\cos x|}{(1 + \sin^2 x)^{3/2}}.$$

Hence $\kappa(0) = 1$ and $\kappa(\pi/2) = 0$. The radius of curvature at $x = 0$ is 1. The radius of curvature at $x = \pi/2$ is infinite.

**4.**
$$\begin{aligned}\mathbf{r} &= t^3\mathbf{i} + t^2\mathbf{j} + t\mathbf{k}\\ \mathbf{v} &= 3t^2\mathbf{i} + 2t\mathbf{j} + \mathbf{k}\\ \mathbf{a} &= 6t\mathbf{i} + 2\mathbf{j}\\ \mathbf{v}(1) &= 3\mathbf{i} + 2\mathbf{j} + \mathbf{k}, \quad \mathbf{a}(1) = 6\mathbf{i} + 2\mathbf{j}\\ \mathbf{v}(1) \times \mathbf{a}(1) &= -2\mathbf{i} + 6\mathbf{j} - 6\mathbf{k}\\ \kappa(1) &= \frac{\sqrt{4+36+36}}{(9+4+1)^{3/2}} = \frac{2\sqrt{19}}{14^{3/2}}\end{aligned}$$

At $t = 1$ the radius of curvature is $14^{3/2}/(2\sqrt{19})$.

**6.**
$$\begin{aligned}\mathbf{r} &= t\mathbf{i} - t^2\mathbf{j}\\ \mathbf{v} &= \mathbf{i} - 2t\mathbf{j}\\ \mathbf{a} &= -2\mathbf{j}\\ \mathbf{v} \times \mathbf{a} &= -2\mathbf{k}\end{aligned}$$

At $(1, -1, 0)$, where $t = 1$, we have

$$\begin{aligned}\hat{\mathbf{T}} &= \mathbf{v}/|\mathbf{v}| = (\mathbf{i} - 2\mathbf{j})/\sqrt{5}\\ \hat{\mathbf{B}} &= (\mathbf{v} \times \mathbf{a})/|\mathbf{v} \times \mathbf{a}| = -\mathbf{k}\\ \hat{\mathbf{N}} &= \hat{\mathbf{B}} \times \hat{\mathbf{T}} = -(2\mathbf{i} + \mathbf{j})/\sqrt{5}.\end{aligned}$$

**8.**
$$\begin{aligned}\mathbf{r} &= t\mathbf{i} + t^2\mathbf{j} + t\mathbf{k}\\ \mathbf{v} &= \mathbf{i} + 2t\mathbf{j} + \mathbf{k}\\ \mathbf{a} &= 2\mathbf{j}\\ \mathbf{v}\times\mathbf{a} &= -2\mathbf{i} + 2\mathbf{k}\end{aligned}$$
At $(1, 1, 1)$, where $t = 1$, we have
$$\begin{aligned}\hat{\mathbf{T}} &= \mathbf{v}/|\mathbf{v}| = (\mathbf{i} + 2\mathbf{j} + \mathbf{k})/\sqrt{6}\\ \hat{\mathbf{B}} &= (\mathbf{v}\times\mathbf{a})/|\mathbf{v}\times\mathbf{a}| = -(\mathbf{i} - \mathbf{k})/\sqrt{2}\\ \hat{\mathbf{N}} &= \hat{\mathbf{B}}\times\hat{\mathbf{T}} = -(\mathbf{i} - \mathbf{j} + \mathbf{k})/\sqrt{3}.\end{aligned}$$

**10.**
$$\begin{aligned}\mathbf{r} &= t\mathbf{i} + 2t\mathbf{j} + \sin t\mathbf{k}\\ \mathbf{v} &= \mathbf{i} + 2\mathbf{j} + \cos t\mathbf{k}\\ \mathbf{a} &= -\sin t\mathbf{k}, \qquad \frac{d\mathbf{a}}{dt} = -\cos t\mathbf{k}\\ \mathbf{v}\times\mathbf{a} &= -2\sin t\mathbf{i} + \sin t\mathbf{j}\\ v &= |\mathbf{v}| = \sqrt{5 + \cos^2 t}, \qquad |\mathbf{v}\times\mathbf{a}| = \sqrt{5}|\sin t|\\ (\mathbf{v}\times\mathbf{a}) \bullet \frac{d\mathbf{a}}{dt} &= 0\\ \hat{\mathbf{T}} &= \frac{\mathbf{v}}{v} = \frac{\mathbf{i} + 2\mathbf{j} + \cos t\mathbf{k}}{\sqrt{5 + \cos^2 t}}\\ \hat{\mathbf{B}} &= \frac{\mathbf{v}\times\mathbf{a}}{|\mathbf{v}\times\mathbf{a}|} = \frac{\operatorname{sgn}(\sin t)}{\sqrt{5}}(-2\mathbf{i} + \mathbf{j})\\ \hat{\mathbf{N}} &= \hat{\mathbf{B}}\times\hat{\mathbf{T}} = \frac{\operatorname{sgn}(\sin t)}{\sqrt{25 + 5\cos^2 t}}(\cos t\mathbf{i} + 2\cos t\mathbf{j} - 5\mathbf{k})\\ \kappa &= \frac{|\mathbf{v}\times\mathbf{a}|}{v^3} = \frac{\sqrt{5}|\sin t|}{(5 + \cos^2 t)^{3/2}},\\ \tau &= \frac{(\mathbf{v}\times\mathbf{a}) \bullet \dfrac{d\mathbf{a}}{dt}}{|\mathbf{v}\times\mathbf{a}|^2} = 0.\end{aligned}$$

**12.**
$$\begin{aligned}\mathbf{r} &= e^t\cos t\mathbf{i} + e^t\sin t\mathbf{j} + e^t\mathbf{k}\\ \mathbf{v} &= e^t(\cos t - \sin t)\mathbf{i} + e^t(\sin t + \cos t)\mathbf{j} + e^t\mathbf{k}\\ \mathbf{a} &= -2e^t\sin t\mathbf{i} + 2e^t\cos t\mathbf{j} + e^t\mathbf{k}\\ \frac{d\mathbf{a}}{dt} &= -2e^t(\cos t + \sin t)\mathbf{i} + 2e^t(\cos t - \sin t)\mathbf{j} + e^t\mathbf{k}\\ \mathbf{v}\times\mathbf{a} &= e^{2t}(\sin t - \cos t)\mathbf{i} - e^{2t}(\cos t + \sin t)\mathbf{j} + 2e^{2t}\mathbf{k}\\ v &= |\mathbf{v}| = \sqrt{3}e^t, \qquad |\mathbf{v}\times\mathbf{a}| = \sqrt{6}e^{2t}\\ (\mathbf{v}\times\mathbf{a}) \bullet \frac{d\mathbf{a}}{dt} &= 2e^{3t}\\ \hat{\mathbf{T}} &= \frac{\mathbf{v}}{v} = \frac{(\cos t - \sin t)\mathbf{i} + (\cos t + \sin t)\mathbf{j} + \mathbf{k}}{\sqrt{3}}\\ \hat{\mathbf{B}} &= \frac{\mathbf{v}\times\mathbf{a}}{|\mathbf{v}\times\mathbf{a}|} = \frac{(\sin t - \cos t)\mathbf{i} - (\cos t + \sin t)\mathbf{j} + 2\mathbf{k}}{\sqrt{6}}\\ \hat{\mathbf{N}} &= \hat{\mathbf{B}}\times\hat{\mathbf{T}} = -\frac{(\cos t + \sin t)\mathbf{i} - (\cos t - \sin t)\mathbf{j}}{\sqrt{2}}\\ \kappa &= \frac{|\mathbf{v}\times\mathbf{a}|}{v^3} = \frac{\sqrt{2}}{3e^t}\\ \tau &= \frac{(\mathbf{v}\times\mathbf{a}) \bullet \dfrac{d\mathbf{a}}{dt}}{|\mathbf{v}\times\mathbf{a}|^2} = \frac{1}{3e^t}.\end{aligned}$$

**14.**
$$\begin{aligned}\mathbf{r} &= x\mathbf{i} + \sin x\mathbf{j}\\ \mathbf{v} &= \frac{dx}{dt}\mathbf{i} + \cos x\frac{dx}{dt}\mathbf{j} = k(\mathbf{i} + \cos x\mathbf{j})\\ v &= k\sqrt{1 + \cos^2 x}\\ \mathbf{a} &= -k\sin x\frac{dx}{dt}\mathbf{j} = -k^2\sin x\mathbf{j}\\ \mathbf{v}\times\mathbf{a} &= -k^3\sin x\mathbf{k}\\ \kappa &= \frac{|\mathbf{v}\times\mathbf{a}|}{v^3} = \frac{|\sin x|}{(1 + \cos^2 x)^{3/2}}.\end{aligned}$$
The tangential and normal components of acceleration are
$$\begin{aligned}\frac{dv}{dt} &= \frac{k}{2\sqrt{1 + \cos^2 x}}2\cos x)(-\sin x)\frac{dx}{dt} = -\frac{k^2\cos x\sin x}{\sqrt{1 + \cos^2 x}}\\ v^2\kappa &= \frac{k^2|\sin x|}{\sqrt{1 + \cos^2 x}}.\end{aligned}$$

**16.**
$$\begin{aligned}\mathbf{r} &= a\cos t\mathbf{i} + b\sin t\mathbf{j}\\ \mathbf{v} &= -a\sin t\mathbf{i} + b\cos t\mathbf{j}\\ \mathbf{a} &= -a\cos t\mathbf{i} - b\sin t\mathbf{j}\\ \mathbf{v}\times\mathbf{a} &= ab\mathbf{k}\\ v &= \sqrt{a^2\sin^2 t + b^2\cos^2 t}.\end{aligned}$$
The tangential component of acceleration is
$$\frac{dv}{dt} = \frac{(a^2 - b^2)\sin t\cos t}{\sqrt{a^2\sin^2 t + b^2\cos^2 t}},$$
which is zero if $t$ is an integer multiple of $\pi/2$, that is, at the ends of the major and minor axes of the ellipse. The normal component of acceleration is
$$v^2\kappa = v^2\frac{|\mathbf{v}\times\mathbf{a}|}{v^3} = \frac{ab}{\sqrt{a^2\sin^2 t + b^2\cos^2 t}}.$$

**18.**

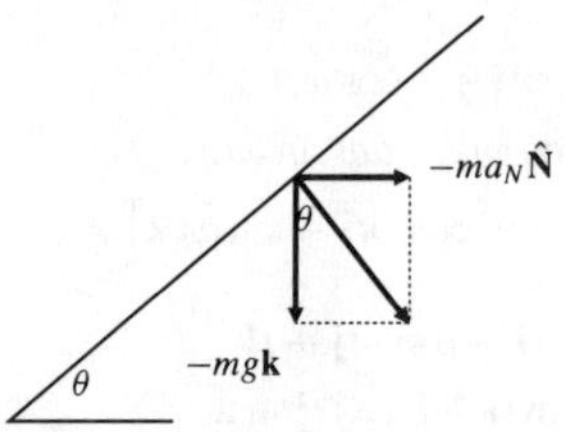

$-ma_N\hat{\mathbf{N}}$
$\theta$
$-mg\mathbf{k}$
$\theta$

Fig. 7.5.18

The path of the road, $y = x^2$, has curvature
$$\kappa = \frac{\left|\dfrac{d^2y}{dx^2}\right|}{\left(1 + \left(\dfrac{dy}{dx}\right)^2\right)^{3/2}} = \frac{2}{(1 + 4x^2)^{3/2}}.$$

The normal component of the acceleration of a vehicle travelling at speed $v_0$ along the road is

$$a_N = v_0^2\kappa = \frac{2v_0^2}{(1+4x^2)^{3/2}}.$$

If the road is banked at angle $\theta$ (see the figure), then the resultant of the centrifugal force $-ma_N\hat{\mathbf{N}}$ and the gravitational force $-mg\mathbf{k}$ is normal to the roadway provided

$$\tan\theta = \frac{ma_N}{mg}, \quad \text{i.e.,} \quad \theta = \tan^{-1}\frac{2v_0^2}{g(1+4x^2)^{3/2}}.$$

**20.** The curve with polar equation $r = f(\theta)$ is given parametrically by

$$\mathbf{r} = f(\theta)\cos\theta\mathbf{i} + f(\theta)\sin\theta\mathbf{j}.$$

Thus we have

$$\begin{aligned}
\mathbf{v} &= \big(f'(\theta)\cos\theta - f(\theta)\sin\theta\big)\mathbf{i} \\
&\quad + \big(f'(\theta)\sin\theta + f(\theta)\cos\theta\big)\mathbf{j} \\
\mathbf{a} &= \big(f''(\theta)\cos\theta - 2f'(\theta)\sin\theta - f(\theta)\cos\theta\big)\mathbf{i} \\
&\quad + \big(f''(\theta)\sin\theta + 2f'(\theta)\cos\theta - f(\theta)\sin\theta\big)\mathbf{j} \\
v &= |\mathbf{v}| = \sqrt{\big(f'(\theta)\big)^2 + \big(f(\theta)\big)^2} \\
\mathbf{v}\times\mathbf{a} &= \Big[2\big(f'(\theta)\big)^2 + \big(f(\theta)\big)^2 - f(\theta)f''(\theta)\Big]\mathbf{k}.
\end{aligned}$$

The curvature is, therefore,

$$\frac{\big|2\big(f'(\theta)\big)^2 + \big(f(\theta)\big)^2 - f(\theta)f''(\theta)\big|}{\Big[\big(f'(\theta)\big)^2 + \big(f(\theta)\big)^2\Big]^{3/2}}.$$

**22.** By Exercise 8 of Section 7.4, the required curve must be a circular helix with parameters $a = 1/2$ (radius), and $b = 1/2$. Its equation will be

$$\mathbf{r} = \frac{1}{2}\cos t\mathbf{i}_1 + \frac{1}{2}\sin t\mathbf{j}_1 + \frac{1}{2}t\mathbf{k}_1 + \mathbf{r}_0$$

for some right-handed basis $\{\mathbf{i}_1, \mathbf{j}_1, \mathbf{k}_1\}$, and some constant vector $\mathbf{r}_0$. Example 3 of Section 7.4 provides values for $\hat{\mathbf{T}}(0)$, $\hat{\mathbf{N}}(0)$, and $\hat{\mathbf{B}}(0)$, which we can equate to the given values of these vectors:

$$\begin{aligned}
\mathbf{i} &= \hat{\mathbf{T}}(0) = \frac{1}{\sqrt{2}}\mathbf{j}_1 + \frac{1}{\sqrt{2}}\mathbf{k}_1 \\
\mathbf{j} &= \hat{\mathbf{N}}(0) = -\mathbf{i}_1 \\
\mathbf{k} &= \hat{\mathbf{B}}(0) = -\frac{1}{\sqrt{2}}\mathbf{j}_1 + \frac{1}{\sqrt{2}}\mathbf{k}_1.
\end{aligned}$$

Solving these equations for $\mathbf{i}_1$, $\mathbf{j}_1$, and $\mathbf{k}_1$ in terms of the given basis vectors, we obtain

$$\begin{aligned}
\mathbf{i}_1 &= -\mathbf{j} \\
\mathbf{j}_1 &= \frac{1}{\sqrt{2}}\mathbf{i} - \frac{1}{\sqrt{2}}\mathbf{k} \\
\mathbf{k}_1 &= \frac{1}{\sqrt{2}}\mathbf{i} + \frac{1}{\sqrt{2}}\mathbf{k}.
\end{aligned}$$

Therefore

$$\mathbf{r}(t) = \frac{t+\sin t}{2\sqrt{2}}\mathbf{i} - \frac{\cos t}{2}\mathbf{j} + \frac{t-\sin t}{2\sqrt{2}}\mathbf{k} + \mathbf{r}_0.$$

We also require that $\mathbf{r}(0) = \mathbf{i}$, so $\mathbf{r}_0 = \mathbf{i} + \frac{1}{2}\mathbf{j}$. The required equation is, therefore,

$$\mathbf{r}(t) = \left(\frac{t+\sin t}{2\sqrt{2}} + 1\right)\mathbf{i} + \frac{1-\cos t}{2}\mathbf{j} + \frac{t-\sin t}{2\sqrt{2}}\mathbf{k}.$$

**24.** For $\mathbf{r} = a\cos t\mathbf{i} + a\sin t\mathbf{j} + bt\mathbf{k}$, we have, by Example 3 of Section 7.4,

$$\hat{\mathbf{N}} = -\cos t\mathbf{i} - \sin t\mathbf{j}, \qquad \kappa = \frac{a}{a^2+b^2}.$$

The centre of curvature $\mathbf{r}_c$ is given by

$$\mathbf{r}_c = \mathbf{r} + \rho\hat{\mathbf{N}} = \mathbf{r} + \frac{1}{\kappa}\hat{\mathbf{N}}.$$

Thus the evolute has equation

$$\begin{aligned}
\mathbf{r} &= a\cos t\mathbf{i} + a\sin t\mathbf{j} + bt\mathbf{k} \\
&\quad - \frac{a^2+b^2}{a}(\cos t\mathbf{i} + \sin t\mathbf{j}) \\
&= -\frac{b^2}{a}\cos t\mathbf{i} - \frac{b^2}{a}\sin t\mathbf{j} + bt\mathbf{k}.
\end{aligned}$$

The evolute is also a circular helix.

**26.** For the ellipse $\mathbf{r} = 2\cos t\mathbf{i} + \sin t\mathbf{j}$, we have

$$\begin{aligned}
\mathbf{v} &= -2\sin t\mathbf{i} + \cos t\mathbf{j} \\
\mathbf{a} &= -2\cos t\mathbf{i} - \sin t\mathbf{j} \\
\mathbf{v}\times\mathbf{a} &= 2\mathbf{k} \\
v &= \sqrt{4\sin^2 t + \cos^2 t} = \sqrt{3\sin^2 t + 1}.
\end{aligned}$$

The curvature is $\kappa = \dfrac{2}{(3\sin^2 t + 1)^{3/2}}$, so the radius of curvature is $\rho = \dfrac{(3\sin^2 t + 1)^{3/2}}{2}$. We have

$$\hat{\mathbf{T}} = \frac{-2\sin t\mathbf{i} + \cos t\mathbf{j}}{\sqrt{3\sin^2 t + 1}}, \qquad \hat{\mathbf{B}} = \mathbf{k}$$

$$\hat{\mathbf{N}} = -\frac{\cos t\mathbf{i} + 2\sin t\mathbf{j}}{\sqrt{3\sin^2 t + 1}}.$$

Therefore the evolute has equation

$$\mathbf{r} = 2\cos t\mathbf{i} + \sin t\mathbf{j} - \frac{3\sin^2 t + 1}{2}(\cos t\mathbf{i} + 2\sin t\mathbf{j})$$
$$= \frac{3}{2}\cos^3 \mathbf{i} - 3\sin^3 t\mathbf{j}.$$

**28.** We require

$$f(0) = 1, \quad f'(0) = 0, \quad f''(0) = -1,$$
$$f(-1) = 1, \quad f'(-1) = 0, \quad f''(-1) = 0.$$

The condition $f''(0) = -1$ follows from the fact that

$$\frac{d^2}{dx^2}\sqrt{1-x^2}\bigg|_{x=0} = -1.$$

As in Example 5, we try

$$f(x) = A + Bx + Cx^2 + Dx^3 + Ex^4 + Fx^5$$
$$f'(x) = B + 2Cx + 3Dx^2 + 4Ex^3 + 5Fx^4$$
$$f'' = 2C + 6Dx + 12Ex^2 + 20Fx^3.$$

The required conditions force the coefficients to satisfy the system of equations

$$A - B + C - D + E - F = 1$$
$$B - 2C + 3D - 4E + 5F = 0$$
$$2C - 6D + 12E - 20F = 0$$
$$A = 1$$
$$B = 0$$
$$2C = -1$$

which has solution $A = 1$, $B = 0$, $C = -1/2$, $D = -3/2$, $E = -3/2$, $F = -1/2$. Thus we can use a track section in the shape of the graph of

$$f(x) = 1 - \frac{1}{2}x^2 - \frac{3}{2}x^3 - \frac{3}{2}x^4 - \frac{1}{2}x^5 = 1 - \frac{1}{2}x^2(1+x)^3.$$

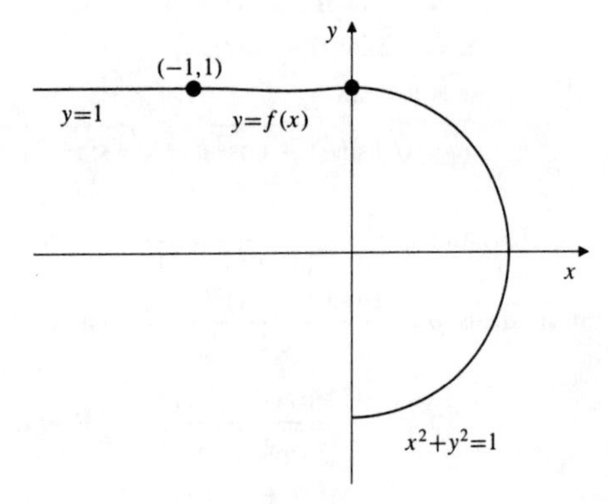


Fig. 7.5.28

## Section 7.6 Kepler's Laws of Planetary Motion (page 365)

**2.** Position: $\mathbf{r} = r\hat{\mathbf{r}} = k\hat{\mathbf{r}}$.
Velocity: $\mathbf{v} = k\dot{\hat{\mathbf{r}}} = k\dot{\theta}\hat{\boldsymbol{\theta}}$; speed: $v = k\dot{\theta}$.
Acceleration: $k\ddot{\theta}\hat{\boldsymbol{\theta}} + k\dot{\theta}\dot{\hat{\boldsymbol{\theta}}} = -k\dot{\theta}^2\hat{\mathbf{r}} + k\ddot{\theta}\hat{\boldsymbol{\theta}}$.
Radial component of acceleration: $-k\dot{\theta}^2$.
Transverse component of acceleration: $k\ddot{\theta} = \dot{v}$ (the rate of change of the speed).

**4.** Path: $r = \theta$. Thus $\dot{r} = \dot{\theta}$, $\ddot{r} = \ddot{\theta}$.
Speed: $v = \sqrt{(\dot{r})^2 + (r\dot{\theta})^2} = \dot{\theta}\sqrt{1+r^2}$.
Transverse acceleration = 0 (central force). Thus $r\ddot{\theta} + 2\dot{r}\dot{\theta} = 0$, or $\ddot{\theta} = -2\dot{\theta}^2/r$.
Radial acceleration:

$$\ddot{r} - r\dot{\theta}^2 = \ddot{\theta} - r\dot{\theta}^2$$
$$= -\left(\frac{2}{r} + r\right)\dot{\theta}^2 = -\frac{(2+r^2)v^2}{r(1+r^2)}.$$

The magnitude of the acceleration is, therefore, $\dfrac{(2+r^2)v^2}{r(1+r^2)}$.

**6.** Let the period and the semi-major axis of the orbit of Halley's comet be $T_H = 76$ years and $a_H$ km respectively. Similar parameters for the earth's orbit are $T_E = 1$ year and $a_E = 150 \times 10^6$ km. By Kepler's third law

$$\frac{T_H^2}{a_H^3} = \frac{T_E^2}{a_E^3}.$$

Thus

$$a_H = 150 \times 10^6 \times 76^{2/3} \approx 2.69 \times 10^9.$$

The major axis of Halley's comet's orbit is $2a_H \approx 5.38 \times 10^9$ km.

**8.** The period $T$ (in years) and radius $R$ (in km) of the asteroid's orbit satisfies

$$\frac{T^2}{R^3} = \frac{T_{\text{earth}}^2}{R_{\text{earth}}^2} = \frac{1^2}{(150 \times 10^6)^3}.$$

Thus the radius of the asteroid's orbit is $R \approx 150 \times 10^6 T^{2/3}$ km.

**10.** At perihelion, $r = a - c = (1-\epsilon)a$.
At aphelion $r = a + c = (1+\epsilon)a$.
Since $\dot{r} = 0$ at perihelion and aphelion, the speed is $v = r\dot{\theta}$ at each point. Since $r^2\dot{\theta} = h$ is constant over the orbit, $v = h/r$. Therefore

$$v_{\text{perihelion}} = \frac{h}{a(1-\epsilon)}, \qquad v_{\text{aphelion}} = \frac{h}{a(1+\epsilon)}.$$

If $v_{\text{perihelion}} = 2v_{\text{aphelion}}$ then

$$\frac{h}{a(1-\epsilon)} = \frac{2h}{a(1+\epsilon)}.$$

Hence $1+\epsilon = 2(1-\epsilon)$, and $\epsilon = 1/3$. The eccentricity of the orbit is 1/3.

**12.** Since $r^2\dot{\theta} = h =$ constant for the planet's orbit, and since the speed is $v = r\dot{\theta}$ at perihelion and at aphelion (the radial velocity is zero at these points), we have

$$r_p v_p = r_a v_a,$$

where the subscripts $p$ and $a$ refer to perihelion and aphelion, respectively. Since $r_p/r_a = 8/10$, we must have $v_p/v_a = 10/8 = 1.25$. Also,

$$r_p = \frac{\ell}{1+\epsilon\cos 0} = \frac{\ell}{1+\epsilon}, \quad r_a = \frac{\ell}{1+\epsilon\cos\pi} = \frac{\ell}{1-\epsilon}.$$

Thus $\ell/(1+\epsilon) = (8/10)\ell/(1-\epsilon)$, and so $10-10\epsilon = 8+8\epsilon$. Hence $2 = 18\epsilon$. The eccentricity of the orbit is $\epsilon = 1/9$.

**14.** As in Exercise 12, $r_P v_P = r_A v_A$, where $r_A = \ell/(1-\epsilon)$ and $r_P = \ell/(1+\epsilon)$, $\epsilon$ being the eccentricity of the orbit. Thus

$$\frac{v_P}{v_A} = \frac{r_A}{r_P} = \frac{1+\epsilon}{1-\epsilon}.$$

Solving this equation for $\epsilon$ in terms of $v_P$ and $v_A$, we get

$$\epsilon = \frac{v_P - v_A}{v_P + v_A}.$$

By conservation of energy the speed $v$ at the ends of the minor axis of the orbit (where $r = a$) satisfies

$$\frac{v^2}{2} - \frac{k}{a} = \frac{v_P^2}{2} - \frac{k}{r_P} = \frac{v_A^2}{2} - \frac{k}{r_A}.$$

The latter equality shows that

$$v_P^2 - v_A^2 = 2k\left(\frac{1}{r_P} - \frac{1}{r_A}\right) = \frac{4k\epsilon}{\ell}.$$

Using this result and the parameters of the orbit from the box on page 363 in the text, we obtain

$$\begin{aligned} v^2 &= v_P^2 + 2k\left(\frac{1}{a} - \frac{1}{r_P}\right) \\ &= v_P^2 + \frac{2k}{\ell}\big(1-\epsilon^2-(1+\epsilon)\big) \\ &= v_P^2 - \frac{2k\epsilon}{\ell}(1+\epsilon) \\ &= v_P^2 - \frac{v_P^2 - v_A^2}{2}\left(1 + \frac{v_P - v_A}{v_P + v_A}\right) \\ &= v_P^2 - \frac{v_P - v_A}{2}(2v_P) = v_P v_A. \end{aligned}$$

Thus $v = \sqrt{v_P v_A}$.

**16.** By conservation of energy, we have

$$\frac{k}{r} - \frac{1}{2}\left(\dot{r}^2 + \frac{h^2}{r^2}\right) = -K$$

where $K$ is a constant for the orbit (the total energy). The term in the parentheses is $v^2$, the square of the speed. Thus

$$\frac{k}{r} - \frac{1}{2}v^2 = -K = \frac{k}{r_0} - \frac{1}{2}v_0^2,$$

where $r_0$ and $v_0$ are the given distance and speed. We evaluate $-K$ at perihelion.

The parameters of the orbit are

$$\ell = \frac{h^2}{k}, \quad a = \frac{h^2}{k(1-\epsilon^2)}, \quad b = \frac{h^2}{k\sqrt{1-\epsilon^2}}, \quad c = \epsilon a.$$

At perihelion $P$ we have

$$r = a - c = (1-\epsilon)a = \frac{h^2}{k(1+\epsilon)}.$$

Since $\dot{r} = 0$ at perihelion, the speed there is $v = r\dot{\theta}$. By Kepler's second law, $r^2\dot{\theta} = h$, so $v = h/r = k(1+\epsilon)/h$. Thus

$$\begin{aligned} -K &= \frac{k}{r} - \frac{v^2}{2} \\ &= \frac{k^2}{h^2}(1+\epsilon) - \frac{1}{2}\frac{k^2}{h^2}(1+\epsilon)^2 \\ &= \frac{k^2}{2h^2}(1+\epsilon)\big[2-(1+\epsilon)\big] \\ &= \frac{k^2}{2h^2}(1-\epsilon^2) = \frac{k}{2a}. \end{aligned}$$

Thus $a = \dfrac{k}{-2K}$. By Kepler's third law,

$$T^2 = \frac{4\pi^2}{k}a^3 = \frac{4\pi^2}{k}\left(\frac{k}{-2K}\right)^3.$$

Thus $T = \dfrac{2\pi}{\sqrt{k}}\left(\dfrac{2}{r_0} - \dfrac{v_0^2}{k}\right)^{-3/2}$.

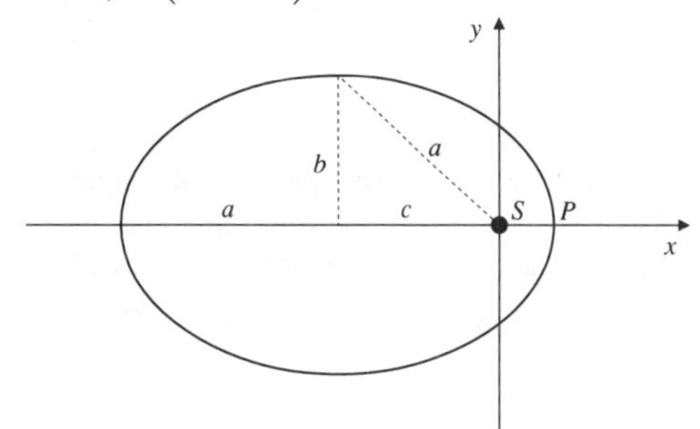


Fig. 7.6.16

**18.** Start with

$$\ddot{r} - \frac{h^2}{r^3} = -\frac{k}{r^2}.$$

Let $r(t) = \dfrac{1}{u(\theta)}$, where $\theta = \theta(t)$. Since $r^2\dot{\theta} = h$ (constant), we have

$$\dot{r} = -\frac{1}{u^2}\frac{du}{d\theta}\dot{\theta} = -r^2\frac{du}{d\theta}\frac{h}{r^2} = -h\frac{du}{d\theta}$$

$$\ddot{r} = -h\frac{d^2u}{d\theta^2}\dot{\theta} = -\frac{h^2}{r^2}\frac{d^2u}{d\theta^2} = -h^2u^2\frac{d^2u}{d\theta^2}.$$

Thus $-h^2u^2\dfrac{d^2u}{d\theta^2} - h^2u^3 = -ku^2$, or

$$\frac{d^2u}{d\theta^2} + u = \frac{k}{h^2}.$$

This is the DE for simple harmonic motion with a constant forcing term (nonhomogeneous term) on the right-hand side. It is easily verified that

$$u = \frac{k}{h^2}\big(1 + \epsilon\cos(\theta - \theta_0)\big)$$

is a solution for any choice of the constants $\epsilon$ and $\theta_0$. Expressing the solution in terms of $r$, we have

$$r = \frac{h^2/k}{1 + \epsilon\cos(\theta - \theta_0)},$$

which is an ellipse if $|\epsilon| < 1$.

**20.** Since $\dfrac{k}{r} = \dfrac{1}{2}v^2 - K$ by conservation of energy, if $K < 0$, then

$$\frac{k}{r} \ge -K > 0,$$

so $r \le -\dfrac{k}{K}$. The orbit is, therefore, bounded.

**22.** By Exercise 17, the asymptotes make angle $\theta = \cos^{-1}(1/\epsilon)$ with the transverse axis, as shown in the figure. The angle of deviation $\delta$ satisfies $2\theta + \delta = \pi$, so $\theta = \dfrac{\pi}{2} - \dfrac{\delta}{2}$, and

$$\cos\theta = \sin\frac{\delta}{2}, \qquad \sin\theta = \cos\frac{\delta}{2}.$$

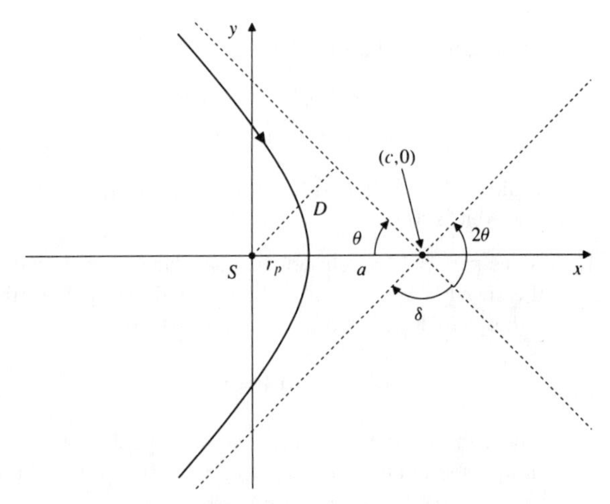


Fig. 7.6.22

By conservation of energy,

$$\frac{v^2}{2} - \frac{k}{r} = \text{constant} = \frac{v_\infty^2}{2}$$

for all points on the orbit. At perihelion,

$$r = r_p = c - a = (\epsilon - 1)a = \frac{\ell}{\epsilon + 1},$$

$$v = v_p = r_p\dot{\theta} = \frac{h}{r_p} = \frac{h(\epsilon + 1)}{\ell}.$$

Since $h^2 = k\ell$, we have

$$\begin{aligned} v_\infty^2 &= v_p^2 - \frac{2k}{r_p} \\ &= \frac{h^2}{\ell^2}(\epsilon + 1)^2 - \frac{2k}{\ell}(\epsilon + 1) \\ &= \frac{k}{\ell}\big[(\epsilon + 1)^2 - 2(\epsilon + 1)\big] \\ &= \frac{k}{\ell}(\epsilon^2 - 1) = \frac{k}{a}. \end{aligned}$$

Thus $av_\infty^2 = k$.
If $D$ is the perpendicular distance from the sun $S$ to an asymptote of the orbit (see the figure) then

$$\begin{aligned} D &= c\sin\theta = \epsilon a\sin\theta = s\frac{\sin\theta}{\cos\theta} \\ &= a\frac{\cos(\delta/2)}{\sin(\delta/2)} = a\cot\frac{\delta}{2}. \end{aligned}$$

Therefore

$$\frac{Dv_\infty^2}{k} = \frac{v_\infty^2 a}{k}\cot\frac{\delta}{2} = \cot\frac{\delta}{2}.$$

## Review Exercises 7 (page 367)

**2.** $\mathbf{r} = t\cos t\mathbf{i} + t\sin t\mathbf{j} + (2\pi - t)\mathbf{k}$, $(0 \le t \le 2\pi)$ is a conical helix wound around the cone $z = 2\pi - \sqrt{x^2 + y^2}$ starting at the vertex $(0, 0, 2\pi)$, and completing one revolution to end up at $(2\pi, 0, 0)$. Since

$$\mathbf{v} = (\cos t - t\sin t)\mathbf{i} + (\sin t + t\cos t)\mathbf{j} - \mathbf{k},$$

the length of the curve is

$$L = \int_0^{2\pi} \sqrt{2+t^2}\,dt = \pi\sqrt{2+4\pi^2} + \ln\left(\frac{2\pi+\sqrt{2+4\pi^2}}{\sqrt{2}}\right)$$

units. (See the solution to Exercise 22 of Section 7.3 for the details of the evaluation of this integral.)

**4.** The position, velocity, speed, and acceleration of the particle are given by

$$\mathbf{r} = x\mathbf{i} + x^2\mathbf{j}$$
$$\mathbf{v} = \frac{dx}{dt}(\mathbf{i} + 2x\mathbf{j}), \quad v = \left|\frac{dx}{dt}\right|\sqrt{1+4x^2}$$
$$\mathbf{a} = \frac{d^2x}{dt^2}(\mathbf{i} + 2x\mathbf{j}) + 2\left(\frac{dx}{dt}\right)^2\mathbf{j}.$$

Let us assume that the particle is moving to the right, so that $dx/dt > 0$. Since the speed is $t$, we have

$$\frac{dx}{dt} = \frac{t}{\sqrt{1+4x^2}}$$
$$\frac{d^2x}{dt^2} = \frac{\sqrt{1+4x^2} - \dfrac{4tx}{\sqrt{1+4x^2}}\dfrac{dx}{dt}}{1+4x^2}.$$

If the particle is at $(\sqrt{2}, 2)$ at $t = 3$, then $dx/dt = 1$ at that time, and

$$\frac{d^2x}{dt^2} = \frac{3-4\sqrt{2}}{9}.$$

Hence the acceleration is

$$\mathbf{a} = \frac{3-4\sqrt{2}}{9}(\mathbf{i} + 2\sqrt{2}\mathbf{j}) + 2\mathbf{j}.$$

If the particle is moving to the left, so that $dx/dt < 0$, a similar calculation shows that at $t = 3$ its acceleration is

$$\mathbf{a} = -\frac{3+4\sqrt{2}}{9}(\mathbf{i} + 2\sqrt{2}\mathbf{j}) + 2\mathbf{j}.$$

**6.** Tangential acceleration: $dv/dt = e^t - e^{-t}$.
Normal acceleration: $v^2\kappa = \sqrt{2}$.
Since $v = 2\cosh t$, the minimum speed is 2 at time $t = 0$.

**8.** If $r = e^{-\theta}$, and $\dot{\theta} = k$, then $\dot{r} = -e^{-\theta}\dot{\theta} = -kr$, and $\ddot{r} = k^2 r$. Since $\mathbf{r} = r\hat{\mathbf{r}}$, we have

$$\mathbf{v} = \dot{r}\hat{\mathbf{r}} + r\dot{\theta}\hat{\boldsymbol{\theta}} = -kr\hat{\mathbf{r}} + kr\hat{\boldsymbol{\theta}}$$
$$\mathbf{a} = (\ddot{r} - r\dot{\theta}^2)\hat{\mathbf{r}} + (r\ddot{\theta} + 2\dot{r}\dot{\theta})\hat{\boldsymbol{\theta}}$$
$$= (k^2 r - k^2 r)\hat{\mathbf{r}} + (0 - 2k^2 r)\hat{\boldsymbol{\theta}} = -2k^2 r\hat{\boldsymbol{\theta}}.$$

**10.** $s = 4a\left(1 - \cos\dfrac{t}{2}\right) \Rightarrow t = 2\cos^{-1}\left(1 - \dfrac{s}{4a}\right) = t(s).$

The required arc length parametrization of the cycloid is

$$\mathbf{r} = a\big(t(s) - \sin t(s)\big)\mathbf{i} + a\big(1 - \cos t(s)\big)\mathbf{j}.$$

**12.**

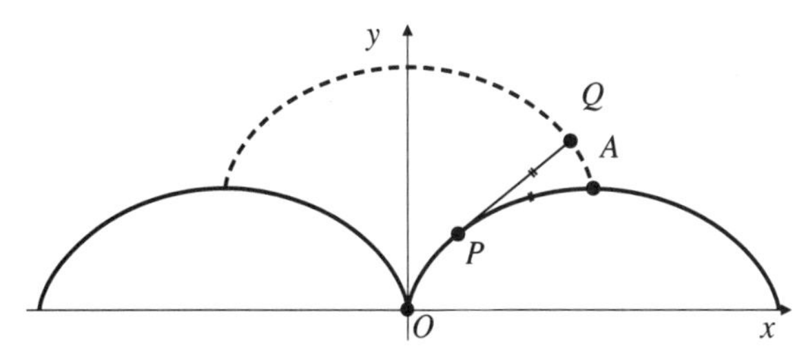


Fig. R-7.12

Let $P$ be the point with position vector $\mathbf{r}(t)$ on the cycloid. By Exercise 9, the arc $OP$ has length $4a - 4a\cos(t/2)$, and so $PQ$ has length $4a$ - arc $OP = 4a\cos(t/2)$ units. Thus

$$\overrightarrow{PQ} = 4a\cos\frac{t}{2}\hat{\mathbf{T}}(t)$$
$$= 4a\cos\frac{t}{2}\left(\sin\frac{t}{2}\mathbf{i} + \cos\frac{t}{2}\mathbf{j}\right)$$
$$= 2a\sin t\mathbf{i} + 2a(1 + \cos t)\mathbf{j}.$$

It follows that $Q$ has position vector

$$\mathbf{r}_Q = \mathbf{r} + \overrightarrow{PQ}$$
$$= a(t - \sin t)\mathbf{i} + a(1 - \cos t)\mathbf{j} + 2a\sin t\mathbf{i} + 2a(1 + \cos t)\mathbf{j}$$
$$= a(t + \sin t)\mathbf{i} + a(1 + \cos t + 2)\mathbf{j} \quad (\text{let } t = u + \pi)$$
$$= a(u - \sin u + \pi)\mathbf{i} + a(1 - \cos u + 2)\mathbf{j}.$$

Thus $\mathbf{r}_Q(t)$ represents the same cycloid as $\mathbf{r}(t)$, but translated $\pi a$ units to the left and $2a$ units upward. From Exercise 11, the given cycloid is the evolute of its involute.

**14.** By Kepler's Second Law the position vector $\mathbf{r}$ from the origin (the sun) to the planet sweeps out area at a constant rate, say $h/2$:

$$\frac{dA}{dt} = \frac{h}{2}.$$

As observed in the text, $dA/dt = r^2\dot{\theta}/2$, so $r^2\dot{\theta} = h$, and

$$\mathbf{r} \times \mathbf{v} = (r\hat{\mathbf{r}}) \times (\dot{r}\hat{\mathbf{r}} + r\dot{\theta}\hat{\boldsymbol{\theta}}) = r^2\dot{\theta}\hat{\mathbf{r}} \times \hat{\boldsymbol{\theta}} = h\mathbf{k} = \mathbf{h}$$

is a constant vector.

**16.** By Exercise 15, $\mathbf{F}(\mathbf{r}) = m(\ddot{r} - r\dot{\theta}^2)\hat{\mathbf{r}} = -f(\mathbf{r})\hat{\mathbf{r}}$. We are given that $r = \ell/(1 + \epsilon\cos\theta)$. Thus

$$\begin{aligned}
\dot{r} &= -\frac{\ell}{(1+\epsilon\cos\theta)^2}(-\epsilon\sin\theta)\dot{\theta}\\
&= \frac{\epsilon\ell\sin\theta}{(1+\epsilon\cos\theta)^2}\dot{\theta}\\
&= \frac{\epsilon\sin\theta}{\ell}r^2\dot{\theta} = \frac{h\epsilon}{\ell}\sin\theta\\
\ddot{r} &= \frac{h\epsilon}{\ell}(\cos\theta)\dot{\theta} = \frac{h^2\epsilon\cos\theta}{\ell r^2}.
\end{aligned}$$

It follows that

$$\begin{aligned}
\ddot{r} - r\dot{\theta}^2 &= \frac{h^2\epsilon\cos\theta}{\ell r^2} - \frac{h^2}{r^3}\\
&= \frac{h^2}{\ell r^2}\left(\epsilon\cos\theta - \frac{\ell}{r}\right) = -\frac{h^2}{\ell r^2},
\end{aligned}$$

(because $(\ell/r) = 1 + \epsilon\cos\theta$). Hence

$$f(\mathbf{r}) = \frac{mh^2}{\ell r^2}.$$

This says that the magnitude of the force on the planet is inversely proportional to the square of its distance from the sun. Thus Newton's law of gravitation follows from Kepler's laws and the second law of motion.

## Challenging Problems 7 (page 368)

**2.** $\begin{cases} \dfrac{d\mathbf{v}}{dt} = \mathbf{k}\times\mathbf{v} - 32\mathbf{k} \\ \mathbf{v}(0) = 70\mathbf{i} \end{cases}$

a) If $\mathbf{v} = v_1\mathbf{i} + v_2\mathbf{j} + v_3\mathbf{k}$, then $\mathbf{k}\times\mathbf{v} = v_1\mathbf{j} - v_2\mathbf{i}$. Thus the initial-value problem breaks down into component equations as

$$\begin{cases} \dfrac{dv_1}{dt} = -v_2 \\ v_1(0) = 70 \end{cases} \qquad \begin{cases} \dfrac{dv_2}{dt} = v_1 \\ v_2(0) = 0 \end{cases} \qquad \begin{cases} \dfrac{dv_3}{dt} = -32 \\ v_3(0) = 0. \end{cases}$$

b) If $\mathbf{r} = x\mathbf{i} + y\mathbf{j} + z\mathbf{k}$ denotes the position of the baseball $t$ s after it is thrown, then $x(0) = y(0) = z(0) = 0$ and we have

$$\frac{dz}{dt} = v_3 = -32t \;\Rightarrow\; z = -16t^2.$$

Also, $\dfrac{d^2v_1}{dt^2} = -\dfrac{dv_2}{dt} = -v_1$ (the equation of simple harmonic motion), so

$$v_1(t) = A\cos t + B\sin t, \quad v_2(t) = A\sin t - B\cos t.$$

Since $v_1(0) = 70$, $v_2(0) = 0$, $x(0) = 0$, and $y(0) = 0$, we have

$$\begin{aligned}
\frac{dx}{dt} &= v_1 = 70\cos t & \frac{dy}{dt} &= v_2 = 70\sin t\\
x(t) &= 70\sin t & y(t) &= 70(1-\cos t).
\end{aligned}$$

At time $t$ seconds after it is thrown, the ball is at position

$$\mathbf{r} = 70\sin t\,\mathbf{i} + 70(1-\cos t)\mathbf{j} - 16t^2\mathbf{k}.$$

c) At $t = 1/5$ s, the ball is at about $(13.9, 1.40, -0.64)$. If it had been thrown without the vertical spin, its position at time $t$ would have been

$$\mathbf{r} = 70t\mathbf{i} - 16t^2\mathbf{k},$$

so its position at $t = 1/5$ s would have been $(14, 0, -0.64)$. Thus the spin has deflected the ball approximately 1.4 ft to the left (as seen from above) of what would have been its parabolic path had it not been given the spin.

**4.**

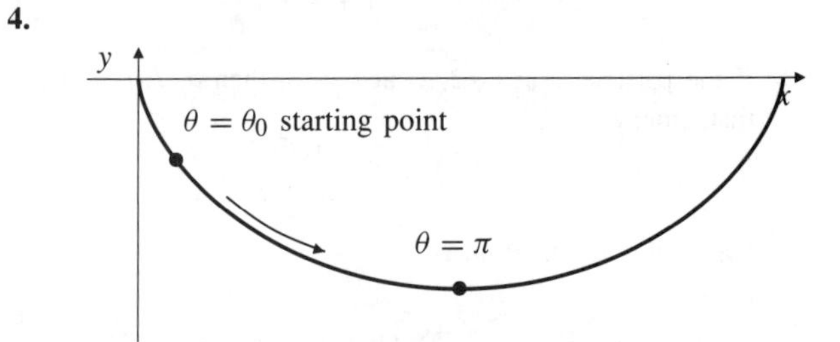


Fig. C-7.4

The arc length element on $x = a(\theta - \sin\theta)$, $y = a(\cos\theta - 1)$ is (for $\theta \le \pi$)

$$\begin{aligned}
ds &= a\sqrt{(1-\cos\theta)^2 + \sin^2\theta}\,d\theta\\
&= a\sqrt{2(1-\cos\theta)}\,d\theta = 2a\sin(\theta/2)\,d\theta.
\end{aligned}$$

If the bead slides downward from rest at height $y(\theta_0)$ to height $y(\theta)$, its gravitational potential energy has decreased by

$$mg\big[y(\theta_0) - y(\theta)\big] = mga(\cos\theta_0 - \cos\theta).$$

Since there is no friction, all this potential energy is converted to kinetic energy, so its speed $v$ at height $y(\theta)$ is given by

$$\frac{1}{2}mv^2 = mga(\cos\theta_0 - \cos\theta),$$

and so $v = \sqrt{2ga(\cos\theta_0 - \cos\theta)}$. The time required for the bead to travel distance $ds$ at speed $v$ is $dt = ds/v$, so the time $T$ required for the bead to slide from its starting position at $\theta = \theta_0$ to the lowest point on the wire, $\theta = \pi$, is

$$\begin{aligned}
T &= \int_{\theta=\theta_0}^{\theta=\pi} \frac{ds}{v} = \int_{\theta_0}^{\pi} \frac{1}{v}\frac{ds}{d\theta}\,d\theta \\
&= \sqrt{\frac{2a}{g}} \int_{\theta_0}^{\pi} \frac{\sin(\theta/2)}{\sqrt{\cos\theta_0 - \cos\theta}}\,d\theta \\
&= \sqrt{\frac{2a}{g}} \int_{\theta_0}^{\pi} \frac{\sin(\theta/2)}{\sqrt{2\cos^2(\theta_0/2) - 2\cos^2(\theta/2)}}\,d\theta \\
&\qquad \text{Let } u = \cos(\theta/2) \\
&\qquad du = -\tfrac{1}{2}\sin(\theta/2)\,d\theta \\
&= 2\sqrt{\frac{a}{g}} \int_0^{\cos(\theta_0/2)} \frac{du}{\sqrt{\cos^2(\theta_0/2) - u^2}} \\
&= 2\sqrt{\frac{a}{g}} \sin^{-1}\left(\frac{u}{\cos(\theta_0/2)}\right)\Bigg|_0^{\cos(\theta_0/2)} \\
&= \pi\sqrt{ag}
\end{aligned}$$

which is independent of $\theta_0$.

**6.** a) At time $t$, the hare is at $P = (0, vt)$ and the fox is at $Q = \big(x(t), y(t)\big)$, where $x$ and $y$ are such that the slope $dy/dx$ of the fox's path is the slope of the line $PQ$:

$$\frac{dy}{dx} = \frac{y - vt}{x}.$$

b) Since $\dfrac{d}{dt}\dfrac{dy}{dx} = \dfrac{d^2y}{dx^2}\dfrac{dx}{dt}$, we have

$$\begin{aligned}
\frac{dx}{dt}\frac{d^2y}{dx^2} &= \frac{d}{dt}\left(\frac{y - vt}{x}\right) \\
&= \frac{x\left(\dfrac{dy}{dt} - v\right) - (y - vt)\dfrac{dx}{dt}}{x^2} \\
&= \frac{1}{x}\left(\frac{dy}{dx}\frac{dx}{dt} - v\right) - \frac{1}{x^2}(y - vt)\frac{dx}{dt} \\
&= \frac{1}{x^2}(y - vt)\frac{dx}{dt} - \frac{v}{x} - \frac{1}{x^2}(y - vt)\frac{dx}{dt} \\
&= -\frac{v}{x}.
\end{aligned}$$

Thus $x\dfrac{d^2y}{dx^2} = -\dfrac{v}{dx/dt}$.
Since the fox's speed is also $v$, we have

$$\left(\frac{dx}{dt}\right)^2 + \left(\frac{dy}{dt}\right)^2 = v^2.$$

Also, the fox is always running to the left (towards the $y$-axis from points where $x > 0$), so $dx/dt < 0$. Hence

$$\frac{v}{-\left(\dfrac{dx}{dt}\right)} = \sqrt{1 + \frac{(dy/dt)^2}{(dx/dt)^2}} = \sqrt{1 + \left(\frac{dy}{dx}\right)^2},$$

and so the fox's path $y = y(x)$ satisfies the DE

$$x\frac{d^2y}{dx^2} = \sqrt{1 + \left(\frac{dy}{dx}\right)^2}.$$

c) If $u = dy/dx$, then $u = 0$ and $y = 0$ when $x = a$, and

$$\begin{aligned}
x\frac{du}{dx} &= \sqrt{1 + u^2} \\
\int \frac{du}{\sqrt{1 + u^2}} &= \int \frac{dx}{x} \qquad \begin{array}{l}\text{Let } u = \tan\theta \\ du = \sec^2\theta\,d\theta\end{array} \\
\int \sec\theta\,d\theta &= \ln x + \ln C \\
\ln(\tan\theta + \sec\theta) &= \ln(Cx) \\
u + \sqrt{1 + u^2} &= Cx.
\end{aligned}$$

Since $u = 0$ when $x = a$, we have $C = 1/a$.

$$\begin{aligned}
\sqrt{1 + u^2} &= \frac{x}{a} - u \\
1 + u^2 &= \frac{x^2}{a^2} - \frac{2xu}{a} + u^2 \\
\frac{2xu}{a} &= \frac{x^2}{a^2} - 1 \\
\frac{dy}{dx} &= u = \frac{x}{2a} - \frac{a}{2x} \\
y &= \frac{x^2}{4a} - \frac{a}{2}\ln x + C_1.
\end{aligned}$$

Since $y = 0$ when $x = a$, we have $C_1 = -\dfrac{a}{4} + \dfrac{a}{2}\ln a$, so

$$y = \frac{x^2 - a^2}{4} - \frac{a}{2}\ln\frac{x}{a}$$

is the path of the fox.

# CHAPTER 8. VECTOR FIELDS

## Section 8.1 Vector and Scalar Fields (page 376)

**2.** $\mathbf{F} = x\mathbf{i} + y\mathbf{j}$.
The field lines satisfy $\dfrac{dx}{x} = \dfrac{dy}{y}$.
Thus $\ln y = \ln x + \ln C$, or $y = Cx$. The field lines are straight half-lines emanating from the origin.

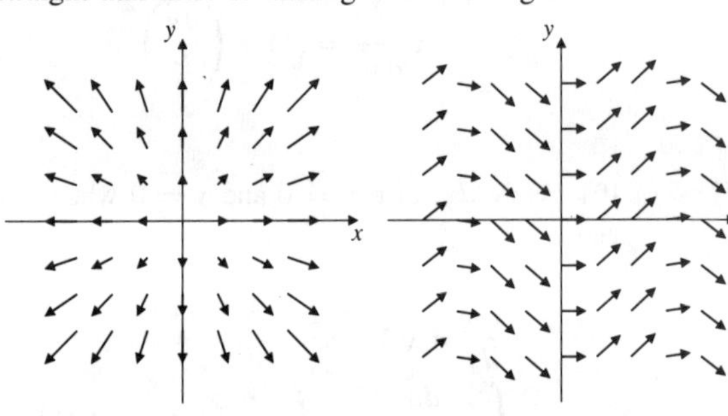


Fig. 8.1.2 Fig. 8.1.4

**4.** $\mathbf{F} = \mathbf{i} + \sin x\mathbf{j}$.
The field lines satisfy $dx = \dfrac{dy}{\sin x}$.
Thus $\dfrac{dy}{dx} = \sin x$. The field lines are the curves $y = -\cos x + C$.

**6.** $\mathbf{F} = \nabla(x^2 - y) = 2x\mathbf{i} - \mathbf{j}$.
The field lines satisfy $\dfrac{dx}{2x} = \dfrac{dy}{-1}$. They are the curves $y = -\dfrac{1}{2}\ln x + C$.

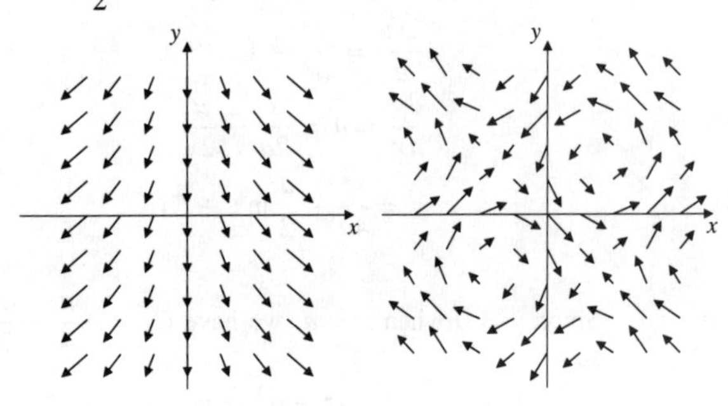


Fig. 8.1.6 Fig. 8.1.8

**8.** $\mathbf{F} = \cos y\mathbf{i} - \cos x\mathbf{j}$.
The field lines satisfy $\dfrac{dx}{\cos y} = -\dfrac{dy}{\cos x}$, that is, $\cos x\,dx + \cos y\,dy = 0$. Thus they are the curves $\sin x + \sin y = C$.

**10.** $\mathbf{v}(x, y, z) = x\mathbf{i} + y\mathbf{j} - x\mathbf{k}$.
The streamlines satisfy $\dfrac{dx}{x} = \dfrac{dy}{y} = -\dfrac{dz}{x}$. Thus $z + x = C_1$, $y = C_2 x$. The streamlines are straight half-lines emanating from the $z$-axis and perpendicular to the vector $\mathbf{i} + \mathbf{k}$.

**12.** $\mathbf{v} = \dfrac{x\mathbf{i} + y\mathbf{j}}{(1 + z^2)(x^2 + y^2)}$.
The streamlines satisfy $dz = 0$ and $\dfrac{dx}{x} = \dfrac{dy}{y}$. Thus $z = C_1$ and $y = C_2 x$. The streamlines are horizontal half-lines emanating from the $z$-axis.

**14.** $\mathbf{v} = e^{xyz}(x\mathbf{i} + y^2\mathbf{j} + z\mathbf{k})$. The field lines satisfy

$$\frac{dx}{x} = \frac{dy}{y^2} = \frac{dz}{z},$$

so they are given by $z = C_1 x$, $\ln|x| = \ln|C_2| - (1/y)$ (or, equivalently, $x = C_2 e^{-1/y}$).

**16.** $\mathbf{v}(x, y) = x\mathbf{i} + (x + y)\mathbf{j}$. The field lines satisfy

$$\frac{dx}{x} = \frac{dy}{x+y}$$
$$\frac{dy}{dx} = \frac{x+y}{x} \qquad \text{Let } y = xv(x)$$
$$\frac{dy}{dx} = v + x\frac{dv}{dx}$$
$$v + x\frac{dv}{dx} = \frac{x(1+v)}{x} = 1 + v.$$

Thus $dv/dx = 1/x$, and so $v(x) = \ln|x| + C$. The field lines have equations $y = x\ln|x| + Cx$.

**18.** $\mathbf{F} = \hat{\mathbf{r}} + \theta\hat{\boldsymbol{\theta}}$. The field lines satisfy $dr = r\,d\theta/\theta$, or $dr/r = d\theta/\theta$, so they are the spirals $r = C\theta$.

**20.** $\mathbf{F} = r\hat{\mathbf{r}} - \hat{\boldsymbol{\theta}}$. The field lines satisfy $dr/r = -r\,d\theta$, or $-dr/r^2 = d\theta$, so they are the spirals $1/r = \theta + C$, or $r = 1/(\theta + C)$.

## Section 8.2 Conservative Fields (page 385)

**2.** $\mathbf{F} = y\mathbf{i} + x\mathbf{j} + z^2\mathbf{k}$, $F_1 = y$, $F_2 = x$, $F_3 = z^2$. We have

$$\frac{\partial F_1}{\partial y} = 1 = \frac{\partial F_2}{\partial x},$$
$$\frac{\partial F_1}{\partial z} = 0 = \frac{\partial F_3}{\partial x},$$
$$\frac{\partial F_2}{\partial z} = 0 = \frac{\partial F_3}{\partial y}.$$

Therefore, $\mathbf{F}$ may be conservative. If $\mathbf{F} = \nabla\phi$, then

$$\frac{\partial \phi}{\partial x} = y, \quad \frac{\partial \phi}{\partial y} = x, \quad \frac{\partial \phi}{\partial z} = z^2.$$

Therefore,

$$\begin{aligned}\phi(x,y,z) &= \int y\,dx = xy + C_1(y,z)\\ x = \frac{\partial\phi}{\partial y} &= x + \frac{\partial C_1}{\partial y} \Rightarrow \frac{\partial C_1}{\partial y} = 0\\ C_1(y,z) = C_2(z), &\qquad \phi(x,y,z) = xy + C_2(z)\\ z^2 = \frac{\partial\phi}{\partial z} &= C_2'(z) \Rightarrow C_2(z) = \frac{z^3}{3}.\end{aligned}$$

Thus $\phi(x,y,z) = xy + \dfrac{z^3}{3}$ is a potential for **F**, and **F** is conservative on $\mathbb{R}^3$.

**4.** $\mathbf{F} = \dfrac{x\mathbf{i} + y\mathbf{j}}{x^2+y^2}$, $F_1 = \dfrac{x}{x^2+y^2}$, $F_2 = \dfrac{y}{x^2+y^2}$. We have

$$\frac{\partial F_1}{\partial y} = -\frac{2xy}{(x^2+y^2)^2} = \frac{\partial F_2}{\partial x}.$$

Therefore, **F** may be conservative. If $\mathbf{F} = \nabla\phi$, then

$$\frac{\partial\phi}{\partial x} = \frac{x}{x^2+y^2}, \qquad \frac{\partial\phi}{\partial y} = \frac{y}{x^2+y^2}.$$

Therefore,

$$\begin{aligned}\phi(x,y) &= \int \frac{x}{x^2+y^2}\,dx = \frac{\ln(x^2+y^2)}{2} + C_1(y)\\ \frac{y}{x^2+y^2} &= \frac{\partial\phi}{\partial y} = \frac{y}{x^2+y^2} + c_1'(y) \Rightarrow c_1'(y) = 0.\end{aligned}$$

Thus we can choose $C_1(y) = 0$, and

$$\phi(x,y) = \frac{1}{2}\ln(x^2+y^2)$$

is a scalar potential for **F**, and **F** is conservative everywhere on $\mathbb{R}^2$ except at the origin.

**6.** $\mathbf{F} = e^{x^2+y^2+z^2}(xz\mathbf{i} + yz\mathbf{j} + xy\mathbf{k})$.
$F_1 = xze^{x^2+y^2+z^2}$, $F_2 = yze^{x^2+y^2+z^2}$,
$F_3 = xye^{x^2+y^2+z^2}$. We have

$$\begin{aligned}\frac{\partial F_1}{\partial y} &= 2xyze^{x^2+y^2+z^2} = \frac{\partial F_2}{\partial x},\\ \frac{\partial F_1}{\partial z} &= (x + 2xz^2)e^{x^2+y^2+z^2},\\ \frac{\partial F_3}{\partial x} &= (y + 2x^2y)e^{x^2+y^2+z^2} \neq \frac{\partial F_1}{\partial z}.\end{aligned}$$

Thus **F** cannot be conservative.

**8.**
$$\begin{aligned}\frac{\partial}{\partial x}\ln|\mathbf{r}| &= \frac{1}{|\mathbf{r}|}\frac{\mathbf{r}\bullet\dfrac{\partial\mathbf{r}}{\partial x}}{|\mathbf{r}|} = \frac{x}{|\mathbf{r}|^2}\\ \nabla\ln|\mathbf{r}| &= \frac{x\mathbf{i}+y\mathbf{j}+z\mathbf{k}}{|\mathbf{r}|^2} = \frac{\mathbf{r}}{|\mathbf{r}|^2}.\end{aligned}$$

**10.** $\mathbf{F} = \dfrac{2x}{z}\mathbf{i} + \dfrac{2y}{z}\mathbf{j} - \dfrac{x^2+y^2}{z^2}\mathbf{k} = \mathbf{G} + \mathbf{k}$,
where **G** is the vector field **F** of Exercise 9. Since **G** is conservative (except on the plane $z = 0$), so is **F**, which has scalar potential

$$\phi(x,y,z) = \frac{x^2+y^2}{z} + z = \frac{x^2+y^2+z^2}{z},$$

since $\dfrac{x^2+y^2}{z}$ is a potential for **G** and $z$ is a potential for the vector **k**.

The equipotential surfaces of **F** are $\phi(x,y,z) = C$, or

$$x^2+y^2+z^2 = Cz$$

which are spheres tangent to the $xy$-plane having centres on the $z$-axis.

The field lines of **F** satisfy

$$\frac{dx}{\dfrac{2x}{z}} = \frac{dy}{\dfrac{2y}{z}} = \frac{dz}{1 - \dfrac{x^2+y^2}{z^2}}.$$

As in Exercise 9, the first equation has solutions $y = Ax$, representing vertical planes containing the $z$-axis. The remaining equations can then be written in the form

$$\frac{dz}{dx} = \frac{z^2-x^2-y^2}{2xz} = \frac{z^2-(1+A^2)x^2}{2zx}.$$

This first order DE is of homogeneous type (see Section 10.2), and can be solved by a change of dependent variable: $z = xv(x)$. We have

$$\begin{aligned}v + x\frac{dv}{dx} &= \frac{dz}{dx} = \frac{x^2v^2-(1+A^2)x^2}{2x^2v}\\ x\frac{dv}{dx} &= \frac{v^2-(1+A^2)}{2v} - v = -\frac{v^2+(1+A^2)}{2v}\\ \frac{2v\,dv}{v^2+(1+A^2)} &= -\frac{dx}{x}\\ \ln\big(v^2+(1+A^2)\big) &= -\ln x + \ln B\\ v^2+1+A^2 &= \frac{B}{x}\\ \frac{z^2}{x^2}+1+A^2 &= \frac{B}{x}\\ z^2+x^2+y^2 &= Bx.\end{aligned}$$

These are spheres centred on the $x$-axis and passing through the origin. The field lines are the intersections of the planes $y = Ax$ with these spheres, so they are vertical circles passing through the origin and having centres in the $xy$-plane. (The technique used to find these circles excludes those circles with centres on the $y$-axis, but they are also field lines of **F**.)

Note: In two dimensions, circles passing through the origin and having centres on the $x$-axis intersect perpendicularly circles passing through the origin and having centres on the $y$-axis. Thus the nature of the field lines of **F** can be determined geometrically from the nature of the equipotential surfaces.

**12.** The scalar potential for the source-sink system is

$$\phi(x, y, z) = \phi(\mathbf{r}) = -\frac{2}{|\mathbf{r}|} + \frac{1}{|\mathbf{r} - \mathbf{k}|}.$$

Thus, the velocity field is

$$\begin{aligned}\mathbf{v} = \nabla\phi &= \frac{2\mathbf{r}}{|\mathbf{r}|^3} - \frac{\mathbf{r} - \mathbf{k}}{|\mathbf{r} - \mathbf{k}|^3} \\ &= \frac{2(x\mathbf{i} + y\mathbf{j} + z\mathbf{k})}{(x^2 + y^2 + z^2)^{3/2}} - \frac{x\mathbf{i} + y\mathbf{j} + (z - 1)\mathbf{k}}{(x^2 + y^2 + (z - 1)^2)^{3/2}}.\end{aligned}$$

For vertical velocity we require

$$\frac{2x}{(x^2 + y^2 + z^2)^{3/2}} = \frac{x}{(x^2 + y^2 + (z - 1)^2)^{3/2}},$$

and a similar equation for $y$. Both equations will be satisfied at all points of the $z$-axis, and also wherever

$$\begin{aligned}2\big(x^2 + y^2 + (z - 1)^2\big)^{3/2} &= \big(x^2 + y^2 + z^2\big)^{3/2} \\ 2^{2/3}\big(x^2 + y^2 + (z - 1)^2\big) &= x^2 + y^2 + z^2 \\ x^2 + y^2 + (z - K)^2 &= K^2 - K,\end{aligned}$$

where $K = 2^{2/3}/(2^{2/3} - 1)$. This latter equation represents a sphere, $S$, since $K^2 - K > 0$. The velocity is vertical at all points on $S$, as well as at all points on the $z$-axis.

Since the source at the origin is twice as strong as the sink at $(0, 0, 1)$, only half the fluid it emits will be sucked into the sink. By symmetry, this half will the half emitted into the half-space $z > 0$. The rest of the fluid emitted at the origin will flow outward to infinity. There is one point where $\mathbf{v} = \mathbf{0}$. This point (which is easily calculated to be $(0, 0, 2 + \sqrt{2})$) lies inside $S$. Streamlines emerging from the origin parallel to the $xy$-plane lead to this point. Streamlines emerging into $z > 0$ cross $S$ and approach the sink. Streamlines emerging into $z < 0$ flow to infinity. Some of these cross $S$ twice, some others are tangent to $S$, some do not intersect $S$ anywhere.

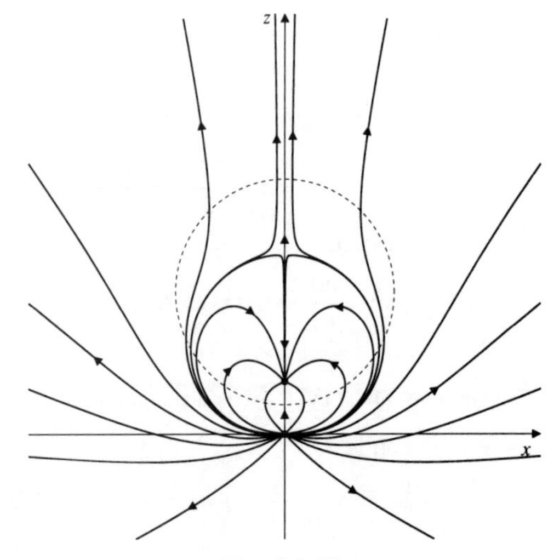

Fig. 8.2.12

**14.** For $\mathbf{v}(x, y) = \dfrac{m(x\mathbf{i} + y\mathbf{j})}{x^2 + y^2}$, we have

$$\frac{\partial v_1}{\partial y} = -\frac{2mxy}{(x^2 + y^2)^2} = \frac{\partial v_2}{\partial x},$$

so **v** may be conservative, except at $(0, 0)$. We have

$$\begin{aligned}\phi(x, y) &= m\int \frac{x\,dx}{x^2 + y^2} = \frac{m}{2}\ln(x^2 + y^2) + C_1(y) \\ \frac{my}{x^2 + y^2} &= \frac{\partial\phi}{\partial y} = \frac{my}{x^2 + y^2} + \frac{dC_1}{dy}.\end{aligned}$$

Thus we may take $C_1(y) = 0$, and obtain

$$\phi(x, y) = \frac{m}{2}\ln(x^2 + y^2) = m\ln|\mathbf{r}|,$$

as a scalar potential for the velocity field **v** of a line source of strength of $m$.

**16.** The equipotential curves for the two-dimensional dipole have equations $y = 0$ or

$$\begin{aligned}-\frac{\mu y}{x^2 + y^2} &= \frac{1}{C} \\ x^2 + y^2 + \mu C y &= 0 \\ x^2 + \left(y + \frac{\mu C}{2}\right)^2 &= \frac{\mu^2 C^2}{4}.\end{aligned}$$

These equipotentials are circles tangent to the $x$-axis at the origin.

**18.** The velocity field for a point source of strength $m\,dt$ at $(0, 0, t)$ is

$$\mathbf{v}_t(x, y, z) = \frac{m\big(x\mathbf{i} + y\mathbf{j} + (z - t)\mathbf{k}\big)}{\big(x^2 + y^2 + (z - t)^2\big)^{3/2}}.$$

Hence we have

$$\begin{aligned}
&\int_{-\infty}^{\infty} \mathbf{v}_t(x, y, z)\,dt \\
&= m\int_{-\infty}^{\infty} \frac{x\mathbf{i} + y\mathbf{j} + (z-t)\mathbf{k}}{\left(x^2+y^2+(z-t)^2\right)^{3/2}}\,dt \\
&= m(x\mathbf{i} + y\mathbf{j})\int_{-\infty}^{\infty} \frac{dt}{\left(x^2+y^2+(z-t)^2\right)^{3/2}} \\
&\qquad \text{Let } z - t = \sqrt{x^2+y^2}\tan\theta \\
&\qquad -dt = \sqrt{x^2+y^2}\sec^2\theta\,d\theta \\
&= \frac{m(x\mathbf{i} + y\mathbf{j})}{x^2+y^2}\int_{-\pi/2}^{\pi/2} \cos\theta\,d\theta \\
&= \frac{2m(x\mathbf{i} + y\mathbf{j})}{x^2+y^2},
\end{aligned}$$

which is the velocity field of a line source of strength $2m$ along the $z$-axis.

The definition of strength of a point source in 3-space was made to ensure that the velocity field of a source of strength 1 had speed 1 at distance 1 from the source. This corresponds to fluid being emitted from the source at a volume rate of $4\pi$. Similarly, the definition of strength of a line source guaranteed that a source of strength 1 gives rise to fluid speed of 1 at unit distance 1 from the line source. This corresponds to a fluid emission at a volume rate $2\pi$ per unit length along the line. Thus, the integral of a 3-dimensional source gives twice the volume rate of a 2-dimensional source, per unit length along the line.

The potential of a point source $m\,dt$ at $(0, 0, t)$ is

$$\phi(x, y, z) = -\frac{m}{\sqrt{x^2+y^2+(x-t)^2}}.$$

This potential cannot be integrated to give the potential for a line source along the $z$-axis because the integral

$$-m\int_{-\infty}^{\infty} \frac{dt}{\sqrt{x^2+y^2+(z-t)^2}}$$

does not converge, in the usual sense in which convergence of improper integrals was defined.

**20.** If $\mathbf{F} = F_r(r,\theta)\hat{\mathbf{r}} + F_\theta(r,\theta)\hat{\boldsymbol{\theta}}$ is conservative, then $\mathbf{F} = \nabla\phi$ for some scalar field $\phi(r,\theta)$, and by Exercise 19,

$$\frac{\partial\phi}{\partial r} = F_r, \quad \frac{1}{r}\frac{\partial\phi}{\partial\theta} = F_\theta.$$

For the equality of the mixed second partial derivatives of $\phi$, we require that

$$\frac{\partial F_r}{\partial\theta} = \frac{\partial}{\partial r}(rF_\theta) = F_\theta + r\frac{\partial F_\theta}{\partial r},$$

that is, $\displaystyle\frac{\partial F_r}{\partial\theta} - r\frac{\partial F_\theta}{\partial r} = F_\theta.$

**22.** If $\mathbf{F} = r^2\cos\theta\hat{\mathbf{r}} + \alpha r^\beta\sin\theta\hat{\boldsymbol{\theta}} = \nabla\phi(r,\theta)$, then we must have

$$\frac{\partial\phi}{\partial r} = r^2\cos\theta, \quad \frac{1}{r}\frac{\partial\phi}{\partial\theta} = \alpha r^\beta\sin\theta.$$

From the first equation

$$\phi(r,\theta) = \frac{r^3}{3}\cos\theta + C(\theta).$$

The second equation then gives

$$C'(\theta) - \frac{r^3}{3}\sin\theta = \frac{\partial\phi}{\partial\theta} = \alpha r^{\beta+1}\sin\theta.$$

This equation can be solved for a function $C(\theta)$ independent of $r$ only if $\alpha = -1/3$ and $\beta = 2$. In this case, $C(\theta) = C$ (a constant). $\mathbf{F}$ is conservative if $\alpha$ and $\beta$ have these values, and a potential for it is $\phi = \frac{1}{3}r^3\cos\theta + C$.

## Section 8.3 Line Integrals (page 390)

**2.** $C$: $x = t\cos t$, $y = t\sin t$, $z = t$, $(0 \le t \le 2\pi)$. We have

$$\begin{aligned}
ds &= \sqrt{(\cos t - t\sin t)^2 + (\sin t + t\cos t)^2 + 1}\,dt \\
&= \sqrt{2+t^2}\,dt.
\end{aligned}$$

Thus

$$\begin{aligned}
\int_C z\,ds &= \int_0^{2\pi} t\sqrt{2+t^2}\,dt \quad \text{Let } u = 2+t^2 \\
&\qquad\qquad\qquad\qquad\quad du = 2t\,dt \\
&= \frac{1}{2}\int_2^{2+4\pi^2} u^{1/2}\,du \\
&= \frac{1}{3}u^{3/2}\Big|_2^{2+4\pi^2} = \frac{(2+4\pi^2)^{3/2} - 2^{3/2}}{3}.
\end{aligned}$$

**4.** The wire of Example 3 lies in the first octant on the surfaces $z = x^2$ and $z = 2 - x^2 - 2y^2$, and, therefore, also on the surface $x^2 = 2 - x^2 - 2y^2$, or $x^2 + y^2 = 1$, a circular cylinder. Since it goes from $(1, 0, 1)$ to $(0, 1, 0)$ it can be parametrized

$$\begin{aligned}
\mathbf{r} &= \cos t\mathbf{i} + \sin t\mathbf{j} + \cos^2\mathbf{k}, \quad (0 \le t \le \pi/2) \\
\mathbf{v} &= -\sin t\mathbf{i} + \cos t\mathbf{j} - 2\cos t\sin t\mathbf{k} \\
v &= \sqrt{1+\sin^2(2t)} = \sqrt{2-\cos^2(2t)}.
\end{aligned}$$

Since the wire has density $\delta = xy = \sin t \cos t = \frac{1}{2}\sin(2t)$, its mass is

$$m = \frac{1}{2}\int_0^{\pi/2} \sqrt{2 - \cos^2(2t)}\sin(2t)\,dt \quad \text{Let } v = \cos(2t),\ dv = -2\sin(2t)\,dt$$
$$= \frac{1}{4}\int_{-1}^{1}\sqrt{2-v^2}\,dv = \frac{1}{2}\int_0^1 \sqrt{2-v^2}\,dv,$$

which is the same integral obtained in Example 3, and has value $(\pi + 2)/8$.

**6.** $C$ is the same curve as in Exercise 5. We have

$$\int_C e^z\,ds = \int_0^{2\pi} e^t\sqrt{1+2e^{2t}}\,dt \quad \text{Let } \sqrt{2}e^t = \tan\theta,\ \sqrt{2}e^t\,dt = \sec^2\theta\,d\theta$$
$$= \frac{1}{\sqrt{2}}\int_{t=0}^{t=2\pi} \sec^3\theta\,d\theta$$
$$= \frac{1}{2\sqrt{2}}\Big[\sec\theta\tan\theta + \ln|\sec\theta + \tan\theta|\Big]\Bigg|_{t=0}^{t=2\pi}$$
$$= \frac{\sqrt{2}e^t\sqrt{1+2e^{2t}} + \ln(\sqrt{2}e^t + \sqrt{1+2e^{2t}})}{2\sqrt{2}}\Bigg|_0^{2\pi}$$
$$= \frac{e^{2\pi}\sqrt{1+2e^{4\pi}} - \sqrt{3}}{2} + \frac{1}{2\sqrt{2}}\ln\frac{\sqrt{2}e^{2\pi} + \sqrt{1+2e^{4\pi}}}{\sqrt{2}+\sqrt{3}}.$$

**8.** The curve $C$ of intersection of $x^2 + z^2 = 1$ and $y = x^2$ can be parametrized

$$\mathbf{r} = \cos t\,\mathbf{i} + \cos^2 t\,\mathbf{j} + \sin t\,\mathbf{k}, \quad (0 \le t \le 2\pi).$$

Thus

$$ds = \sqrt{\sin^2 t + 4\sin^2 t\cos^2 t + \cos^2 t}\,dt = \sqrt{1+\sin^2 2t}\,dt.$$

We have

$$\int_C \sqrt{1+4x^2z^2}\,ds$$
$$= \int_0^{2\pi} \sqrt{1+4\cos^2 t\sin^2 t}\sqrt{1+\sin^2 2t}\,dt$$
$$= \int_0^{2\pi} (1+\sin^2 2t)\,dt$$
$$= \int_0^{2\pi}\left(1 + \frac{1-\cos 4t}{2}\right)dt$$
$$= \frac{3}{2}(2\pi) = 3\pi.$$

**10.** Here the wire of Exercise 9 extends only from $t = 0$ to $t = \pi$:

$$m = \sqrt{2}\int_0^{\pi} t\,dt = \frac{\pi^2\sqrt{2}}{2}$$
$$M_{x=0} = \sqrt{2}\int_0^{\pi} t\cos t\,dt = -2\sqrt{2}$$
$$M_{y=0} = \sqrt{2}\int_0^{\pi} t\sin t\,dt = \pi\sqrt{2}$$
$$M_{z=0} = \sqrt{2}\int_0^{\pi} t^2\,dt = \frac{\pi^3\sqrt{2}}{3}.$$

The centre of mass is $\left(-\frac{4}{\pi^2}, \frac{2}{\pi}, \frac{2\pi}{3}\right)$.

**12.**
$$m = \int_0^1 (e^t + e^{-t})\,dt = \frac{e^2-1}{e}$$
$$M_{x=0} = \int_0^1 e^t(e^t + e^{-t})\,dt = \frac{e^2+1}{2}$$
$$M_{y=0} = \int_0^1 \sqrt{2}t(e^t + e^{-t})\,dt = \frac{2\sqrt{2}(e-1)}{e}$$
$$M_{z=0} = \int_0^1 e^{-t}(e^t + e^{-t})\,dt = \frac{3e^2-1}{2e^2}$$

The centroid is $\left(\frac{e^3+e}{2e^2-2}, \frac{2\sqrt{2}}{e+1}, \frac{3e^2-1}{2e^3-2e}\right)$.

**14.** On $C$, we have

$$z = \sqrt{1-x^2-y^2} = \sqrt{1-x^2-(1-x)^2} = \sqrt{2(x-x^2)}.$$

Thus $C$ can be parametrized

$$\mathbf{r} = t\mathbf{i} + (1-t)\mathbf{j} + \sqrt{2(t-t^2)}\mathbf{k}, \quad (0 \le t \le 1).$$

Hence

$$ds = \sqrt{1+1+\frac{(1-2t)^2}{2(t-t^2)}}\,dt = \frac{dt}{\sqrt{2(t-t^2)}}.$$

We have

$$\int_C z\,ds = \int_0^1 \sqrt{2(t-t^2)}\frac{dt}{\sqrt{2(t-t^2)}} = 1.$$

**16.** $C$: $y = x^2$, $z = y^2$, from $(0,0,0)$ to $(2,4,16)$. Parametrize $C$ by

$$\mathbf{r} = t\mathbf{i} + t^2\mathbf{j} + t^4\mathbf{k}, \quad (0 \le t \le 2).$$

Since $ds = \sqrt{1+4t^2+16t^6}\,dt$, we have

$$\int_C xyz\,ds = \int_0^2 t^7\sqrt{1+4t^2+16t^6}\,dt.$$

**18.** The straight line $\mathcal{L}$ with equation $Ax + By = C$, $(C \neq 0)$, lies at distance $D = \sqrt{|C|}/\sqrt{A^2+B^2}$ from the origin. So does the line $\mathcal{L}_1$ with equation $y = D$. Since $x^2 + y^2$ depends only on distance from the origin, we have, by symmetry,

$$\begin{aligned}\int_{\mathcal{L}} \frac{ds}{x^2+y^2} &= \int_{\mathcal{L}_1} \frac{ds}{x^2+y^2} \\ &= \int_{-\infty}^{\infty} \frac{dx}{x^2+D^2} \\ &= \frac{2}{D}\tan^{-1}\frac{x}{D}\bigg|_0^{\infty} = \frac{2}{D}\left(\frac{\pi}{2} - 0\right) \\ &= \frac{\pi}{D} = \frac{\pi\sqrt{A^2+B^2}}{|C|}.\end{aligned}$$

## Section 8.4 Line Integrals of Vector Fields (page 397)

**2.** $\mathbf{F} = \cos x\mathbf{i} - y\mathbf{j} = \nabla\left(\sin x - \frac{y^2}{2}\right)$.

$\mathcal{C}$: $y = \sin x$ from (0,0) to $(\pi, 0)$.

$$\int_{\mathcal{C}} \mathbf{F} \bullet d\mathbf{r} = \left(\sin x - \frac{y^2}{2}\right)\bigg|_{(0,0)}^{(\pi,0)} = 0.$$

**4.** $\mathbf{F} = z\mathbf{i} - y\mathbf{j} + 2x\mathbf{k}$.
$\mathcal{C}$: $\mathbf{r} = t\mathbf{i} + t^2\mathbf{j} + t^3\mathbf{k}$, $(0 \le t \le 1)$.

$$\begin{aligned}\int_{\mathcal{C}} \mathbf{F} \bullet d\mathbf{r} &= \int_0^1 [t^3 - t^2(2t) + 2t(3t^2)]\,dt \\ &= \int_0^1 5t^3\,dt = \frac{5t^4}{4}\bigg|_0^1 = \frac{5}{4}.\end{aligned}$$

**6.** $\mathbf{F} = (x-z)\mathbf{i} + (y-z)\mathbf{j} - (x+y)\mathbf{k}$
$= \nabla\left(\frac{x^2+y^2}{2} - (x+y)z\right)$.

$\mathcal{C}$ is a given polygonal path from (0,0,0) to (1,1,1) (but any other piecewise smooth path from the first point to the second would do as well).

$$\int_{\mathcal{C}} \mathbf{F} \bullet d\mathbf{r} = \left(\frac{x^2+y^2}{2} - (x+y)z\right)\bigg|_{(0,0,0)}^{(1,1,1)} = 1 - 2 = -1.$$

**8.**

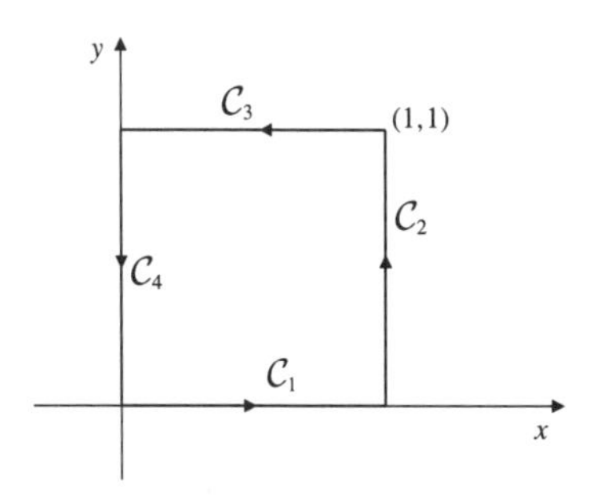


Fig. 8.4.8

$\mathcal{C}$ is made up of four segments as shown in the figure.
On $\mathcal{C}_1$, $y = 0$, $dy = 0$, and $x$ goes from 0 to 1.
On $\mathcal{C}_2$, $x = 1$, $dx = 0$, and $y$ goes from 0 to 1.
On $\mathcal{C}_3$, $y = 1$, $dy = 0$, and $x$ goes from 1 to 0.
On $\mathcal{C}_4$, $x = 0$, $dx = 0$, and $y$ goes from 1 to 0.
Thus

$$\int_{\mathcal{C}_1} x^2y^2\,dx + x^3y\,dy = 0$$
$$\int_{\mathcal{C}_2} x^2y^2\,dx + x^3y\,dy = \int_0^1 y\,dy = \frac{1}{2}$$
$$\int_{\mathcal{C}_3} x^2y^2\,dx + x^3y\,dy = \int_1^0 x^2\,dx = -\frac{1}{3}$$
$$\int_{\mathcal{C}_4} x^2y^2\,dx + x^3y\,dy = 0.$$

Finally, therefore,

$$\int_{\mathcal{C}} x^2y^2\,dx + x^3y\,dy = 0 + \frac{1}{2} - \frac{1}{3} + 0 = \frac{1}{6}.$$

**10.** $\mathbf{F} = (axy+z)\mathbf{i} + x^2\mathbf{j} + (bx+2z)\mathbf{k}$ is conservative if

$$\begin{aligned}\frac{\partial F_1}{\partial y} &= \frac{\partial F_2}{\partial x} &&\Leftrightarrow\quad a = 2 \\ \frac{\partial F_1}{\partial z} &= \frac{\partial F_3}{\partial x} &&\Leftrightarrow\quad b = 1 \\ \frac{\partial F_2}{\partial z} &= \frac{\partial F_3}{\partial y} &&\Leftrightarrow\quad 0 = 0.\end{aligned}$$

If $a = 2$ and $b = 1$, then $\mathbf{F} = \nabla\phi$ where

$$\begin{aligned}\phi &= \int (2xy+z)\,dx = x^2y + xz + C_2(y,z) \\ \frac{\partial C_1}{\partial y} + x^2 &= F_2 = x^2 \quad\Rightarrow\quad C_1(y,z) = C_2(z) \\ \frac{dC_2}{dz} + x &= F_3 = x + 2z \quad\Rightarrow\quad C_2(z) = z^2 + C.\end{aligned}$$

Thus $\phi = x^2y + xz + z^2 + C$ is a potential for $\mathbf{F}$.

**12.** $\mathbf{F} = (y^2 \cos x + z^3)\mathbf{i} + (2y \sin x - 4)\mathbf{j} + (3xz^2 + 2)\mathbf{k}$
$= \nabla(y^2 \sin x + xz^3 - 4y + 2z)$.
The curve $\mathcal{C}$: $x = \sin^{-1} t$, $y = 1 - 2t$, $z = 3t - 1$, $(0 \le t \le 1)$, goes from $(0, 1, -1)$ to $(\pi/2, -1, 2)$. The work done by $\mathbf{F}$ in moving a particle along $\mathcal{C}$ is

$$\begin{aligned} W &= \int_{\mathcal{C}} \mathbf{F} \bullet d\mathbf{r} \\ &= (y^2 \sin x + xz^3 - 4y + 2z)\Big|_{(0,1,-1)}^{(\pi/2,-1,2)} \\ &= 1 + 4\pi + 4 + 4 - 0 - 0 + 4 + 2 = 15 + 4\pi. \end{aligned}$$

**14.** a) $S = \{(x, y) : x > 0, y \ge 0\}$ is a simply connected domain.

b) $S = \{(x, y) : x = 0, y \ge 0\}$ is not a domain. (It has empty interior.)

c) $S = \{(x, y) : x \ne 0, y > 0\}$ is a domain but is not connected. There is no path in $S$ from $(-1, 1)$ to $(1, 1)$.

d) $S = \{(x, y, z) : x^2 > 1\}$ is a domain but is not connected. There is no path in $S$ from $(-2, 0, 0)$ to $(2, 0, 0)$.

e) $S = \{(x, y, z) : x^2 + y^2 > 1\}$ is a connected domain but is not simply connected. The circle $x^2 + y^2 = 2$, $z = 0$ lies in $S$, but cannot be shrunk through $S$ to a point since it surrounds the cylinder $x^2 + y^2 \le 1$ which is outside $S$.

f) $S = \{(x, y, z) : x^2 + y^2 + z^2 > 1\}$ is a simply connected domain even though it has a ball-shaped "hole" in it.

**16.** $\mathcal{C}$ is the curve $\mathbf{r} = a \cos t\mathbf{i} + b \sin t\mathbf{j}$, $(0 \le t \le 2\pi)$.

$$\oint_{\mathcal{C}} x\,dy = \int_0^{2\pi} a \cos t\, b \cos t\, dt = \pi ab$$
$$\oint_{\mathcal{C}} y\,dx = \int_0^{2\pi} b \sin t(-a \sin t)\, dt = -\pi ab.$$

**18.** $\mathcal{C}$ is made up of four segments as shown in the figure.
On $\mathcal{C}_1$, $y = 0$, $dy = 0$, and $x$ goes from 0 to 1.
On $\mathcal{C}_2$, $x = 1$, $dx = 0$, and $y$ goes from 0 to 1.
On $\mathcal{C}_3$, $y = 1$, $dy = 0$, and $x$ goes from 1 to 0.
On $\mathcal{C}_4$, $x = 0$, $dx = 0$, and $y$ goes from 1 to 0.

$$\begin{aligned} \oint_{\mathcal{C}} x\,dy &= \int_{\mathcal{C}_1} + \int_{\mathcal{C}_2} + \int_{\mathcal{C}_3} + \int_{\mathcal{C}_4} \\ &= 0 + \int_0^1 dy + 0 + 0 = 1 \end{aligned}$$
$$\begin{aligned} \oint_{\mathcal{C}} y\,dx &= \int_{\mathcal{C}_1} + \int_{\mathcal{C}_2} + \int_{\mathcal{C}_3} + \int_{\mathcal{C}_4} \\ &= 0 + 0 + \int_1^0 dx + 0 = -1. \end{aligned}$$

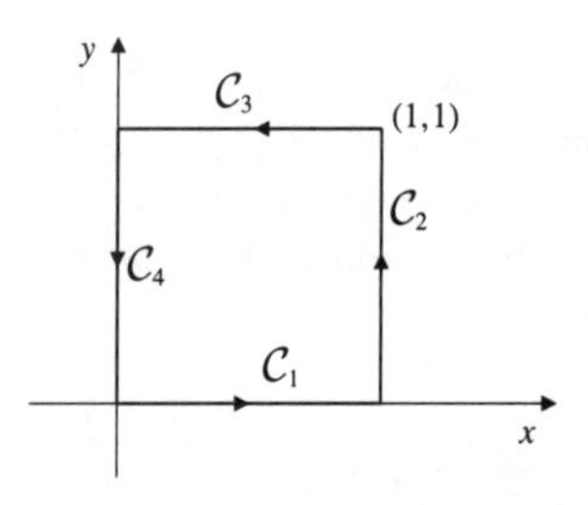


Fig. 8.4.18

**20.** Conjecture: If $D$ is a domain in $\mathbb{R}^2$ whose boundary is a closed, non-self-intersecting curve $\mathcal{C}$, oriented counter-clockwise, then

$$\oint_{\mathcal{C}} x\,dy = \text{area of } D,$$
$$\oint_{\mathcal{C}} y\,dx = -\text{ area of } D.$$

Proof for a domain $D$ that is $x$-simple and $y$-simple: Since $D$ is $x$-simple, it can be specified by the inequalities

$$c \le y \le d, \qquad f(y) \le x \le g(y).$$

Let $\mathcal{C}$ consist of the four parts shown in the figure. On $\mathcal{C}_1$ and $\mathcal{C}_3$, $dy = 0$.
On $\mathcal{C}_2$, $x = g(y)$, where $y$ goes from $c$ to $d$.
On $\mathcal{C}_2$, $x = f(y)$, where $y$ goes from $d$ to $c$. Thus

$$\begin{aligned} \oint_{\mathcal{C}} x\,dy &= \int_{\mathcal{C}_1} + \int_{\mathcal{C}_2} + \int_{\mathcal{C}_3} + \int_{\mathcal{C}_4} \\ &= 0 + \int_c^d g(y)\,dy + 0 + \int_d^c f(y)\,dy \\ &= \big(g(y) - f(y)\big)\,dy = \text{area of } D. \end{aligned}$$

The proof that $\oint_{\mathcal{C}} y\,dx = -(\text{area of } D)$ is similar, and uses the fact that $D$ is $y$-simple.

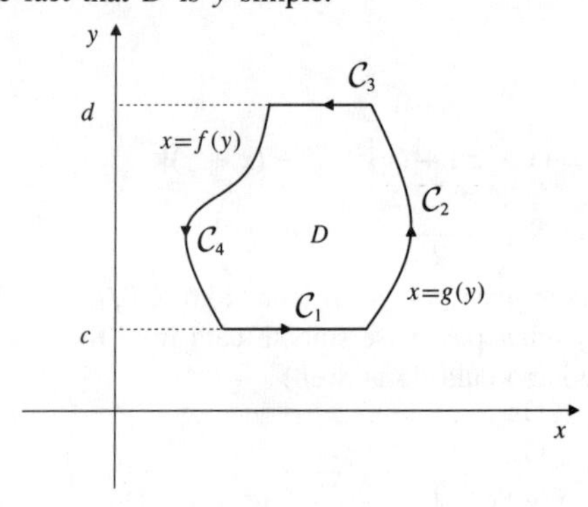


Fig. 8.4.20

**22.** a) $\mathcal{C}$: $x = a\cos t$, $x = a\sin t$, $0 \le t \le 2\pi$.

$$\frac{1}{2\pi}\oint_{\mathcal{C}} \frac{x\,dy - y\,dx}{x^2+y^2}$$
$$= \frac{1}{2\pi}\int_0^{2\pi} \frac{a^2\cos^2 t + a^2\sin^2 t}{a^2\cos^2 t + a^2\sin^2 t}\,dt = 1.$$

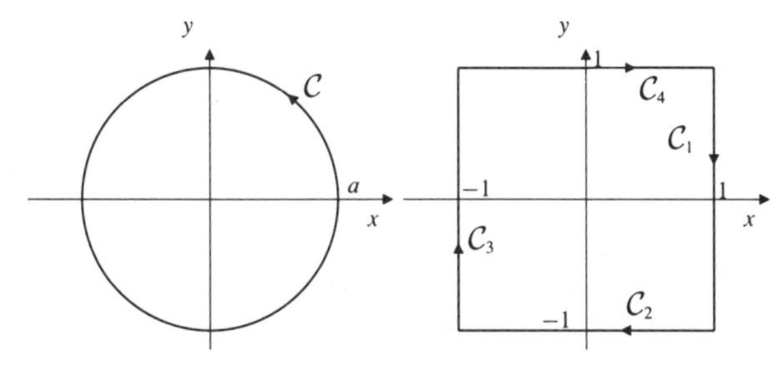


Fig. 8.4.22(a) Fig. 8.4.22(b)

b) See the figure. $\mathcal{C}$ has four parts.
On $\mathcal{C}_1$, $x = 1$, $dx = 0$, $y$ goes from 1 to $-1$.
On $\mathcal{C}_2$, $y = -1$, $dy = 0$, $x$ goes from 1 to $-1$.
On $\mathcal{C}_3$, $x = -1$, $dx = 0$, $y$ goes from $-1$ to 1.
On $\mathcal{C}_4$, $x = 1$, $dx = 0$, $y$ goes from 1 to $-1$.

$$\frac{1}{2\pi}\oint_{\mathcal{C}} \frac{x\,dy - y\,dx}{x^2+y^2}$$
$$= \frac{1}{2\pi}\left[\int_1^{-1}\frac{dy}{1+y^2} + \int_1^{-1}\frac{dx}{x^2+1}\right.$$
$$\left.\int_{-1}^{1}\frac{-dy}{1+y^2} + \int_{-1}^{1}\frac{-dx}{x^2+1}\right]$$
$$= -\frac{2}{\pi}\int_{-1}^{1}\frac{dt}{1+t^2}$$
$$= -\frac{2}{\pi}\tan^{-1}t\Big|_{-1}^{1} = -\frac{2}{\pi}\left(\frac{\pi}{4}+\frac{\pi}{4}\right) = -1.$$

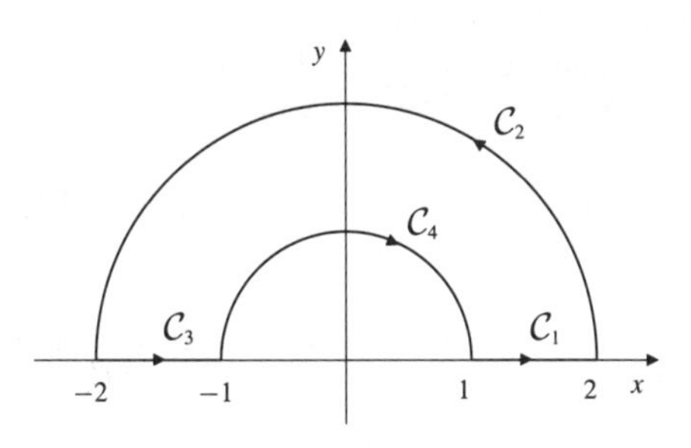


Fig. 8.4.22

c) See the figure. $\mathcal{C}$ has four parts.
On $\mathcal{C}_1$, $y = 0$, $dy = 0$, $x$ goes from 1 to 2.
On $\mathcal{C}_2$, $x = 2\cos t$, $y = 2\sin t$, $t$ goes from 0 to $\pi$.
On $\mathcal{C}_3$, $y = 0$, $dy = 0$, $x$ goes from $-2$ to $-1$.
On $\mathcal{C}_4$, $x = \cos t$, $y = \sin t$, $t$ goes from $\pi$ to 0.

$$\frac{1}{2\pi}\oint_{\mathcal{C}} \frac{x\,dy - y\,dx}{x^2+y^2}$$
$$= \frac{1}{2\pi}\left[0 + \int_0^{\pi}\frac{4\cos^2 t + 4\sin^2 t}{4\cos^2 t + 4\sin^2 t}\,dt\right.$$
$$\left. + 0 + \int_{\pi}^{0}\frac{\cos^2 t + \sin^2 t}{\cos^2 t + \sin^2 t}\,dt\right]$$
$$= \frac{1}{2\pi}(\pi - \pi) = 0.$$

**24.** If $\mathcal{C}$ is a closed, piecewise smooth curve in $\mathbb{R}^2$ having equation $\mathbf{r} = \mathbf{r}(t)$, $a \le t \le b$, and if $\mathcal{C}$ does not pass through the origin, then the polar angle function $\theta = \theta\big(x(t), y(t)\big) = \theta(t)$ can be defined so as to vary continuously on $\mathcal{C}$. Therefore,

$$\theta(x,y)\Big|_{t=a}^{t=b} = 2\pi \times w(\mathcal{C}),$$

where $w(\mathcal{C})$ is the number of times $\mathcal{C}$ winds around the origin in a counterclockwise direction. For example, $w(\mathcal{C})$ equals 1, $-1$ and 0 respectively, for the curves $\mathcal{C}$ in parts (a), (b) and (c) of Exercise 22. Since

$$\nabla\theta = \frac{\partial\theta}{\partial x}\mathbf{i} + \frac{\partial\theta}{\partial y}\mathbf{j}$$
$$= \frac{-y\mathbf{i} + x\mathbf{j}}{x^2+y^2},$$

we have

$$\frac{1}{2\pi}\oint_{\mathcal{C}} \frac{x\,dy - y\,dx}{x^2+y^2} = \frac{1}{2\pi}\oint_{\mathcal{C}} \nabla\theta \bullet d\mathbf{r}$$
$$= \frac{1}{2\pi}\theta(x,y)\Big|_{t=a}^{t=b} = w(\mathcal{C}).$$

## Section 8.5 Surfaces and Surface Integrals (page 409)

**2.**

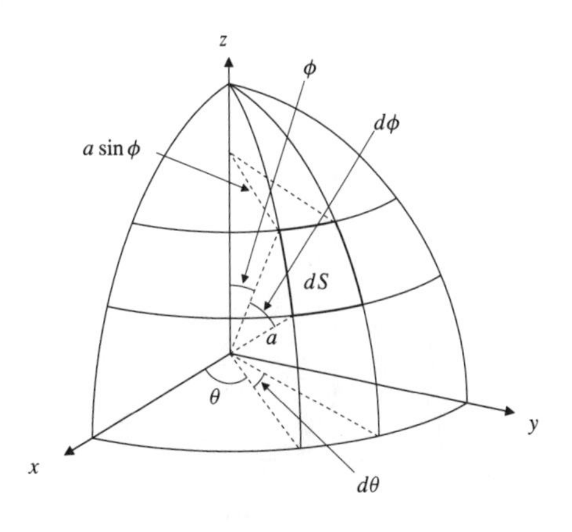


Fig. 8.5.2

The area element $dS$ is bounded by the curves in which the coordinate planes at $\theta$ and $\theta + d\theta$ and the coordinate cones at $\phi$ and $\phi + d\phi$ intersect the sphere $R = a$. (See the figure.) The element is rectangular with sides $a\,d\phi$ and $a\sin\phi\,d\theta$. Thus

$$dS = a^2 \sin\phi\,d\phi\,d\theta.$$

**4.** One-quarter of the required area is shown in the figure. It lies above the semicircular disk $R$ bounded by $x^2 + y^2 = 2ay$, or, in terms of polar coordinates, $r = 2a\sin\theta$. On the sphere $x^2 + y^2 + z^2 = 4a^2$, we have

$$2z\frac{\partial z}{\partial x} = -2x, \qquad \text{or} \qquad \frac{\partial z}{\partial x} = -\frac{x}{z}.$$

Similarly, $\dfrac{\partial z}{\partial y} = -\dfrac{y}{z}$, so the surface area element on the sphere can be written

$$dS = \sqrt{1 + \frac{x^2 + y^2}{z^2}}\,dx\,dy = \frac{2a\,dx\,dy}{\sqrt{4a^2 - x^2 - y^2}}.$$

The required area is

$$\begin{aligned}
S &= 4\iint_R \frac{2a}{\sqrt{4a^2 - x^2 - y^2}}\,dx\,dy \\
&= 8a\int_0^{\pi/2} d\theta \int_0^{2a\sin\theta} \frac{r\,dr}{\sqrt{4a^2 - r^2}} \qquad \begin{array}{l}\text{Let } u = 4a^2 - r^2 \\ du = -2r\,dr\end{array} \\
&= 4a\int_0^{\pi/2} d\theta \int_{4a^2\cos^2\theta}^{4a^2} u^{-1/2}\,du \\
&= 8a\int_0^{\pi/2} (2a - 2a\cos\theta)\,d\theta \\
&= 16a^2(\theta - \sin\theta)\Big|_0^{\pi/2} = 8a^2(\pi - 2) \text{ sq. units.}
\end{aligned}$$

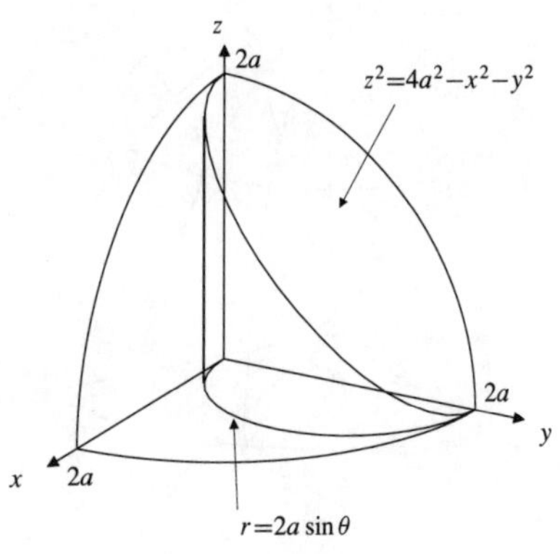


Fig. 8.5.4

**6.** The cylinder $x^2 + y^2 = 2ay$ intersects the sphere $x^2+y^2+z^2 = 4a^2$ on the parabolic cylinder $2ay+z^2 = 4a^2$. By Exercise 5, the area element on $x^2 + y^2 - 2ay = 0$ is

$$\begin{aligned}
dS &= \left|\frac{2x\mathbf{i} + (2y - 2a)\mathbf{j}}{2x}\right| dy\,dz \\
&= \sqrt{1 + \frac{(y-a)^2}{2ay - y^2}}\,dy\,dz \\
&= \sqrt{\frac{2ay - y^2 + y^2 - 2ay + a^2}{2ay - y^2}}\,dy\,dz = \frac{a}{\sqrt{2ay - y^2}}\,dy\,dz.
\end{aligned}$$

The area of the part of the cylinder inside the sphere is 4 times the part shown in Figure 8.23 in the text, that is, 4 times the double integral of $dS$ over the region $0 \le y \le 2a$, $0 \le z \le \sqrt{4a^2 - 2ay}$, or

$$\begin{aligned}
S &= 4\int_0^{2a} \frac{a\,dy}{\sqrt{2ay - y^2}} \int_0^{\sqrt{4a^2 - 2ay}} dz \\
&= 4a\int_0^{2a} \frac{\sqrt{2a(2a - y)}}{\sqrt{y(2a - y)}}\,dy = 4\sqrt{2}a^{3/2}\int_0^{2a} \frac{dy}{\sqrt{y}} \\
&= 4\sqrt{2}a^{3/2}(2\sqrt{y})\Big|_0^{2a} = 16a^2 \text{ sq. units.}
\end{aligned}$$

**8.** The normal to the cone $z^2 = x^2 + y^2$ makes a 45° angle with the vertical, so $dS = \sqrt{2}\,dx\,dy$ is a surface area element for the cone. Both *nappes* (halves) of the cone pass through the interior of the cylinder $x^2 + y^2 = 2ay$, so the area of that part of the cone inside the cylinder is $2\sqrt{2}\pi a^2$ square units, since the cylinder has a circular cross-section of radius $a$.

**10.** One-eighth of the required area lies in the first octant, above the triangle $T$ with vertices $(0, 0, 0)$, $(a, 0, 0)$ and $(a, a, 0)$. (See the figure.)
The surface $x^2 + z^2 = a^2$ has normal $\mathbf{n} = x\mathbf{i} + z\mathbf{k}$, so an area element on it can be written

$$dS = \frac{|\mathbf{n}|}{|\mathbf{n} \bullet \mathbf{k}|}\,dx\,dy = \frac{a}{z}\,dx\,dy = \frac{a\,dx\,dy}{\sqrt{a^2 - x^2}}.$$

The area of the part of that cylinder lying inside the cylinder $y^2 + z^2 = a^2$ is

$$\begin{aligned}
S &= 8\iint_T \frac{a\,dx\,dy}{\sqrt{a^2 - x^2}} = 8a\int_0^a \frac{dx}{\sqrt{a^2 - x^2}}\int_0^x dy \\
&= 8a\int_0^a \frac{x\,dx}{\sqrt{a^2 - x^2}} \\
&= -8a\sqrt{a^2 - x^2}\Big|_0^a = 8a^2 \text{ sq. units.}
\end{aligned}$$

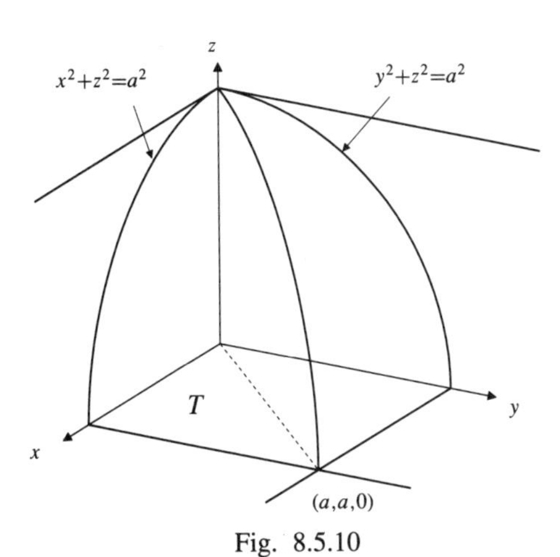

Fig. 8.5.10

**12.** We want to find $A_1$, the area of that part of the cylinder $x^2 + z^2 = a^2$ inside the cylinder $y^2 + z^2 = b^2$, and $A_2$, the area of that part of $y^2 + z^2 = b^2$ inside $x^2 + z^2 = a^2$. We have

$$A_1 = 8 \times (\text{area of } \mathcal{S}_1),$$
$$A_2 = 8 \times (\text{area of } \mathcal{S}_2),$$

where $\mathcal{S}_1$ and $\mathcal{S}_2$ are the parts of these surfaces lying in the first octant, as shown in the figure.
A normal to $\mathcal{S}_1$ is $\mathbf{n}_1 = x\mathbf{i} + z\mathbf{k}$, and the area element on $\mathcal{S}_1$ is

$$dS_1 = \frac{|\mathbf{n}_1|}{|\mathbf{n}_1 \bullet \mathbf{i}|}\,dy\,dz = \frac{a\,dy\,dz}{\sqrt{a^2 - z^2}}.$$

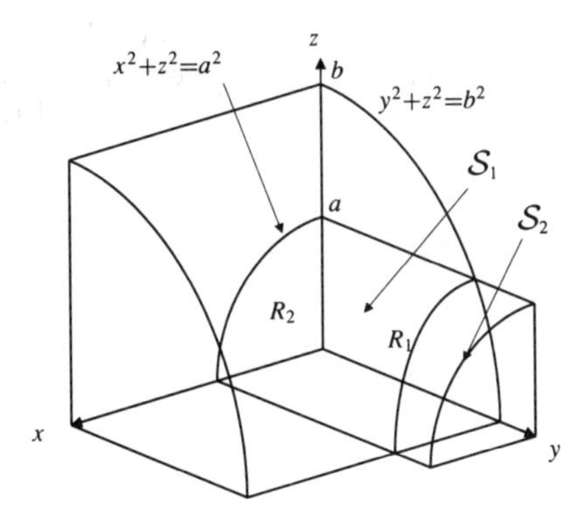

Fig. 8.5.12

A normal to $\mathcal{S}_2$ is $\mathbf{n}_2 = x\mathbf{j} + z\mathbf{k}$, and the area element on $\mathcal{S}_2$ is

$$dS_2 = \frac{|\mathbf{n}_2|}{|\mathbf{n}_2 \bullet \mathbf{j}|}\,dx\,dz = \frac{b\,dx\,dz}{\sqrt{b^2 - z^2}}.$$

Let $R_1$ be the region of the first quadrant of the $yz$-plane bounded by $y^2 + z^2 = b^2$, $y = 0$, $z = 0$, and $z = a$.
Let $R_2$ be the quarter-disk in the first quadrant of the $xz$-plane bounded by $x^2 + z^2 = a^2$, $x = 0$, and $z = 0$. Then

$$\begin{aligned}
A_1 &= 8\iint_{R_1} dS_1 = 8a\int_0^a \frac{dz}{\sqrt{a^2 - z^2}}\int_0^{\sqrt{b^2-z^2}} dy \\
&= 8a\int_0^a \frac{\sqrt{b^2 - z^2}}{\sqrt{a^2 - z^2}}\,dz \quad \text{Let } z = a\sin t \\
&\qquad\qquad\qquad\qquad\qquad dz = a\cos t\,dt \\
&= 8a\int_0^{\pi/2} \sqrt{b^2 - a^2\sin^2 t}\,dt \\
&= 8ab\int_0^{\pi/2} \sqrt{1 - \frac{a^2}{b^2}\sin^2 t}\,dt \\
&= 8abE\left(\frac{a}{b}\right) \text{ sq. units.}
\end{aligned}$$

$$\begin{aligned}
A_2 &= 8\iint_{R_2} dS_2 = 8b\int_0^a \frac{dz}{\sqrt{b^2 - z^2}}\int_0^{\sqrt{a^2-z^2}} dx \\
&= 8b\int_0^a \frac{\sqrt{a^2 - z^2}}{\sqrt{b^2 - z^2}}\,dz \quad \text{Let } z = b\sin t \\
&\qquad\qquad\qquad\qquad\qquad dz = b\cos t\,dt \\
&= 8b\int_0^{\sin^{-1}(a/b)} \sqrt{a^2 - b^2\sin^2 t}\,dt \\
&= 8ab\int_0^{\sin^{-1}(a/b)} \sqrt{1 - \frac{b^2}{a^2}\sin^2 t}\,dt \\
&= 8abE\left(\frac{b}{a}, \sin^{-1}\frac{a}{b}\right) \text{ sq. units.}
\end{aligned}$$

**14.** The cone $z = \sqrt{2(x^2 + y^2)}$ has area element

$$\begin{aligned}
dS &= \sqrt{1 + \left(\frac{\partial z}{\partial x}\right)^2 + \left(\frac{\partial z}{\partial y}\right)^2}\,dx\,dy \\
&= \sqrt{1 + \frac{4(x^2 + y^2)}{z^2}}\,dx\,dy = \sqrt{3}\,dx\,dy.
\end{aligned}$$

The projection of $\mathcal{S}$, the part of the cone lying below the plane $z = 1 + y$, on the $xy$-plane is an elliptic disk $E$ bounded by $x^2 + (y - 1)^2/2 = 1$. Thus

$$\iint_{\mathcal{S}} y\,dS = \sqrt{3}\iint_E y\,dx\,dy = \sqrt{3}A\bar{y} = \sqrt{6}\pi.$$

**16.** The surface $z = \sqrt{2xy}$ has area element

$$\begin{aligned}
dS &= \sqrt{1 + \frac{y}{2x} + \frac{x}{2y}}\,dx\,dy \\
&= \sqrt{\frac{2xy + y^2 + x^2}{2xy}}\,dx\,dy = \frac{|x + y|}{\sqrt{2xy}}\,dx\,dy.
\end{aligned}$$

If its density is $kz$, the mass of the specified part of the surface is

$$\begin{aligned} m &= \int_0^5 dx \int_0^2 k\sqrt{2xy}\,\frac{x+y}{\sqrt{2xy}}\,dy \\ &= k\int_0^5 dx \int_0^2 (x+y)\,dy \\ &= k\int_0^5 (2x+2)\,dx = 35k \text{ units.} \end{aligned}$$

**18.** The upper half of the spheroid $\dfrac{x^2}{a^2}+\dfrac{y^2}{a^2}+\dfrac{z^2}{c^2}=1$ has a circular disk of radius $a$ as projection onto the $xy$-plane. Since

$$\frac{2x}{a^2}+\frac{2z}{c^2}\frac{\partial z}{\partial x}=0 \quad\Rightarrow\quad \frac{\partial z}{\partial x}=-\frac{c^2x}{a^2z},$$

and, similarly, $\dfrac{\partial z}{\partial y}=-\dfrac{c^2y}{a^2z}$, the area element on the spheroid is

$$\begin{aligned} dS &= \sqrt{1+\frac{c^4}{a^4}\frac{x^2+y^2}{z^2}}\,dx\,dy \\ &= \sqrt{1+\frac{c^2}{a^2}\frac{x^2+y^2}{a^2-x^2-y^2}}\,dx\,dy \\ &= \sqrt{\frac{a^4+(c^2-a^2)r^2}{a^2(a^2-r^2)}}\,r\,dr\,d\theta \end{aligned}$$

in polar coordinates. Thus the area of the spheroid is

$$\begin{aligned} S &= \frac{2}{a}\int_0^{2\pi} d\theta \int_0^a \sqrt{\frac{a^4+(c^2-a^2)r^2}{a^2-r^2}}\,r\,dr \\ &\qquad \text{Let } u^2 = a^2-r^2 \\ &\qquad u\,du = -r\,dr \\ &= \frac{4\pi}{a}\int_0^a \sqrt{a^4+(c^2-a^2)(a^2-u^2)}\,du \\ &= \frac{4\pi}{a}\int_0^a \sqrt{a^2c^2-(c^2-a^2)u^2}\,du \\ &= 4\pi c\int_0^a \sqrt{1-\frac{c^2-a^2}{a^2c^2}u^2}\,du. \end{aligned}$$

For the case of a prolate spheroid $0 < a < c$, let $k^2=\dfrac{c^2-a^2}{a^2c^2}$. Then

$$\begin{aligned} S &= 4\pi c\int_0^a \sqrt{1-k^2u^2}\,du \quad \text{Let } ku=\sin v \\ &\qquad\qquad\qquad\qquad\qquad k\,du=\cos v\,dv \\ &= \frac{4\pi c}{k}\int_0^{\sin^{-1}(ka)} \cos^2 v\,dv \\ &= \frac{2\pi c}{k}(v+\sin v\cos v)\Big|_0^{\sin^{-1}(ka)} \\ &= \frac{2\pi ac^2}{\sqrt{c^2-a^2}}\sin^{-1}\frac{\sqrt{c^2-a^2}}{c}+2\pi a^2 \text{ sq. units.} \end{aligned}$$

**20.** $x = au\cos v,\quad y = au\sin v,\quad z = bv,$ $(0\le u\le 1,\quad 0\le v\le 2\pi)$. This surface is a spiral (helical) ramp of radius $a$ and height $2\pi b$, wound around the $z$-axis. (It's like a circular staircase with a ramp instead of stairs.) We have

$$\begin{aligned} \frac{\partial(x,y)}{\partial(u,v)} &= \begin{vmatrix} a\cos v & -au\sin v \\ a\sin v & au\cos v \end{vmatrix} = a^2u \\ \frac{\partial(y,z)}{\partial(u,v)} &= \begin{vmatrix} a\sin v & au\cos v \\ 0 & b \end{vmatrix} = ab\sin v \\ \frac{\partial(z,x)}{\partial(u,v)} &= \begin{vmatrix} 0 & b \\ a\cos v & -au\sin v \end{vmatrix} = -ab\cos v \\ dS &= \sqrt{a^4u^2+a^2b^2\sin^2 v+a^2b^2\cos^2 v}\,du\,dv \\ &= a\sqrt{a^2u^2+b^2}\,du\,dv. \end{aligned}$$

The area of the ramp is

$$\begin{aligned} A &= a\int_0^1 \sqrt{a^2u^2+b^2}\,du\int_0^{2\pi} dv \\ &= 2\pi a\int_0^1 \sqrt{a^2u^2+b^2}\,du \quad \text{Let } au=b\tan\theta \\ &\qquad\qquad\qquad\qquad\qquad a\,du=b\sec^2\theta\,d\theta \\ &= 2\pi b^2\int_{u=0}^{u=1} \sec^3\theta\,d\theta \\ &= \pi b^2\big(\sec\theta\tan\theta+\ln|\sec\theta+\tan\theta|\big)\Big|_{u=0}^{u=1} \\ &= \pi b^2\left(\frac{au\sqrt{a^2u^2+b^2}}{b^2}+\ln\left|\frac{au+\sqrt{a^2u^2+b^2}}{b}\right|\right)\Bigg|_0^1 \\ &= \pi a\sqrt{a^2+b^2}+\pi b^2\ln\left(\frac{a+\sqrt{a^2+b^2}}{b}\right) \text{ sq. units.} \end{aligned}$$

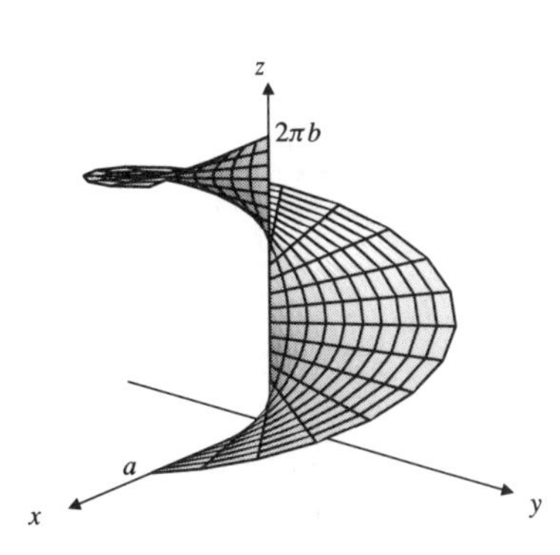


Fig. 8.5.20

**22.** Use spherical coordinates. The area of the eighth-sphere $\mathcal{S}$ is

$$A = \frac{1}{8}(4\pi a^2) = \frac{\pi a^2}{2} \text{ sq. units.}$$

The moment about $z = 0$ is

$$\begin{aligned} M_{z=0} &= \iint_{\mathcal{S}} z\, dS \\ &= \int_0^{\pi/2} d\theta \int_0^{\pi/2} a\cos\phi\, a^2 \sin\phi\, d\phi \\ &= \frac{\pi a^3}{2}\int_0^{\pi/2} \frac{\sin 2\phi}{2}\, d\phi = \frac{\pi a^3}{4}. \end{aligned}$$

Thus $\bar{z} = \dfrac{M_{z=0}}{A} = \dfrac{a}{2}$. By symmetry, $\bar{x} = \bar{y} = \bar{z}$, so the centroid of that part of the surface of the sphere $x^2 + y^2 + z^2 = a^2$ lying in the first octant is $\left(\dfrac{a}{2}, \dfrac{a}{2}, \dfrac{a}{2}\right)$.

**24.** By symmetry, the force of attraction of the hemisphere shown in the figure on the mass $m$ at the origin is vertical. The vertical component of the force exerted by area element $dS = a^2 \sin\phi\, d\phi\, d\theta$ at the position with spherical coordinates $(a, \phi, \theta)$ is

$$dF = \frac{km\sigma\, dS}{a^2}\cos\phi = km\sigma \sin\phi\cos\phi\, d\phi\, d\theta.$$

Thus, the total force on $m$ is

$$F = km\sigma \int_0^{2\pi} d\theta \int_0^{\pi/2} \sin\phi\cos\phi\, d\phi = \pi km\sigma \text{ units.}$$

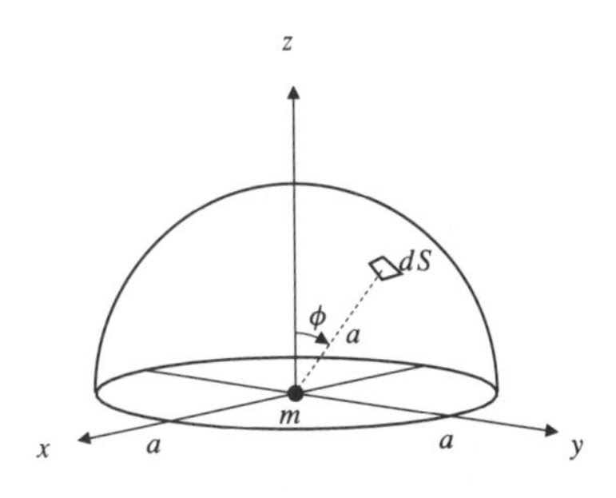


Fig. 8.5.24

**26.** $\mathcal{S}$ is the cylindrical surface $x^2 + y^2 = a^2$, $0 \le z \le h$, with areal density $\sigma$. Its mass is $m = 2\pi a h\sigma$. Since all surface elements are at distance $a$ from the $z$-axis, the radius of gyration of the cylindrical surface about the $z$-axis is $\bar{D} = a$. Therefore the moment of inertia about that axis is

$$I = m\bar{D}^2 = ma^2 = 2\pi\sigma a^3 h.$$

**28.** The surface area element for a conical surface $\mathcal{S}$,

$$z = h\left(1 - \frac{\sqrt{x^2 + y^2}}{a}\right),$$

having base radius $a$ and height $h$, was determined in the solution to Exercise 23 to be

$$dS = \frac{\sqrt{a^2 + h^2}}{a}\, dx\, dy.$$

The mass of $\mathcal{S}$, which has areal density $\sigma$, was also determined in that exercise: $m = \pi\sigma a\sqrt{a^2 + h^2}$. The moment of inertia of $\mathcal{S}$ about the $z$-axis is

$$\begin{aligned} I &= \sigma \iint_{\mathcal{S}} (x^2 + y^2)\, dS \\ &= \frac{\sigma\sqrt{a^2 + h^2}}{a}\int_0^{2\pi} d\theta \int_0^a r^2\, r\, dr \\ &= \frac{2\pi\sigma\sqrt{a^2 + h^2}}{a}\,\frac{a^4}{4} = \frac{\pi\sigma a^3\sqrt{a^2 + h^2}}{2}. \end{aligned}$$

The radius of gyration is $\bar{D} = \sqrt{I/m} = \dfrac{a}{\sqrt{2}}$.

## Section 8.6 Surface Integrals of Vector Fields (page 416)

**2.** On the sphere $\mathcal{S}$ with equation $x^2 + y^2 + z^2 = a^2$ we have

$$\hat{\mathbf{N}} = \frac{x\mathbf{i} + y\mathbf{j} + z\mathbf{k}}{a}.$$

If $\mathbf{F} = x\mathbf{i} + y\mathbf{j} + z\mathbf{k}$, then $\mathbf{F} \bullet \hat{\mathbf{N}} = a$ on $\mathcal{S}$. Thus the flux of $\mathbf{F}$ out of $\mathcal{S}$ is

$$\iint_{\mathcal{S}} \mathbf{F} \bullet \hat{\mathbf{N}}\, dS = a \times 4\pi a^2 = 4\pi a^3.$$

**4.** $\mathbf{F} = y\mathbf{i} + z\mathbf{k}$. Let $\mathcal{S}_1$ be the conical surface and $\mathcal{S}_2$ be the base disk. The flux of $\mathbf{F}$ outward through the surface of the cone is

$$\iint_{\mathcal{S}} \mathbf{F} \bullet \hat{\mathbf{N}} = \iint_{\mathcal{S}_1} + \iint_{\mathcal{S}_2}.$$

On $\mathcal{S}_1$: $\hat{\mathbf{N}} = \dfrac{1}{\sqrt{2}}\left(\dfrac{x\mathbf{i} + y\mathbf{j}}{\sqrt{x^2+y^2}} + \mathbf{k}\right)$, $dS = \sqrt{2}\,dx\,dy$.

Thus

$$\begin{aligned}
&\iint_{\mathcal{S}_1} \mathbf{F} \bullet \hat{\mathbf{N}}\, dS \\
&= \iint_{x^2+y^2\le 1} \left(\frac{xy}{\sqrt{x^2+y^2}} + 1 - \sqrt{x^2+y^2}\right) dx\,dy \\
&= 0 + \pi \times 1^2 - \int_0^{2\pi} d\theta \int_0^1 r^2\, dr \\
&= \pi - \frac{2\pi}{3} = \frac{\pi}{3}.
\end{aligned}$$

On $\mathcal{S}_2$: $\hat{\mathbf{N}} = -\mathbf{k}$ and $z = 0$, so $\mathbf{F} \bullet \hat{\mathbf{N}} = 0$. Thus, the total flux of $\mathbf{F}$ out of the cone is $\pi/3$.

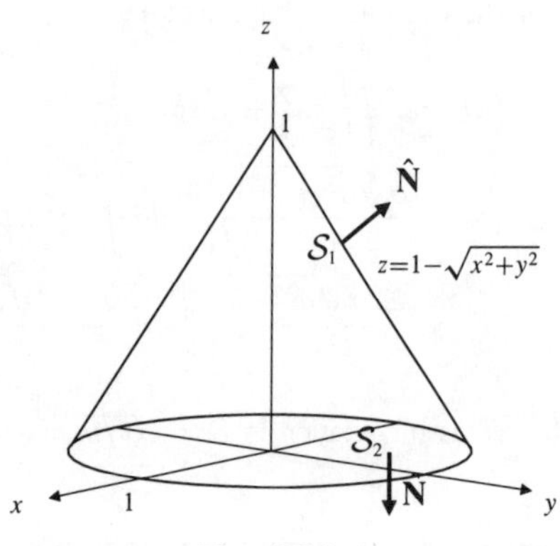


Fig. 8.6.4

**6.** For $z = x^2 - y^2$ the upward surface element is

$$\hat{\mathbf{N}}\, dS = \frac{-2x\mathbf{i} + 2y\mathbf{j} + \mathbf{k}}{1}\, dx\,dy.$$

The flux of $\mathbf{F} = x\mathbf{i} + x\mathbf{j} + \mathbf{k}$ upward through $\mathcal{S}$, the part of $z = x^2 - y^2$ inside $x^2 + y^2 = a^2$ is

$$\begin{aligned}
\iint_{\mathcal{S}} \mathbf{F} \bullet \hat{\mathbf{N}}\, dS &= \iint_{x^2+y^2\le a^2} (-2x^2 + 2xy + 1)\, dx\,dy \\
&= -2\int_0^{2\pi} \cos^2\theta\, d\theta \int_0^a r^3\, dr + 0 + \pi a^2 \\
&= \pi a^2 - 2(\pi)\frac{a^4}{4} = \frac{\pi}{2}a^2(2 - a^2).
\end{aligned}$$

**8.** The upward vector surface element on the top half of $x^2 + y^2 + z^2 = a^2$ is

$$\hat{\mathbf{N}}\, dS = \frac{2x\mathbf{i} + 2y\mathbf{j} + 2z\mathbf{k}}{2z}\, dx\,dy = \left(\frac{x\mathbf{i} + y\mathbf{j}}{z} + \mathbf{k}\right) dx\,dy.$$

The flux of $\mathbf{F} = z^2\mathbf{k}$ upward through the first octant part $\mathcal{S}$ of the sphere is

$$\iint_{\mathcal{S}} \mathbf{F} \bullet \hat{\mathbf{N}}\, dS = \int_0^{\pi/2} d\theta \int_0^a (a^2 - r^2) r\, dr = \frac{\pi a^4}{8}.$$

**10.** $\mathcal{S}$: $\mathbf{r} = u^2 v\mathbf{i} + uv^2\mathbf{j} + v^3\mathbf{k}$, $(0 \le u \le 1,\ 0 \le v \le 1)$, has upward surface element

$$\begin{aligned}
\hat{\mathbf{N}}\, dS &= \frac{\partial \mathbf{r}}{\partial u} \times \frac{\partial \mathbf{r}}{\partial v}\, du\,dv \\
&= (2uv\mathbf{i} + v^2\mathbf{j}) \times (u^2\mathbf{i} + 2uv\mathbf{j} + 3v^2\mathbf{k})\, du\,dv \\
&= (3v^4\mathbf{i} - 6uv^3\mathbf{j} + 3u^2v^2\mathbf{k})\, du\,dv.
\end{aligned}$$

The flux of $\mathbf{F} = 2x\mathbf{i} + y\mathbf{j} + z\mathbf{k}$ upward through $\mathcal{S}$ is

$$\begin{aligned}
&\iint_{\mathcal{S}} \mathbf{F} \bullet \hat{\mathbf{N}}\, dS \\
&= \int_0^1 du \int_0^1 (6u^2v^5 - 6u^2v^5 + 3u^2v^5)\, dv \\
&= \frac{1}{2}\int_0^1 u^2\, du = \frac{1}{6}.
\end{aligned}$$

**12.** $\mathcal{S}$: $\mathbf{r} = e^u \cos v\mathbf{i} + e^u \sin v\mathbf{j} + u\mathbf{k}$, $(0 \le u \le 1,\ 0 \le v \le \pi)$, has upward surface element

$$\begin{aligned}
\hat{\mathbf{N}}\, dS &= \frac{\partial \mathbf{r}}{\partial u} \times \frac{\partial \mathbf{r}}{\partial v}\, du\,dv \\
&= (-e^u \cos v\mathbf{i} - e^u \sin v\mathbf{j} + e^{2u}\mathbf{k})\, du\,dv.
\end{aligned}$$

The flux of $\mathbf{F} = yz\mathbf{i} - xz\mathbf{j} + (x^2 + y^2)\mathbf{k}$ upward through $\mathcal{S}$ is

$$\begin{aligned}
&\iint_{\mathcal{S}} \mathbf{F} \bullet \hat{\mathbf{N}}\, dS \\
&= \int_0^1 du \int_0^\pi (-ue^{2u} \sin v \cos v + ue^{2u} \sin v \cos v + e^{4u})\, dv \\
&= \int_0^1 e^{4u}\, du \int_0^\pi dv = \frac{\pi e^4}{4}.
\end{aligned}$$

**14.** The flux of $\mathbf{F} = \dfrac{m\mathbf{r}}{|\mathbf{r}|^3}$ out of the cube $1 \le x, y, z \le 2$ is equal to three times the total flux out of the pair of opposite faces $z = 1$ and $z = 2$, which have outward normals $-\mathbf{k}$ and $\mathbf{k}$ respectively. This latter flux is $2mI_2 - mI_1$, where

$$\begin{aligned}
I_k &= \int_1^2 dx \int_1^2 \frac{dy}{(x^2+y^2+k^2)^{3/2}} \qquad \text{Let } y = \sqrt{x^2+k^2}\tan u, \quad dy = \sqrt{x^2+k^2}\sec^2 u\, du \\
&= \int_1^2 \frac{dx}{x^2+k^2}\int_{y=1}^{y=2} \cos u\, du \\
&= \int_1^2 \frac{dx}{x^2+k^2}(\sin u)\Bigg|_{y=1}^{y=2} \\
&= \int_1^2 \frac{dx}{x^2+k^2}\left(\frac{y}{\sqrt{x^2+y^2+k^2}}\Bigg|_1^2\right) = J_{k2} - J_{k1},
\end{aligned}$$

where

$$\begin{aligned}
J_{kn} &= n\int_1^2 \frac{dx}{(x^2+k^2)\sqrt{x^2+n^2+k^2}} \qquad \text{Let } x = \sqrt{n^2+k^2}\tan v, \quad dx = \sqrt{n^2+k^2}\sec^2 v\, dv \\
&= n\int_{x=1}^{x=2} \frac{\sec^2 v\, dv}{\left[(n^2+k^2)\tan^2 v + k^2\right]\sec v} \\
&= n\int_{x=1}^{x=2} \frac{\cos v\, dv}{(n^2+k^2)\sin^2 v + k^2\cos^2 v} \\
&= n\int_{x=1}^{x=2} \frac{\cos v\, dv}{k^2+n^2\sin^2 v} \qquad \text{Let } w = n\sin v, \quad dw = n\cos v\, dv \\
&= \int_{x=1}^{x=2} \frac{dw}{k^2+w^2} = \frac{1}{k}\tan^{-1}\frac{w}{k}\Bigg|_{x=1}^{x=2} \\
&= \frac{1}{k}\tan^{-1}\frac{n\sin v}{k}\Bigg|_{x=1}^{x=2} \\
&= \frac{1}{k}\tan^{-1}\frac{nx}{k\sqrt{x^2+n^2+k^2}}\Bigg|_1^2 \\
&= \frac{1}{k}\left(\tan^{-1}\frac{2n}{k\sqrt{4+n^2+k^2}} - \tan^{-1}\frac{n}{k\sqrt{1+n^2+k^2}}\right).
\end{aligned}$$

Thus

$$I_k = \frac{1}{k}\left[\tan^{-1}\frac{4}{k\sqrt{8+k^2}} - 2\tan^{-1}\frac{2}{k\sqrt{5+k^2}} + \tan^{-1}\frac{1}{k\sqrt{2+k^2}}\right].$$

The contribution to the total flux from the pair of surfaces $z = 1$ and $z = 2$ of the cube is

$$\begin{aligned}
&2mI_2 - mI_1 \\
&= m\left[\tan^{-1}\frac{1}{\sqrt{3}} - 2\tan^{-1}\frac{1}{3} + \tan^{-1}\frac{1}{2\sqrt{6}} - \tan^{-1}\frac{4}{3} + 2\tan^{-1}\frac{2}{\sqrt{6}} - \tan^{-1}\frac{1}{\sqrt{3}}\right].
\end{aligned}$$

Using the identities

$$2\tan^{-1} a = \tan^{-1}\frac{2a}{1-a^2}, \quad \text{and}$$
$$\tan^{-1} a = \frac{\pi}{2} - \tan^{-1}\frac{1}{a},$$

we calculate

$$\begin{aligned}
-2\tan^{-1}\frac{1}{3} &= -\tan^{-1}\frac{3}{4} = -\frac{\pi}{2} + \tan^{-1}\frac{4}{3} \\
2\tan^{-1}\frac{2}{\sqrt{6}} &= \tan^{-1}\frac{12}{\sqrt{6}} = \frac{\pi}{2} - \tan^{-1}\frac{1}{2\sqrt{6}}.
\end{aligned}$$

Thus the net flux out of the pair of opposite faces is 0. By symmetry this holds for each pair, and the total flux out of the cube is 0. (You were warned this would be a difficult calculation!)

**16.** $\mathbf{F} = -\dfrac{x\mathbf{i}+y\mathbf{j}}{x^2+y^2}$.

a) The flux of $\mathbf{F}$ inward across the circle of Exercise 7(a) is

$$\begin{aligned}
&-\oint_{\mathcal{C}}\left(-\frac{x\mathbf{i}+y\mathbf{j}}{a^2}\right)\bullet\frac{x\mathbf{i}+y\mathbf{j}}{a}\,ds \\
&= \oint_{\mathcal{C}}\frac{a^2}{a^3}\,ds = \frac{1}{a}\times 2\pi a = 2\pi.
\end{aligned}$$

b) The flux of $\mathbf{F}$ inward across the boundary of the square of Exercise 7(b) is four times the flux inward across the edge $x = 1$, $-1 \le y \le 1$. Thus it is

$$\begin{aligned}
-4\int_{-1}^1\left(-\frac{\mathbf{i}+y\mathbf{j}}{1+y^2}\right)\bullet\mathbf{i}\,dy &= 4\int_{-1}^1\frac{dy}{1+y^2} \\
&= 4\tan^{-1} y\Bigg|_{-1}^1 = 2\pi.
\end{aligned}$$

**18.** Let $\mathbf{F} = F_1\mathbf{i} + F_2\mathbf{j} + F_3\mathbf{k}$ be a constant vector field.

a) If $R$ is a rectangular box, we can choose the origin and coordinate axes in such a way that the box is $0 \le x \le a$, $0 \le y \le b$, $0 \le z \le c$. On the faces $x = 0$ and $x = a$ we have $\hat{\mathbf{N}} = -\mathbf{i}$ and $\hat{\mathbf{N}} = \mathbf{i}$ respectively. Since $F_1$ is constant, the total flux out of the box through these two faces is

$$\iint_{\substack{0\le y\le b \\ 0\le z\le c}} (F_1 - F_1)\,dy\,dz = 0.$$

The flux out of the other two pairs of opposite faces is also 0. Thus the total flux of $\mathbf{F}$ out of the box is 0.

b) If $\mathcal{S}$ is a sphere of radius $a$ we can choose the origin so that $\mathcal{S}$ has equation $x^2+y^2+z^2=a^2$, and so its outward normal is

$$\hat{\mathbf{N}}=\frac{x\mathbf{i}+y\mathbf{j}+z\mathbf{k}}{a}.$$

Thus the flux out of $\mathcal{S}$ is

$$\frac{1}{a}\iint_{\mathcal{S}}(F_1x+F_2y+F_3z)\,ds=0,$$

since the sphere $\mathcal{S}$ is symmetric about the origin.

## Review Exercises 8 (page 417)

**2.** $C$ can be parametrized $x=t$, $y=2t$, $z=t+4t^2$, $(0\le t\le 2)$. Thus

$$\begin{aligned}\int_C 2y\,dx+x\,dy+2\,dz&=\int_0^2[4t(1)+t(2)+2(1+8t)]\,dt\\&=\int_0^2(22t+2)\,dt=48.\end{aligned}$$

**4.** The plane $x+y+z=1$ has area element $dS=\sqrt{3}\,dx\,dy$. If $\mathcal{S}$ is the part of the plane in the first octant, then the projection of $\mathcal{S}$ on the $xy$-plane is the triangle $0\le x\le 1$, $0\le y\le 1-x$. Thus

$$\begin{aligned}\iint_{\mathcal{S}}xyz\,dS&=\sqrt{3}\int_0^1 x\,dx\int_0^{1-x}y(1-x-y)\,dy\\&=\sqrt{3}\int_0^1\frac{x(1-x)^3}{6}\,dx\quad\text{Let } u=1-x\\&\qquad\qquad\qquad\qquad\quad du=-dx\\&=\frac{\sqrt{3}}{6}\int_0^1u^3(1-u)\,du=\frac{\sqrt{3}}{6}\left(\frac14-\frac15\right)=\frac{\sqrt{3}}{120}.\end{aligned}$$

**6.** The plane $x+2y+3z=6$ has downward vector surface element

$$\hat{\mathbf{N}}\,dS=\frac{-\mathbf{i}-2\mathbf{j}-3\mathbf{k}}{3}\,dx\,dy.$$

If $\mathcal{S}$ is the part of the plane in the first octant, then the projection of $\mathcal{S}$ on the $xy$-plane is the triangle $0\le y\le 3$, $0\le x\le 6-2y$. Thus

$$\begin{aligned}&\iint_{\mathcal{S}}(x\mathbf{i}+y\mathbf{j}+z\mathbf{k})\bullet\hat{\mathbf{N}}\,dS\\&=-\frac13\int_0^3dy\int_0^{6-2y}(x+2y+6-x-2y)\,dx\\&=-2\int_0^3(6-2y)=-36+18=-18.\end{aligned}$$

**8.** $\int_C\mathbf{F}\bullet d\mathbf{r}$ can be determined using only the endpoints of $C$, provided

$$\mathbf{F}=(axy+3yz)\mathbf{i}+(x^2+3xz+by^2z)\mathbf{j}+(bxy+cy^3)\mathbf{k}$$

is conservative, that is, if

$$\begin{aligned}ax+3z&=\frac{\partial F_1}{\partial y}=\frac{\partial F_2}{\partial x}=2x+3z\\3y&=\frac{\partial F_1}{\partial z}=\frac{\partial F_3}{\partial x}=by\\3x+by^2&=\frac{\partial F_2}{\partial z}=\frac{\partial F_3}{\partial y}=bx+3cy^2.\end{aligned}$$

Thus we need $a=2$, $b=3$, and $c=1$.
With these values, $\mathbf{F}=\nabla(x^2y+3xyz+y^3z)$. Thus

$$\int_C\mathbf{F}\bullet d\mathbf{r}=(x^2y+3xyz+y^3z)\Big|_{(0,1,-1)}^{(2,1,1,)}=11-(-1)=12.$$

**10.** a) $\mathbf{F}=(1+x)e^{x+y}\mathbf{i}+(xe^{x+y}+2y)\mathbf{j}-2z\mathbf{k}$
$=\nabla(xe^{x+y}+y^2-z^2)$.
Thus $\mathbf{F}$ is conservative.

b) $\mathbf{G}=(1+x)e^{x+y}\mathbf{i}+(xe^{x+y}+2z)\mathbf{j}-2y\mathbf{k}$
$=\mathbf{F}+2(z-y)(\mathbf{j}+\mathbf{k})$.
$C:\mathbf{r}=(1-t)e^t\mathbf{i}+t\mathbf{j}+2t\mathbf{k},\quad(0\le t\le 1)$.
$\mathbf{r}(0)=(1,0,0)$, $\mathbf{r}(1)=(0,1,2)$. Thus

$$\begin{aligned}\int_C\mathbf{G}\bullet d\mathbf{r}&=\int_C\mathbf{F}\bullet d\mathbf{r}+\int_C2(z-y)(\mathbf{j}+\mathbf{k})\bullet d\mathbf{r}\\&=(xe^{x+y}+y^2-z^2)\Big|_{(1,0,0)}^{(0,1,2)}\\&\quad+2\int_0^1(2t-t)(1+2)\,dt\\&=-3-e+3t^2\Big|_0^1=-e.\end{aligned}$$

**12.** The first octant part of the cylinder $y^2+z^2=16$ has outward vector surface element

$$\hat{\mathbf{N}}\,dS=\frac{2y\mathbf{j}+2z\mathbf{k}}{2z}\,dx\,dy=\left(\frac{y}{\sqrt{16-y^2}}\mathbf{j}+\mathbf{k}\right)dx\,dy.$$

The flux of $3z^2x\mathbf{i} - x\mathbf{j} - y\mathbf{k}$ outward through the specified surface $\mathcal{S}$ is

$$\begin{aligned}
\mathbf{F} \bullet \hat{\mathbf{N}}\,dS &= \int_0^5 dx \int_0^4 \left(0 - \frac{xy}{\sqrt{16 - y^2}} - y\right) dy \\
&= \int_0^5 \left(x\sqrt{16 - y^2} - \frac{y^2}{2}\right)\Bigg|_{y=0}^{y=4} dx \\
&= -\int_0^5 (4x + 8)\,dx = -90.
\end{aligned}$$

## Challenging Problems 8 (page 418)

**2.** This is a trick question. Observe that the given parametrization $\mathbf{r}(u, v)$ satisfies

$$\mathbf{r}(u + \pi, v) = \mathbf{r}(u, -v).$$

Therefore the surface $\mathcal{S}$ is traced out twice as $u$ goes from 0 to $2\pi$. (It is a Möbius band. See Figure 8.28 in the text.) If $\mathcal{S}_1$ is the part of the surface corresponding to $0 \le u \le \pi$, and $\mathcal{S}_2$ is the part corresponding to $\pi \le u \le 2\pi$, then $\mathcal{S}_1$ and $\mathcal{S}_2$ coincide as point sets, but their normals are oppositely oriented: $\hat{\mathbf{N}}_2 = -\hat{\mathbf{N}}_1$ at corresponding points on the two surfaces. Hence

$$\iint_{\mathcal{S}_1} \mathbf{F} \bullet \hat{\mathbf{N}}_1\,dS = -\iint_{\mathcal{S}_2} \mathbf{F} \bullet \hat{\mathbf{N}}_2\,dS,$$

for any smooth vector field, and

$$\iint_{\mathcal{S}} \mathbf{F} \bullet \hat{\mathbf{N}}\,dS = \iint_{\mathcal{S}_1} \mathbf{F} \bullet \hat{\mathbf{N}}_1\,dS + \iint_{\mathcal{S}_2} \mathbf{F} \bullet \hat{\mathbf{N}}_2\,dS = 0.$$

# CHAPTER 9. VECTOR CALCULUS

## Section 9.1 Gradient, Divergence, and Curl (page 427)

**2.** $\mathbf{F} = y\mathbf{i} + x\mathbf{j}$

$$\text{div}\,\mathbf{F} = \frac{\partial}{\partial x}(y) + \frac{\partial}{\partial y}(x) + \frac{\partial}{\partial z}(0) = 0 + 0 = 0$$

$$\text{curl}\,\mathbf{F} = \begin{vmatrix} \mathbf{i} & \mathbf{j} & \mathbf{k} \\ \frac{\partial}{\partial x} & \frac{\partial}{\partial y} & \frac{\partial}{\partial z} \\ y & x & 0 \end{vmatrix} = (1-1)\mathbf{k} = \mathbf{0}$$

**4.** $\mathbf{F} = yz\mathbf{i} + xz\mathbf{j} + xy\mathbf{k}$

$$\text{div}\,\mathbf{F} = \frac{\partial}{\partial x}(yz) + \frac{\partial}{\partial y}(xz) + \frac{\partial}{\partial z}(xy) = 0$$

$$\text{curl}\,\mathbf{F} = \begin{vmatrix} \mathbf{i} & \mathbf{j} & \mathbf{k} \\ \frac{\partial}{\partial x} & \frac{\partial}{\partial y} & \frac{\partial}{\partial z} \\ yz & xz & xy \end{vmatrix}$$
$$= (x-x)\mathbf{i} + (y-y)\mathbf{j} + (z-z)\mathbf{k} = \mathbf{0}$$

**6.** $\mathbf{F} = xy^2\mathbf{i} - yz^2\mathbf{j} + zx^2\mathbf{k}$

$$\text{div}\,\mathbf{F} = \frac{\partial}{\partial x}\left(xy^2\right) + \frac{\partial}{\partial y}\left(-yz^2\right) + \frac{\partial}{\partial z}\left(zx^2\right)$$
$$= y^2 - z^2 + x^2$$

$$\text{curl}\,\mathbf{F} = \begin{vmatrix} \mathbf{i} & \mathbf{j} & \mathbf{k} \\ \frac{\partial}{\partial x} & \frac{\partial}{\partial y} & \frac{\partial}{\partial z} \\ xy^2 & -yz^2 & zx^2 \end{vmatrix}$$
$$= 2yz\mathbf{i} - 2xz\mathbf{j} - 2xy\mathbf{k}$$

**8.** $\mathbf{F} = f(z)\mathbf{i} - f(z)\mathbf{j}$

$$\text{div}\,\mathbf{F} = \frac{\partial}{\partial x}f(z) + \frac{\partial}{\partial y}\big(-f(z)\big) = 0$$

$$\text{curl}\,\mathbf{F} = \begin{vmatrix} \mathbf{i} & \mathbf{j} & \mathbf{k} \\ \frac{\partial}{\partial x} & \frac{\partial}{\partial y} & \frac{\partial}{\partial z} \\ f(z) & -f(z) & 0 \end{vmatrix} = f'(z)(\mathbf{i}+\mathbf{j})$$

**10.** $\mathbf{F} = \hat{\mathbf{r}} = \cos\theta\mathbf{i} + \sin\theta\mathbf{j}$

$$\text{div}\,\mathbf{F} = \frac{\sin^2\theta}{r} + \frac{\cos^2\theta}{r} = \frac{1}{r} = \frac{1}{\sqrt{x^2+y^2}}$$

$$\text{curl}\,\mathbf{F} = \begin{vmatrix} \mathbf{i} & \mathbf{j} & \mathbf{k} \\ \frac{\partial}{\partial x} & \frac{\partial}{\partial y} & \frac{\partial}{\partial z} \\ \cos\theta & \sin\theta & 0 \end{vmatrix}$$
$$= -\left(\frac{\cos\theta\sin\theta}{r} - \frac{\cos\theta\sin\theta}{r}\right)\mathbf{k} = \mathbf{0}$$

**12.** We use the Maclaurin expansion of $\mathbf{F}$, as presented in the proof of Theorem 1:

$$\mathbf{F} = \mathbf{F}_0 + \mathbf{F}_1 x + \mathbf{F}_2 y + \mathbf{F}_3 z + \cdots,$$

where

$$\mathbf{F}_0 = \mathbf{F}(0,0,0)$$
$$\mathbf{F}_1 = \frac{\partial}{\partial x}\mathbf{F}(x,y,z)\Big|_{(0,0,0)} = \left(\frac{\partial F_1}{\partial x}\mathbf{i} + \frac{\partial F_2}{\partial x}\mathbf{j} + \frac{\partial F_3}{\partial x}\mathbf{k}\right)\Bigg|_{(0,0,0)}$$
$$\mathbf{F}_2 = \frac{\partial}{\partial y}\mathbf{F}(x,y,z)\Big|_{(0,0,0)} = \left(\frac{\partial F_1}{\partial y}\mathbf{i} + \frac{\partial F_2}{\partial y}\mathbf{j} + \frac{\partial F_3}{\partial y}\mathbf{k}\right)\Bigg|_{(0,0,0)}$$
$$\mathbf{F}_3 = \frac{\partial}{\partial z}\mathbf{F}(x,y,z)\Big|_{(0,0,0)} = \left(\frac{\partial F_1}{\partial z}\mathbf{i} + \frac{\partial F_2}{\partial z}\mathbf{j} + \frac{\partial F_3}{\partial z}\mathbf{k}\right)\Bigg|_{(0,0,0)}$$

and where $\cdots$ represents terms of degree 2 and higher in $x$, $y$, and $z$.
On the top of the box $B_{a,b,c}$, we have $z = c$ and $\hat{\mathbf{N}} = \mathbf{k}$. On the bottom of the box, we have $z = -c$ and $\hat{\mathbf{N}} = -\mathbf{k}$. On both surfaces $dS = dx\,dy$. Thus

$$\left(\iint_{\text{top}} + \iint_{\text{bottom}}\right)\mathbf{F}\bullet\hat{\mathbf{N}}\,dS$$
$$= \int_{-a}^{a} dx \int_{-b}^{b} dy\big(c\mathbf{F}_3\bullet\mathbf{k} - c\mathbf{F}_3\bullet(-\mathbf{k})\big) + \cdots$$
$$= 8abc\mathbf{F}_3\bullet\mathbf{k} + \cdots = 8abc\frac{\partial}{\partial z}F_3(x,y,z)\Big|_{(0,0,0)} + \cdots,$$

where $\cdots$ represents terms of degree 4 and higher in $a$, $b$, and $c$.
Similar formulas obtain for the two other pairs of faces, and the three formulas combine into

$$\oiint_{B_{a,b,c}} \mathbf{F}\bullet\hat{\mathbf{N}}\,dS = 8abc\,\textbf{div}\,\mathbf{F}(0,0,0) + \cdots.$$

It follows that

$$\lim_{a,b,c\to 0+} \frac{1}{8abc}\oiint_{B_{a,b,c}} \mathbf{F}\bullet\hat{\mathbf{N}}\,dS = \textbf{div}\,\mathbf{F}(0,0,0).$$

**14.** We use the same Maclaurin expansion for $\mathbf{F}$ as in Exercise 13. On $\mathcal{C}_\epsilon$ we have

$$\mathbf{r} = \epsilon\cos\theta\mathbf{i} + \epsilon\sin\theta\mathbf{j}, \quad (0 \le \theta \le 2\pi)$$
$$d\mathbf{r} = -\epsilon\sin\theta\mathbf{i} + \epsilon\cos\theta\mathbf{j}$$
$$\mathbf{F}\bullet d\mathbf{r} = \big(-\epsilon\sin\theta\mathbf{F}_0\bullet\mathbf{i} + \epsilon\cos\theta\mathbf{F}_0\bullet\mathbf{j}$$
$$- \epsilon^2\sin\theta\cos\theta\mathbf{F}_1\bullet\mathbf{i} + \epsilon^2\cos^2\theta\mathbf{F}_1\bullet\mathbf{j}$$
$$- \epsilon^2\sin^2\theta\mathbf{F}_2\bullet\mathbf{i} + \epsilon^2\sin\theta\cos\theta\mathbf{F}_2\bullet\mathbf{j} + \cdots\big)\,ds,$$

where $\cdots$ represents terms of degree 3 or higher in $\epsilon$.
Since

$$\int_0^{2\pi}\sin\theta\,d\theta = \int_0^{2\pi}\cos\theta\,d\theta = \int_0^{2\pi}\sin\theta\cos\theta\,d\theta = 0$$
$$\int_0^{2\pi}\cos^2\theta\,d\theta = \int_0^{2\pi}\sin^2\theta\,d\theta = \pi,$$

we have

$$\frac{1}{\pi\epsilon^2}\oint_{\mathcal{C}_\epsilon} \mathbf{F} \bullet d\mathbf{r} = \mathbf{F}_1 \bullet \mathbf{j} - \mathbf{F}_2 \bullet \mathbf{i} + \cdots,$$

where $\cdots$ represents terms of degree at least 1 in $\epsilon$. Hence

$$\begin{aligned}\lim_{\epsilon\to 0+}\frac{1}{\pi\epsilon^2}\oint_{\mathcal{C}_\epsilon} \mathbf{F} \bullet d\mathbf{r} &= \mathbf{F}_1 \bullet \mathbf{j} - \mathbf{F}_2 \bullet \mathbf{i}\\ &= \frac{\partial F_2}{\partial x} - \frac{\partial F_1}{\partial y}\\ &= \mathbf{curl}\,\mathbf{F} \bullet \mathbf{k} = \mathbf{curl}\,\mathbf{F} \bullet \hat{\mathbf{N}}.\end{aligned}$$

## Section 9.2 Some Identities Involving Grad, Div, and Curl (page 433)

**2.** Theorem 3(b):

$$\begin{aligned}\nabla \bullet (\phi\mathbf{F}) &= \frac{\partial}{\partial x}(\phi F_1) + \frac{\partial}{\partial y}(\phi F_2) + \frac{\partial}{\partial z}(\phi F_3)\\ &= \frac{\partial \phi}{\partial x}F_1 + \phi\frac{\partial F_1}{\partial x} + \cdots + \frac{\partial \phi}{\partial z}F_3 + \phi\frac{\partial F_3}{\partial z} + \cdots\\ &= \nabla\phi \bullet \mathbf{F} + \phi\nabla \bullet \mathbf{F}.\end{aligned}$$

**4.** Theorem 3(f). The first component of $\nabla(\mathbf{F} \bullet \mathbf{G})$ is

$$\frac{\partial F_1}{\partial x}G_1 + F_1\frac{\partial G_1}{\partial x} + \frac{\partial F_2}{\partial x}G_2 + F_2\frac{\partial G_2}{\partial x} + \frac{\partial F_3}{\partial x}G_3 + F_3\frac{\partial G_3}{\partial x}.$$

We calculate the first components of the four terms on the right side of the identity to be proved.
The first component of $\mathbf{F} \times (\nabla \times \mathbf{G})$ is

$$F_2\left(\frac{\partial G_2}{\partial x} - \frac{\partial G_1}{\partial y}\right) - F_3\left(\frac{\partial G_1}{\partial z} - \frac{\partial G_3}{\partial x}\right).$$

The first component of $\mathbf{G} \times (\nabla \times \mathbf{F})$ is

$$G_2\left(\frac{\partial F_2}{\partial x} - \frac{\partial F_1}{\partial y}\right) - G_3\left(\frac{\partial F_1}{\partial z} - \frac{\partial F_3}{\partial x}\right).$$

The first component of $(\mathbf{F} \bullet \nabla)\mathbf{G}$ is

$$F_1\frac{\partial G_1}{\partial x} + F_2\frac{\partial G_1}{\partial y} + F_3\frac{\partial G_1}{\partial z}.$$

The first component of $(\mathbf{G} \bullet \nabla)\mathbf{F}$ is

$$G_1\frac{\partial F_1}{\partial x} + G_2\frac{\partial F_1}{\partial y} + G_3\frac{\partial F_1}{\partial z}.$$

When we add these four first components, eight of the fourteen terms cancel out and the six remaining terms are the six terms of the first component of $\nabla(\mathbf{F} \bullet \mathbf{G})$, as calculated above. Similar calculations show that the second and third components of both sides of the identity agree. Thus

$$\nabla(\mathbf{F}\bullet\mathbf{G}) = \mathbf{F}\times(\nabla\times\mathbf{G}) + \mathbf{G}\times(\nabla\times\mathbf{F}) + (\mathbf{F}\bullet\nabla)\mathbf{G} + (\mathbf{G}\bullet\nabla)\mathbf{F}.$$

**6.** Theorem 3(i). We examine the first components of the terms on both sides of the identity

$$\nabla \times (\nabla \times \mathbf{F}) = \nabla(\nabla \bullet \mathbf{F}) - \nabla^2\mathbf{F}.$$

The first component of $\nabla \times (\nabla \times \mathbf{F})$ is

$$\begin{aligned}&\frac{\partial}{\partial y}\left(\frac{\partial F_2}{\partial x} - \frac{\partial F_1}{\partial y}\right) - \frac{\partial}{\partial z}\left(\frac{\partial F_1}{\partial z} - \frac{\partial F_3}{\partial x}\right)\\ &= \frac{\partial^2 F_2}{\partial y\partial x} - \frac{\partial^2 F_1}{\partial y^2} - \frac{\partial^2 F_1}{\partial z^2} + \frac{\partial^2 F_3}{\partial z\partial x}.\end{aligned}$$

The first component of $\nabla(\nabla \bullet \mathbf{F})$ is

$$\frac{\partial}{\partial x}\nabla \bullet \mathbf{F} = \frac{\partial^2 F_1}{\partial x^2} + \frac{\partial^2 F_2}{\partial x\partial y} + \frac{\partial^2 F_3}{\partial x\partial z}.$$

The first component of $-\nabla^2\mathbf{F}$ is

$$-\nabla^2 F_1 = -\frac{\partial^2 F_1}{\partial x^2} - \frac{\partial^2 F_1}{\partial y^2} - \frac{\partial^2 F_1}{\partial z^2}.$$

Evidently the first components of both sides of the given identity agree. By symmetry, so do the other components.

**8.** If $\mathbf{r} = x\mathbf{i} + y\mathbf{j} + z\mathbf{k}$ and $r = |\mathbf{r}|$, then

$$\nabla \bullet \mathbf{r} = 3, \qquad \nabla \times \mathbf{r} = \mathbf{0}, \qquad \nabla r = \frac{\mathbf{r}}{r}.$$

If $\mathbf{c}$ is a constant vector, then its divergence and curl are both zero. By Theorem 3(d), (e), and (f) we have

$$\begin{aligned}\nabla \bullet (\mathbf{c} \times \mathbf{r}) &= (\nabla \times \mathbf{c}) \bullet \mathbf{r} - \mathbf{c} \bullet (\nabla \times \mathbf{r}) = 0\\ \nabla \times (\mathbf{c} \times \mathbf{r}) &= (\nabla \bullet \mathbf{r})\mathbf{c} + (\mathbf{r} \bullet \nabla)\mathbf{c} - (\nabla \bullet \mathbf{c})\mathbf{r} - (\mathbf{c} \bullet \nabla)\mathbf{r}\\ &= 3\mathbf{c} + \mathbf{0} - \mathbf{0} - \mathbf{c} = 2\mathbf{c}\end{aligned}$$

$$\begin{aligned}\nabla(\mathbf{c} \bullet \mathbf{r}) &= \mathbf{c} \times (\nabla \times \mathbf{r}) + \mathbf{r} \times (\nabla \times \mathbf{c}) + (\mathbf{c} \bullet \nabla)\mathbf{r} + (\mathbf{r} \bullet \nabla)\mathbf{c}\\ &= \mathbf{0} + \mathbf{0} + \mathbf{c} + \mathbf{0} = \mathbf{c}.\end{aligned}$$

**10.** Given that $\mathbf{div}\,\mathbf{F} = 0$ and $\mathbf{curl}\,\mathbf{F} = \mathbf{0}$, Theorem 3(i) implies that $\nabla^2\mathbf{F} = 0$ too. Hence the components of $\mathbf{F}$ are harmonic functions.
If $\mathbf{F} = \nabla\phi$, then

$$\nabla^2\phi = \nabla \bullet \nabla\phi = \nabla \bullet \mathbf{F} = 0,$$

so $\phi$ is also harmonic.

**12.** If $\nabla^2\phi = 0$ and $\nabla^2\psi = 0$, then

$$\begin{aligned}
&\nabla \bullet (\phi\nabla\psi - \psi\nabla\phi) \\
&= \nabla\phi \bullet \nabla\psi + \phi\nabla^2\psi - \nabla\psi \bullet \nabla\phi - \psi\nabla^2\phi = 0,
\end{aligned}$$

so $\phi\nabla\psi - \psi\nabla\phi$ is solenoidal.

**14.** By Theorem 3(b), (d), and (h), we have

$$\begin{aligned}
&\nabla \bullet \big(f(\nabla g \times \nabla h)\big) \\
&= \nabla f \bullet (\nabla g \times \nabla h) + f\nabla \bullet (\nabla g \times \nabla h) \\
&= \nabla f \bullet (\nabla g \times \nabla h) + f\big((\nabla \times \nabla g) \bullet \nabla h - \nabla g \bullet (\nabla \times \nabla h)\big) \\
&= \nabla f \bullet (\nabla g \times \nabla h) + \mathbf{0} - \mathbf{0} = \nabla f \bullet (\nabla g \times \nabla h).
\end{aligned}$$

**16.** If $\nabla \times \mathbf{G} = \mathbf{F} = -y\mathbf{i} + x\mathbf{j}$, then

$$\begin{aligned}
\frac{\partial G_3}{\partial y} - \frac{\partial G_2}{\partial z} &= -y \\
\frac{\partial G_1}{\partial z} - \frac{\partial G_3}{\partial x} &= x \\
\frac{\partial G_2}{\partial x} - \frac{\partial G_1}{\partial y} &= 0.
\end{aligned}$$

As in Example 1, we try to find a solution with $G_2 = 0$. Then

$$G_3 = -\int y\,dy = -\frac{y^2}{2} + M(x, z).$$

Again we try $M(x, z) = 0$, so $G_3 = -\dfrac{y^2}{2}$. Thus $\dfrac{\partial G_3}{\partial x} = 0$ and

$$G_1 = \int x\,dz = xz + N(x, y).$$

Since $\dfrac{\partial G_1}{\partial y} = 0$ we may take $N(x, y) = 0$.

$\mathbf{G} = xz\mathbf{i} - \dfrac{1}{2}y^2\mathbf{k}$ is a vector potential for $\mathbf{F}$. (Of course, this answer is not unique.)

**18.** For $(x, y, z)$ in $D$ let $\mathbf{v} = x\mathbf{i} + y\mathbf{j} + z\mathbf{k}$. The line segment $\mathbf{r}(t) = t\mathbf{v}$, $(0 \le t \le 1)$, lies in $D$, so $\mathbf{div\,F} = 0$ on the path. We have

$$\begin{aligned}
\mathbf{G}(x, y, z) &= \int_0^1 t\mathbf{F}\big(\mathbf{r}(t)\big) \times \mathbf{v}\,dt \\
&= \int_0^1 t\mathbf{F}\big(\xi(t), \eta(t), \zeta(t)\big) \times \mathbf{v}\,dt
\end{aligned}$$

where $\xi = tx$, $\eta = ty$, $\zeta = tz$. The first component of $\mathbf{curl\,G}$ is

$$\begin{aligned}
&(\mathbf{curl\,G})_1 \\
&= \int_0^1 t\big(\mathbf{curl}\,(\mathbf{F} \times \mathbf{v})\big)_1\,dt \\
&= \int_0^1 t\left(\frac{\partial}{\partial y}(\mathbf{F} \times \mathbf{v})_3 - \frac{\partial}{\partial z}(\mathbf{F} \times \mathbf{v})_2\right) dt \\
&= \int_0^1 t\left(\frac{\partial}{\partial y}(F_1 y - F_2 x) - \frac{\partial}{\partial z}(F_3 x - F_1 z)\right) dt \\
&= \int_0^1 \left(tF_1 + t^2 y\frac{\partial F_1}{\partial \eta} - t^2 x\frac{\partial F_2}{\partial \eta} - t^2 x\frac{\partial F_3}{\partial \zeta}\right. \\
&\qquad \left. + tF_1 + t^2 z\frac{\partial F_1}{\partial \zeta}\right) dt \\
&= \int_0^1 \left(2tF_1 + t^2 x\frac{\partial F_1}{\partial \xi} + t^2 y\frac{\partial F_2}{\partial \eta} + t^2 z\frac{\partial F_3}{\partial \zeta}\right) dt.
\end{aligned}$$

To get the last line we used the fact that $div\mathbf{F} = 0$ to replace $-t^2 x\dfrac{\partial F_2}{\partial \eta} - t^2 x\dfrac{\partial F_3}{\partial \zeta}$ with $t^2 x\dfrac{\partial F_1}{\partial \xi}$. Continuing the calculation, we have

$$\begin{aligned}
(\mathbf{curl\,G})_1 &= \int_0^1 \frac{d}{dt}\big(t^2 F_1(\xi, \eta, \zeta)\big)\,dt \\
&= t^2 F_1(tx, ty, tz)\Big|_0^1 = F_1(x, y, z).
\end{aligned}$$

Similarly, $(\mathbf{curl\,G})_2 = F_2$ and $(\mathbf{curl\,G})_3 = F_3$. Thus $\mathbf{curl\,G} = \mathbf{F}$, as required.

## Section 9.3 The Divergence Theorem (page 440)

**2.** If $\mathbf{F} = ye^z\mathbf{i} + x^2e^z\mathbf{j} + xy\mathbf{k}$, then $\mathbf{div\,F} = 0$, and

$$\oiint_{\mathcal{S}} \mathbf{F} \bullet \hat{\mathbf{N}}\,dS = \iiint_B 0\,dV = 0.$$

**4.** If $\mathbf{F} = x^3\mathbf{i} + 3yz^2\mathbf{j} + (3y^2z + x^2)\mathbf{k}$, then $\mathbf{div\,F} = 3x^2 + 3z^2 + 3y^2$, and

$$\begin{aligned}
\oiint_{\mathcal{S}} \mathbf{F} \bullet \hat{\mathbf{N}}\,dS &= 3\iiint_B (x^2 + y^2 + z^2)\,dV \\
&= 3\int_0^{2\pi} d\theta \int_0^{\pi} \sin\phi\,d\phi \int_0^a \rho^4\,d\rho \\
&= \frac{12}{5}\pi a^5.
\end{aligned}$$

**6.** If $R$ is the ellipsoid $x^2 + y^2 + 4(z-1)^2 \le 4$, then $\overline{x} = 0$, $\overline{y} = 0$, $\overline{z} = 1$, and $V = (4\pi/3)(2)(2)(1) = 16\pi/3$. The flux of $\mathbf{F}$ out of $R$ is $2(0 + 0 + 1)(16\pi/3) = 32\pi/3$.

**8.** If $R$ is the cylinder $x^2 + y^2 \le 2y$ (or, equivalently, $x^2 + (y-1)^2 \le 1$), $0 \le z \le 4$, then $\overline{x} = 0$, $\overline{y} = 1$, $\overline{z} = 2$, and $V = (\pi 1^2)(4) = 4\pi$. The flux of $\mathbf{F}$ out of $R$ is $2(0 + 1 + 2)(4\pi) = 24\pi$.

**10.** The required surface integral,

$$I = \iint_{\mathcal{S}} \nabla\phi \bullet \hat{\mathbf{N}}\, dS,$$

can be calculated directly by the methods of Section 8.6. We will do it here by using the Divergence Theorem instead. $\mathcal{S}$ is one face of a tetrahedral domain $D$ whose other faces are in the coordinate planes, as shown in the figure. Since $\phi = xy + z^2$, we have

$$\nabla\phi = y\mathbf{i} + x\mathbf{j} + 2z\mathbf{k}, \qquad \nabla\bullet\nabla\phi = \nabla^2\phi = 2.$$

Thus

$$\iiint_D \nabla\bullet\nabla\phi\, dV = 2\times\frac{abc}{6} = \frac{abc}{3},$$

the volume of the tetrahedron $D$ being $abc/6$ cubic units.

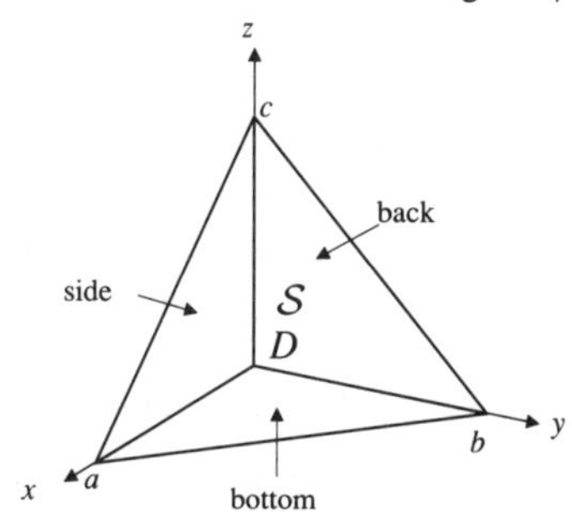


Fig. 9.3.10

The flux of $\nabla\phi$ out of $D$ is the sum of its fluxes out of the four faces of the tetrahedron.

On the bottom, $\hat{\mathbf{N}} = -\mathbf{k}$ and $z = 0$, so $\nabla\phi\bullet\hat{\mathbf{N}} = 0$, and the flux out of the bottom face is 0.

On the side, $y = 0$ and $\hat{\mathbf{N}} = -\mathbf{j}$, so $\nabla\phi\bullet\hat{\mathbf{N}} = -x$. The flux out of the side face is

$$\iint_{\text{side}} \nabla\phi\bullet\hat{\mathbf{N}}\, dS = -\iint_{\text{side}} x\, dx\, dz = -\frac{ac}{2}\times\frac{a}{3} = -\frac{a^2c}{6}.$$

(We used the fact that $M_{x=0} = \text{area}\times\bar{x}$ and $\bar{x} = a/3$ for that face.)

On the back face, $x = 0$ and $\hat{\mathbf{N}} = -\mathbf{i}$, so the flux out of that face is

$$\iint_{\text{back}} \nabla\phi\bullet\hat{\mathbf{N}}\, dS = -\iint_{\text{back}} y\, dy\, dz = -\frac{bc}{2}\times\frac{b}{3} = -\frac{b^2c}{6}.$$

Therefore, by the Divergence Theorem

$$I - \frac{a^2c}{6} - \frac{b^2c}{6} + 0 = \frac{abc}{3},$$

so $\displaystyle\iint_{\mathcal{S}} \nabla\phi\bullet\hat{\mathbf{N}}\, dS = I = \frac{abc}{3} + \frac{c(a^2+b^2)}{6}.$

**12.** $\mathbf{F} = (y+xz)\mathbf{i} + (y+yz)\mathbf{j} - (2x+z^2)\mathbf{k}$
$\mathbf{div}\,\mathbf{F} = z + (1+z) - 2z = 1$. Thus

$$\iiint_D \mathbf{div}\,\mathbf{F}\, dV = \text{volume of } D = \frac{\pi a^3}{6},$$

where $D$ is the region in the first octant bounded by the sphere and the coordinate planes. The boundary of $D$ consists of the spherical part $\mathcal{S}$ and the four planar parts, called the bottom, side, and back in the figure.

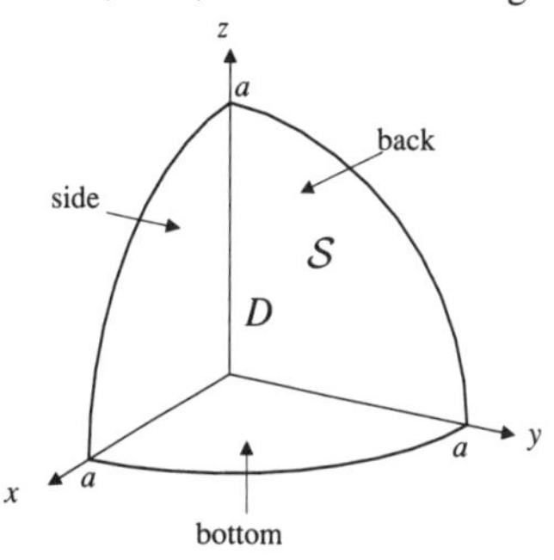


Fig. 9.3.12

On the side, $y = 0$, $\hat{\mathbf{N}} = -\mathbf{j}$, $\mathbf{F}\bullet\hat{\mathbf{N}} = 0$, so

$$\iint_{\text{side}} \mathbf{F}\bullet\hat{\mathbf{N}}\, dS = 0.$$

On the back, $x = 0$, $\hat{\mathbf{N}} = -\mathbf{i}$, $\mathbf{F}\bullet\hat{\mathbf{N}} = -y$, so

$$\begin{aligned}\iint_{\text{back}} \mathbf{F}\bullet\hat{\mathbf{N}}\, dS &= -\int_0^{\pi/2} d\theta \int_0^a r\cos\theta\, r\, dr\\ &= -\sin\theta\Big|_0^{\pi/2}\times\frac{a^3}{3} = -\frac{a^3}{3}.\end{aligned}$$

On the bottom, $z = 0$, $\hat{\mathbf{N}} = -\mathbf{k}$, $\mathbf{F}\bullet\hat{\mathbf{N}} = 2x$, so

$$\iint_{\text{bottom}} \mathbf{F}\bullet\hat{\mathbf{N}}\, dS = 2\int_0^{\pi/2} d\theta \int_0^a r\cos\theta\, r\, dr = \frac{2a^3}{3}.$$

By the Divergence Theorem

$$\iint_{\mathcal{S}} \mathbf{F}\bullet\hat{\mathbf{N}}\, dS + 0 - \frac{a^3}{3} + \frac{2a^3}{3} = \frac{\pi a^3}{6}.$$

Hence the flux of $\mathbf{F}$ upward through $\mathcal{S}$ is

$$\iint_{\mathcal{S}} \mathbf{F}\bullet\hat{\mathbf{N}}\, dS = \frac{\pi a^3}{6} - \frac{a^3}{3}.$$

**14.** Let $D$ be the domain bounded by $\mathcal{S}$, the coordinate planes, and the plane $x = 1$. If

$$\mathbf{F} = 3xz^2\mathbf{i} - x\mathbf{j} - y\mathbf{k},$$

then $\mathbf{div}\,\mathbf{F} = 3z^2$, so the total flux of $\mathbf{F}$ out of $D$ is

$$\oiint_{\text{bdry of } D} \mathbf{F} \bullet \hat{\mathbf{N}}\, dS = \iiint_D 3z^2\, dV$$
$$= 3\int_0^1 dx \int_0^{\pi/2} d\theta \int_0^1 r^2 \cos^2\theta\, r\, dr$$
$$= 3 \times \frac{1}{4} \times \frac{\pi}{4} = \frac{3\pi}{16}.$$

The boundary of $D$ consists of the cylindrical surface $\mathcal{S}$ and four planar surfaces, the side, bottom, back, and front.

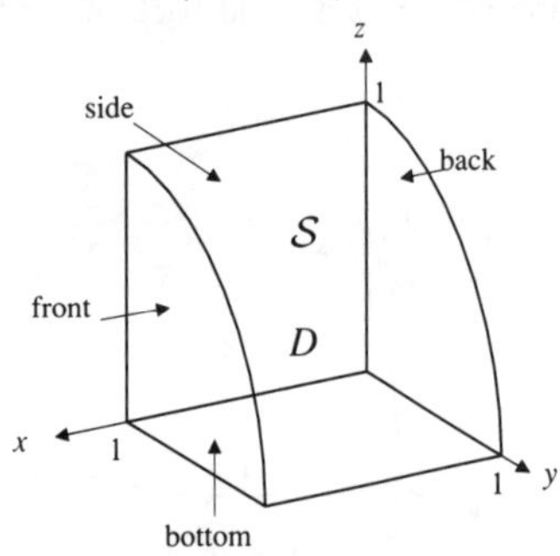


Fig. 9.3.14

On the side, $y = 0$, $\hat{\mathbf{N}} = -\mathbf{j}$, $\mathbf{F} \bullet \hat{\mathbf{N}} = x$, so

$$\iint_{\text{side}} \mathbf{F} \bullet \hat{\mathbf{N}}\, dS = \int_0^1 x\, dx \int_0^1 dz = \frac{1}{2}.$$

On the bottom, $z = 0$, $\hat{\mathbf{N}} = -\mathbf{k}$, $\mathbf{F} \bullet \hat{\mathbf{N}} = y$, so

$$\iint_{\text{bottom}} \mathbf{F} \bullet \hat{\mathbf{N}}\, dS = \int_0^1 y\, dy \int_0^1 dx = \frac{1}{2}.$$

On the back, $x = 0$, $\hat{\mathbf{N}} = -\mathbf{i}$, $\mathbf{F} \bullet \hat{\mathbf{N}} = 0$, so

$$\iint_{\text{back}} \mathbf{F} \bullet \hat{\mathbf{N}}\, dS = 0.$$

On the front, $x = 1$, $\hat{\mathbf{N}} = \mathbf{i}$, $\mathbf{F} \bullet \hat{\mathbf{N}} = 3z^2$, so

$$\iint_{\text{front}} \mathbf{F} \bullet \hat{\mathbf{N}}\, dS = 3\int_0^{\pi/2} d\theta \int_0^1 r^2 \cos^2\theta\, r\, dr = \frac{3\pi}{16}.$$

Hence,

$$\iint_{\mathcal{S}} (3xz^2\mathbf{i} - x\mathbf{j} - y\mathbf{k}) \bullet \hat{\mathbf{N}}\, dS = \frac{3\pi}{16} - \frac{1}{2} - \frac{1}{2} - 0 - \frac{3\pi}{16} = -1.$$

**16.** $\mathbf{F} = x\mathbf{i} + y\mathbf{j} + z\mathbf{k}$ implies that $\mathbf{div}\,\mathbf{F} = 3$. The total flux of $\mathbf{F}$ out of $D$ is

$$\oiint_{\text{bdry of } D} \mathbf{F} \bullet \hat{\mathbf{N}}\, dS = 3\iiint_D dV = 12,$$

since the volume of $D$ is half that of a cube of side 2, that is, 4 square units.

$D$ has three triangular faces, three pentagonal faces, and a hexagonal face. By symmetry, the flux of $\mathbf{F}$ out of each triangular face is equal to that out of the triangular face $T$ in the plane $z = 1$. Since $\mathbf{F} \bullet \hat{\mathbf{N}} = \mathbf{k} \bullet \mathbf{k} = 1$ on that face, these fluxes are

$$\iint_T dx\, dy = \text{area of } T = \frac{1}{2}.$$

Similarly, the flux of $\mathbf{F}$ out of each pentagonal face is equal to the flux out of the pentagonal face $P$ in the plane $z = -1$, where $\mathbf{F} \bullet \hat{\mathbf{N}} = -\mathbf{k} \bullet (-\mathbf{k}) = 1$; that flux is

$$\iint_P dx\, dy = \text{area of } P = 4 - \frac{1}{2} = \frac{7}{2}.$$

Thus the flux of $\mathbf{F}$ out of the remaining hexagonal face $H$ is

$$12 - 3 \times \left(\frac{1}{2} + \frac{7}{2}\right) = 0.$$

(This can also be seen directly, since $\mathbf{F}$ radiates from the origin, so is everywhere tangent to the plane of the hexagonal face, the plane $x + y + z = 0$.)

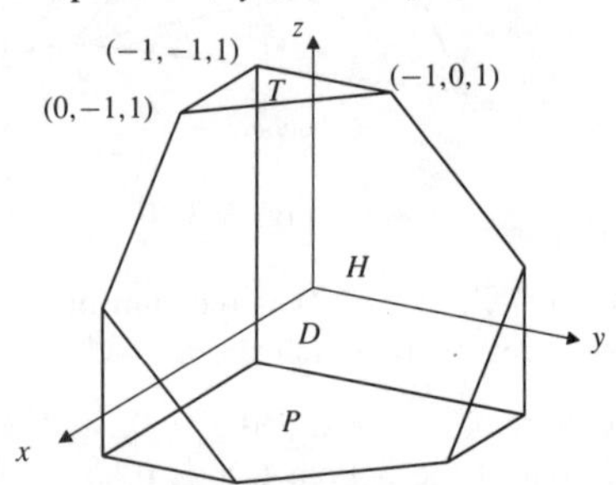


Fig. 9.3.16

**18.** $\phi = x^2 - y^2 + z^2$, $\mathbf{G} = \frac{1}{3}(-y^3\mathbf{i} + x^3\mathbf{j} + z^3\mathbf{k})$.
$\mathbf{F} = \nabla\phi + \mu\,\mathbf{curl}\,\mathbf{G}$.

Let $R$ be the region of 3-space occupied by the sandpile. Then $R$ is bounded by the upper surface $\mathcal{S}$ of the sandpile and by the disk $D$: $x^2 + y^2 \le 1$ in the plane $z = 0$. The outward (from $R$) normal on $D$ is $-\mathbf{k}$. The flux of $\mathbf{F}$ out of $R$ is given by

$$\iint_{\mathcal{S}} \mathbf{F} \bullet \hat{\mathbf{N}}\, dS + \iint_D \mathbf{F} \bullet (-\mathbf{k})\, dA = \iiint_R \mathbf{div}\,\mathbf{F}\, dV.$$

Now $\mathbf{div}\,\mathbf{curl}\,\mathbf{G} = 0$ by Theorem 3(g). Also
$\mathbf{div}\,\nabla\phi = \mathbf{div}\,(2x\mathbf{i} - 2y\mathbf{j} + 2z\mathbf{k}) = 2 - 2 + 2 = 2$. Therefore

$$\iiint_R \mathbf{div}\,\mathbf{F}\, dV = \iiint_R (2 + \mu \times 0)\, dV = 2(5\pi) = 10\pi.$$

In addition,

$$\mathbf{curl\,G} = \frac{1}{3}\begin{vmatrix} \mathbf{i} & \mathbf{j} & \mathbf{k} \\ \frac{\partial}{\partial x} & \frac{\partial}{\partial y} & \frac{\partial}{\partial z} \\ -y^3 & x^3 & z^3 \end{vmatrix} = 3(x^2+y^2)\mathbf{k},$$

and $\nabla\phi \bullet \mathbf{k} = 2z = 0$ on $D$, so

$$\iint_D \mathbf{F}\bullet\mathbf{k}\,dA = 3\mu\int_0^{2\pi} d\theta \int_0^1 r^3\,dr = \frac{3\pi\mu}{2}.$$

The flux of $\mathbf{F}$ out of $\mathcal{S}$ is $10\pi + (3\pi\mu)/2$.

**20.** If $\mathbf{r} = x\mathbf{i} + y\mathbf{j} + z\mathbf{k}$, then $\mathbf{div\,r} = 3$ and

$$\frac{1}{3}\oiint_{\mathcal{S}} \mathbf{r}\bullet\hat{\mathbf{N}}\,dS = \frac{1}{3}\iiint_D 3\,dV = V.$$

**22.** Taking $\mathbf{F} = \nabla\phi$ in the first identity in Theorem 7(a), we have

$$\oiint_{\mathcal{S}} \nabla\phi \times \hat{\mathbf{N}}\,dS = -\iiint_D \mathbf{curl}\,\nabla\phi\,dV = 0,$$

since $\nabla\times\nabla\phi = 0$ by Theorem 3(h).

**24.** If $\mathbf{F} = \nabla\phi$ in the previous exercise, then $\mathbf{div\,F} = \nabla^2\phi$ and

$$\iiint_D \phi\nabla^2\phi\,dV + \iiint_D |\nabla\phi|^2\,dV = \oiint_{\mathcal{S}} \phi\nabla\phi\bullet\hat{\mathbf{N}}\,dS.$$

If $\nabla^2\phi = 0$ in $D$ and $\phi = 0$ on $\mathcal{S}$, then

$$\iiint_D |\nabla\phi|^2\,dV = 0.$$

Since $\phi$ is assumed to be smooth, $\nabla\phi = 0$ throughout $D$, and therefore $\phi$ is constant on each connected component of $D$. Since $\phi = 0$ on $\mathcal{S}$, these constants must all be 0, and $\phi = 0$ on $D$.

**26.** Re-examine the solution to Exercise 24 above. If $\nabla^2\phi = 0$ in $D$ and $\partial\phi/\partial n = \nabla\phi\bullet\hat{\mathbf{N}} = 0$ on $\mathcal{S}$, then we can again conclude that

$$\iiint_D |\nabla\phi|\,dV = 0$$

and $\nabla\phi = 0$ throughout $D$. Thus $\phi$ is constant on the connected components of $D$. (We can't conclude the constant is 0 because we don't know the value of $\phi$ on $\mathcal{S}$.) If $u$ and $v$ are solutions of the given Neumann problem, then $\phi = u - v$ satisfies

$$\nabla^2\phi = \nabla^2 u - \nabla^2 v = f - f = 0 \text{ on } D$$
$$\frac{\partial\phi}{\partial n} = \frac{\partial u}{\partial n} - \frac{\partial v}{\partial n} = g - g = 0 \text{ on } \mathcal{S},$$

so $\phi$ is constant on any connected component of $\mathcal{S}$, and $u$ and $v$ can only differ by a constant on $\mathcal{S}$.

**28.** By Theorem 3(b),

$$\begin{aligned}\mathbf{div}\,(\phi\nabla\psi - \psi\nabla\phi) &= \nabla\phi\bullet\nabla\psi + \phi\nabla^2\psi - \nabla\psi\bullet\nabla\phi - \psi\nabla^2\phi \\ &= \phi\nabla^2\psi - \psi\nabla^2\phi.\end{aligned}$$

Hence, by the Divergence Theorem,

$$\begin{aligned}\iiint_D (\phi\nabla^2\psi - \psi\nabla^2\phi)\,dV &= \iiint_D \mathbf{div}\,(\phi\nabla\psi - \psi\nabla\phi)\,dV \\ &= \oiint_{\mathcal{S}} (\phi\nabla\psi - \psi\nabla\phi)\bullet\hat{\mathbf{N}}\,dS \\ &= \oiint_{\mathcal{S}} \left(\phi\frac{\partial\psi}{\partial n} - \psi\frac{\partial\phi}{\partial n}\right) dS.\end{aligned}$$

**30.**

$$\begin{aligned}\frac{1}{\text{vol}(D_\epsilon)}\oiint_{\mathcal{S}_\epsilon} \mathbf{F}\bullet\hat{\mathbf{N}}\,dS &= \frac{1}{\text{vol}(D_\epsilon)}\iiint_{D_\epsilon} \mathbf{div\,F}\,dV \\ &= \frac{1}{\text{vol}(D_\epsilon)}\left[\iiint_{D_\epsilon} \mathbf{div\,F}(P_0)\,dV \right.\\ &\quad \left. + \iiint_{D_\epsilon} \big(\mathbf{div\,F} - \mathbf{div\,F}(P_0)\big)\,dV\right] \\ &= \mathbf{div\,F}(P_0) + \frac{1}{\text{vol}(D_\epsilon)}\iiint_{D_\epsilon}\big(\mathbf{div\,F} - \mathbf{div\,F}(P_0)\big)\,dV.\end{aligned}$$

Thus

$$\begin{aligned}&\left|\frac{1}{\text{vol}(D_\epsilon)}\oiint_{\mathcal{S}_\epsilon} \mathbf{F}\bullet\hat{\mathbf{N}}\,dS - \mathbf{div\,F}(P_0)\right| \\ &\le \frac{1}{\text{vol}(D_\epsilon)}\iiint_{D_\epsilon} |\mathbf{div\,F} - \mathbf{div\,F}(P_0)|\,dV \\ &\le \max_{P \text{ in } D_\epsilon} |\mathbf{div\,F} - \mathbf{div\,F}(P_0)| \\ &\to 0 \text{ as } \epsilon \to 0+ \text{ assuming } \mathbf{div\,F} \text{ is continuous.}\end{aligned}$$

$$\lim_{\epsilon\to 0+} \frac{1}{\text{vol}(D_\epsilon)}\oiint_{\mathcal{S}_\epsilon} \mathbf{F}\bullet\hat{\mathbf{N}}\,dS = \mathbf{div\,F}(P_0).$$

## Section 9.4 Green's Theorem and Stokes's Theorem (page 447)

**2.**

$$\begin{aligned}&\oint_C (x^2 - xy)\,dx + (xy - y^2)\,dy \\ &= -\iint_T \left[\frac{\partial}{\partial x}(xy - y^2) - \frac{\partial}{\partial y}(x^2 - xy)\right] dA \\ &= -\iint_T (y + x)\,dA \\ &= -(\overline{y} + \overline{x}) \times (\text{area of } T) = -\left(\frac{1}{3} + 1\right) \times 1 = -\frac{4}{3}.\end{aligned}$$

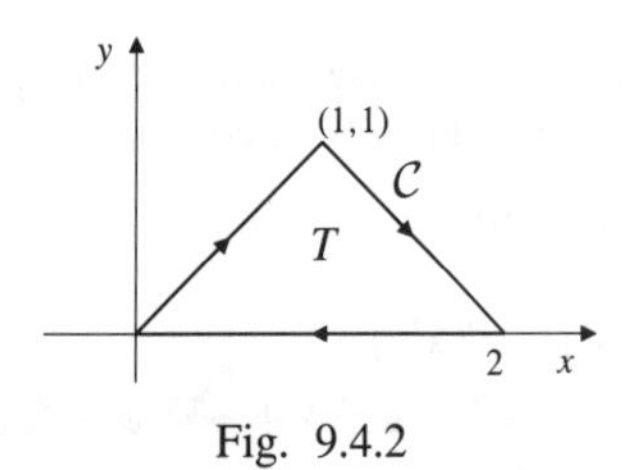

Fig. 9.4.2

**4.** Let $D$ be the region $x^2 + y^2 \le 9$, $y \ge 0$. Since $\mathcal{C}$ is the clockwise boundary of $D$,

$$\begin{aligned}&\oint_{\mathcal{C}} x^2 y\,dx - xy^2\,dy\\ &= -\iint_D \left[\frac{\partial}{\partial x}(-xy^2) - \frac{\partial}{\partial y}(x^2 y)\right] dx\,dy\\ &= \iint_D (y^2 + x^2)\,dA = \int_0^{\pi} d\theta \int_0^3 r^3\,dr = \frac{81\pi}{4}.\end{aligned}$$

**6.** $\mathbf{F} = ye^x\mathbf{i} + (x + e^x)\mathbf{j} + z^2\mathbf{k} = x\mathbf{j} + \nabla\phi$, where $\phi = ye^x + (z^3/3)$. The curve $\mathcal{C}$:

$$\mathbf{r} = (1+\cos t)\mathbf{i} + (1+\sin t)\mathbf{j} + (1-\sin t-\cos t)\mathbf{k}, \quad (0 \le t \le 2\pi)$$

is a closed curve, so

$$\begin{aligned}\oint_{\mathcal{C}} \mathbf{F} \bullet d\mathbf{r} &= \oint_{\mathcal{C}} x\mathbf{j} \bullet d\mathbf{r} + \oint_{\mathcal{C}} \nabla\phi \bullet d\mathbf{r}\\ &= \oint_{\mathcal{C}} x\,dy\\ &= \int_0^{2\pi} (1 + \cos t)\cos t\,dt = \int_0^{2\pi} \cos^2 t\,dt = \pi.\end{aligned}$$

Remark: this problem can be done by using Stokes's Theorem: $\mathcal{C}$ is the boundary of an elliptic disk in the plane $x + y + z = 3$ lying inside the cylinder $(x - 1)^2 + (y - 1)^2 = 1$. However, the above direct solution is simpler.

**8.** a) $\mathbf{F} = x^2\mathbf{j}$

$$\oint_{\mathcal{C}} \mathbf{F} \bullet d\mathbf{r} = \oint_{\mathcal{C}} x^2\,dy = \iint_R 2x\,dA = 2A\bar{x}.$$

b) $\mathbf{F} = xy\mathbf{i}$

$$\oint_{\mathcal{C}} \mathbf{F} \bullet d\mathbf{r} = \oint_{\mathcal{C}} xy\,dx = -\iint_R x\,dA = -A\bar{x}.$$

c) $\mathbf{F} = y^2\mathbf{i} + 3xy\mathbf{j}$

$$\begin{aligned}&\oint_{\mathcal{C}} \mathbf{F} \bullet d\mathbf{r} = \oint_{\mathcal{C}} y^2\,dx + 3xy\,dy\\ &= \iint_R (3y - 2y)\,dA = A\bar{y}.\end{aligned}$$

**10.** Let $\mathcal{S}$ be the part of the surface $z = y^2$ lying inside the cylinder $x^2 + y^2 = 4$, and having upward normal $\hat{\mathbf{N}}$. Then $\mathcal{C}$ is the oriented boundary of $\mathcal{S}$. Let $D$ be the disk $x^2 + y^2 \le 4$, $z = 0$, that is, the projection of $\mathcal{S}$ onto the $xy$-plane.

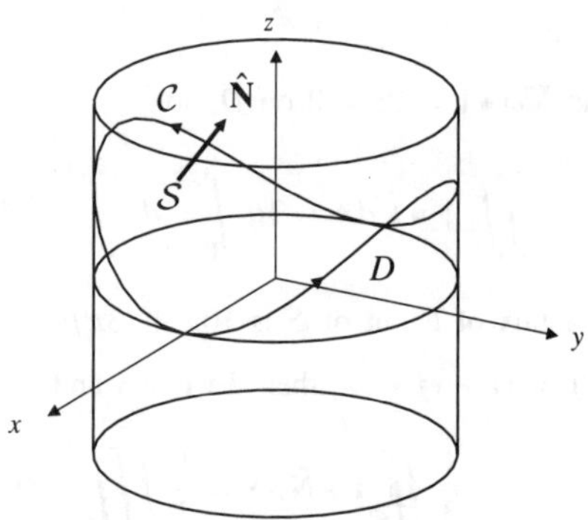

Fig. 9.4.10

If $\mathbf{F} = y\mathbf{i} - x\mathbf{j} + z^2\mathbf{k}$, then

$$\mathbf{curl\,F} = \begin{vmatrix} \mathbf{i} & \mathbf{j} & \mathbf{k}\\ \frac{\partial}{\partial x} & \frac{\partial}{\partial y} & \frac{\partial}{\partial z}\\ y & -x & z^2 \end{vmatrix} = -2\mathbf{k}.$$

Since $dS = \dfrac{dx\,dy}{\mathbf{k} \bullet \hat{\mathbf{N}}}$ on $\mathcal{S}$, we have

$$\begin{aligned}\oint_{\mathcal{C}} y\,dx - x\,dy + z^2\,dz &= \oint_{\mathcal{C}} \mathbf{F} \bullet d\mathbf{r} = \iint_{\mathcal{S}} \mathbf{curl\,F} \bullet \hat{\mathbf{N}}\,dS\\ &= \iint_D -2\mathbf{k} \bullet \hat{\mathbf{N}}\,\frac{dx\,dy}{\mathbf{k} \bullet \hat{\mathbf{N}}} = -8\pi.\end{aligned}$$

**12.** The surface $\mathcal{S}$ with equation

$$x^2 + y^2 + 2(z - 1)^2 = 6, \qquad z \ge 0,$$

with outward normal $\hat{\mathbf{N}}$, is that part of an ellipsoid of revolution about the $z$-axis, centred at $(0, 0, 1)$, and lying above the $xy$-plane. The boundary of $\mathcal{S}$ is the circle $\mathcal{C}$: $x^2 + y^2 = 4$, $z = 0$, oriented counterclockwise as seen from the positive $z$-axis. $\mathcal{C}$ is also the oriented boundary of the disk $x^2 + y^2 \le 4$, $z = 0$, with normal $\hat{\mathbf{N}} = \mathbf{k}$. If $\mathbf{F} = (xz - y^3\cos z)\mathbf{i} + x^3 e^z\mathbf{j} + xyze^{x^2+y^2+z^2}\mathbf{k}$, then, on $z = 0$, we have

$$\begin{aligned}\mathbf{curl\,F} \bullet \mathbf{k} &= \left(\frac{\partial}{\partial x}x^3 e^z - \frac{\partial}{\partial y}(xz - y^3\cos z)\right)\Big|_{z=0}\\ &= \left(3x^2 e^z + 3y^2\cos z\right)\Big|_{z=0} = 3(x^2 + y^2).\end{aligned}$$

Thus

$$\begin{aligned}\iint_{\mathcal{S}} \mathbf{curl\,F} \bullet \hat{\mathbf{N}}\,dS &= \oint_{\mathcal{C}} \mathbf{F} \bullet d\mathbf{r} = \iint_D \mathbf{curl\,F} \bullet \mathbf{k}\,dA\\ &= \int_0^{2\pi} d\theta \int_0^2 3r^2\,r\,dr = 24\pi.\end{aligned}$$

**14.** The curve $\mathcal{C}$:

$$\mathbf{r} = \cos t\mathbf{i} + \sin t\mathbf{j} + \sin 2t\mathbf{k}, \qquad 0 \le t \le 2\pi,$$

lies on the surface $z = 2xy$, since $\sin 2t = 2\cos t \sin t$. It also lies on the cylinder $x^2 + y^2 = 1$, so it is the boundary of that part of $z = 2xy$ lying inside that cylinder. Since $\mathcal{C}$ is oriented counterclockwise as seen from high on the $z$-axis, $\mathcal{S}$ should be oriented with upward normal,

$$\hat{\mathbf{N}} = \frac{-2y\mathbf{i} - 2x\mathbf{j} + \mathbf{k}}{\sqrt{1 + 4(x^2 + y^2)}},$$

and has area element

$$dS = \sqrt{1 + 4(x^2 + y^2)}\,dx\,dy.$$

If $\mathbf{F} = (e^x - y^3)\mathbf{i} + (e^y + x^3)\mathbf{j} + e^z\mathbf{k}$, then

$$\mathbf{curl\,F} = \begin{vmatrix} \mathbf{i} & \mathbf{j} & \mathbf{k} \\ \dfrac{\partial}{\partial x} & \dfrac{\partial}{\partial y} & \dfrac{\partial}{\partial z} \\ e^x - y^3 & e^y + x^3 & e^z \end{vmatrix} = 3(x^2 + y^2)\mathbf{k}.$$

If $D$ is the disk $x^2 + y^2 \le 1$ in the $xy$-plane, then

$$\begin{aligned} \oint_{\mathcal{C}} \mathbf{F} \bullet d\mathbf{r} &= \iint_{\mathcal{S}} \mathbf{curl\,F} \bullet \hat{\mathbf{N}}\,dS = \iint_D 3(x^2 + y^2)\,dx\,dy \\ &= 3\int_0^{2\pi} d\theta \int_0^1 r^2\,r\,dr = \frac{3\pi}{2}. \end{aligned}$$

**16.** The closed curve

$$\mathbf{r} = (1 + \cos t)\mathbf{i} + (1 + \sin t)\mathbf{j} + (1 - \cos t - \sin t)\mathbf{k},$$

$(0 \le t \le 2\pi)$, lies in the plane $x + y + z = 3$ and is oriented counterclockwise as seen from above. Therefore it is the boundary of a region $\mathcal{S}$ in that plane with normal field $\hat{\mathbf{N}} = (\mathbf{i} + \mathbf{j} + \mathbf{k})/\sqrt{3}$. The projection of $\mathcal{S}$ onto the $xy$-plane is the circular disk $D$ of radius 1 with centre at $(1, 1)$.

If $\mathbf{F} = ye^x\mathbf{i} + (x^2 + e^x)\mathbf{j} + z^2e^z\mathbf{k}$, then

$$\mathbf{curl\,F} = \begin{vmatrix} \mathbf{i} & \mathbf{j} & \mathbf{k} \\ \dfrac{\partial}{\partial x} & \dfrac{\partial}{\partial y} & \dfrac{\partial}{\partial z} \\ ye^x & x^2 + e^x & z^2 + e^z \end{vmatrix} = 2x\mathbf{k}.$$

By Stokes's Theorem,

$$\begin{aligned} \oint_{\mathcal{C}} \mathbf{F} \bullet d\mathbf{r} &= \iint_{\mathcal{S}} \mathbf{curl\,F} \bullet \hat{\mathbf{N}}\,dS \\ &= \iint_{\mathcal{S}} \frac{2x}{\sqrt{3}}\,dS = \iint_D \frac{2x}{\sqrt{3}}(\sqrt{3})\,dx\,dy \\ &= 2\bar{x}A = 2\pi, \end{aligned}$$

where $\bar{x} = 1$ is the $x$-coordinate of the boundary of $D$, and $A = \pi 1^2 = \pi$ is the area of $D$.

**18.** The curve $\mathcal{C}$: $(x - 1)^2 + 4y^2 = 16$, $2x + y + z = 3$, oriented counterclockwise as seen from above, bounds an elliptic disk $\mathcal{S}$ on the plane $2x + y + z = 3$. $\mathcal{S}$ has normal $\hat{\mathbf{N}} = (2\mathbf{i} + \mathbf{j} + \mathbf{k})/\sqrt{6}$. Since its projection onto the $xy$-plane is an elliptic disk with centre at $(1, 0, 0)$ and area $\pi(4)(2) = 8\pi$, therefore $\mathcal{S}$ has area $8\sqrt{6}\pi$ and centroid $(1, 0, 1)$. If

$$\mathbf{F} = (z^2 + y^2 + \sin x^2)\mathbf{i} + (2xy + z)\mathbf{j} + (xz + 2yz)\mathbf{k},$$

then

$$\begin{aligned} \mathbf{curl\,F} &= \begin{vmatrix} \mathbf{i} & \mathbf{j} & \mathbf{k} \\ \dfrac{\partial}{\partial x} & \dfrac{\partial}{\partial y} & \dfrac{\partial}{\partial z} \\ z^2 + y^2 + \sin x^2 & 2xy + z & xz + 2yz \end{vmatrix} \\ &= (2z - 1)\mathbf{i} + z\mathbf{j}. \end{aligned}$$

By Stokes's Theorem,

$$\begin{aligned} \oint_{\mathcal{C}} \mathbf{F} \bullet d\mathbf{r} &= \iint_{\mathcal{S}} \mathbf{curl\,F} \bullet \hat{\mathbf{N}}\,dS \\ &= \frac{1}{\sqrt{6}} \iint_{\mathcal{S}} (2(2z - 1) + z)\,dS \\ &= \frac{5\bar{z} - 2}{\sqrt{6}}(8\sqrt{6}\pi) = 24\pi. \end{aligned}$$

**20.** We are given that $\mathcal{C}$ bounds a region $R$ in a plane $P$ with unit normal $\hat{\mathbf{N}} = a\mathbf{i} + b\mathbf{j} + c\mathbf{k}$. Therefore, $a^2 + b^2 + c^2 = 1$. If $\mathbf{F} = (bz - cy)\mathbf{i} + (cx - az)\mathbf{j} + (ay - bx)\mathbf{k}$, then

$$\begin{aligned} \mathbf{curl\,F} &= \begin{vmatrix} \mathbf{i} & \mathbf{j} & \mathbf{k} \\ \dfrac{\partial}{\partial x} & \dfrac{\partial}{\partial y} & \dfrac{\partial}{\partial z} \\ bz - cy & cx - az & ay - bx \end{vmatrix} \\ &= 2a\mathbf{i} + 2b\mathbf{j} + 2c\mathbf{k}. \end{aligned}$$

Hence $\mathbf{curl\,F} \bullet \hat{\mathbf{N}} = 2(a^2 + b^2 + c^2) = 2$. We have

$$\begin{aligned} \frac{1}{2}\oint_{\mathcal{C}} &(bz - cy)\,dx + (cx - az)\,dy + (ay - bx)\,dz \\ &= \frac{1}{2}\oint_{\mathcal{C}} \mathbf{F} \bullet d\mathbf{r} = \frac{1}{2}\iint_R \mathbf{curl\,F} \bullet \hat{\mathbf{N}}\,dS \\ &= \frac{1}{2}\iint_R 2\,dS = \text{area of } R. \end{aligned}$$

**22.** The circle $\mathcal{C}_\epsilon$ of radius $\epsilon$ centred at $P$ is the oriented boundary of the disk $\mathcal{S}_\epsilon$ of area $\pi\epsilon^2$ having constant normal field $\hat{\mathbf{N}}$. By Stokes's Theorem,

$$\begin{aligned}\oint_{\mathcal{C}_\epsilon} \mathbf{F}\bullet d\mathbf{r} &= \iint_{\mathcal{S}_\epsilon} \mathbf{curl\,F}\bullet\hat{\mathbf{N}}\,dS \\ &= \iint_{\mathcal{S}_\epsilon} \mathbf{curl\,F}(P)\bullet\hat{\mathbf{N}}\,dS \\ &\quad + \iint_{\mathcal{S}_\epsilon}\big(\mathbf{curl\,F}-\mathbf{curl\,F}(P)\big)\bullet\hat{\mathbf{N}}\,dS \\ &= \pi\epsilon^2\mathbf{curl\,F}(P)\bullet\hat{\mathbf{N}} \\ &\quad + \iint_{\mathcal{S}_\epsilon}\big(\mathbf{curl\,F}-\mathbf{curl\,F}(P)\big)\bullet\hat{\mathbf{N}}\,dS.\end{aligned}$$

Since $\mathbf{F}$ is assumed smooth, its curl is continuous at $P$. Therefore

$$\begin{aligned}&\left|\frac{1}{\pi\epsilon^2}\oint_{\mathcal{C}_\epsilon}\mathbf{F}\bullet d\mathbf{r} - \mathbf{curl\,F}(P)\bullet\hat{\mathbf{N}}\right| \\ &\le \frac{1}{\pi\epsilon^2}\iint_{\mathcal{S}_\epsilon}\left|\big(\mathbf{curl\,F}-\mathbf{curl\,F}(P)\big)\bullet\hat{\mathbf{N}}\right|\,dS \\ &\le \max_{Q \text{ on } \mathcal{S}_\epsilon}|\mathbf{curl\,F}(Q)-\mathbf{curl\,F}(P)| \\ &\to 0 \text{ as } \epsilon\to 0+.\end{aligned}$$

Thus $\displaystyle\lim_{\epsilon\to 0+}\oint_{\mathcal{C}_\epsilon}\mathbf{F}\bullet d\mathbf{r} = \mathbf{curl\,F}(P)\bullet\hat{\mathbf{N}}$.

## Section 9.5 Some Physical Applications of Vector Calculus (page 455)

**2.** The first component of $\mathbf{F}(\mathbf{G}\bullet\hat{\mathbf{N}})$ is $(F_1\mathbf{G})\bullet\hat{\mathbf{N}}$. Applying the Divergence Theorem and Theorem 3(b), we obtain

$$\begin{aligned}\oiint_{\mathcal{S}}(F_1\mathbf{G})\bullet\hat{\mathbf{N}}\,dS &= \iiint_D \mathbf{div}\,(F_1\mathbf{G})\,dV \\ &= \iiint_D \big(\nabla F_1\bullet\mathbf{G} + F_1\nabla\bullet\mathbf{G}\big)\,dS.\end{aligned}$$

But $\nabla F_1 \bullet \mathbf{G}$ is the first component of $(\mathbf{G}\bullet\nabla)\mathbf{F}$, and $F_1\nabla\bullet\mathbf{G}$ is the first component of $\mathbf{F}\mathbf{div}\,\mathbf{G}$. Similar results obtain for the other components, so

$$\oiint_{\mathcal{S}}\mathbf{F}(\mathbf{G}\bullet\hat{\mathbf{N}})\,dS = \iiint_D\big(\mathbf{F}\mathbf{div}\,\mathbf{G} + (\mathbf{G}\bullet\nabla)\mathbf{F}\big)\,dV.$$

**4.** Since $\mathbf{r} = x\mathbf{i}+y\mathbf{j}+z\mathbf{k}$ and $\mathbf{b} = b_1\mathbf{i}+b_2\mathbf{j}+b_3\mathbf{k}$, we have

$$\begin{aligned}|\mathbf{r}-\mathbf{b}|^2 &= (x-b_1)^2+(y-b_2)^2+(z-b_3)^2 \\ 2|\mathbf{r}-\mathbf{b}|\frac{\partial}{\partial x}|\mathbf{r}-\mathbf{b}| &= 2(x-b_1) \\ \frac{\partial}{\partial x}|\mathbf{r}-\mathbf{b}| &= \frac{x-b_1}{|\mathbf{r}-\mathbf{b}|}.\end{aligned}$$

Similar formulas hold for the other first partials of $|\mathbf{r}-\mathbf{b}|$, so

$$\begin{aligned}&\nabla\left(\frac{1}{|\mathbf{r}-\mathbf{b}|}\right) \\ &= \frac{-1}{|\mathbf{r}-\mathbf{b}|^2}\left(\frac{\partial}{\partial x}|\mathbf{r}-\mathbf{b}|\mathbf{i}+\cdots+\frac{\partial}{\partial z}|\mathbf{r}-\mathbf{b}|\mathbf{k}\right) \\ &= \frac{-1}{|\mathbf{r}-\mathbf{b}|^2}\,\frac{(x-b_1)\mathbf{i}+(y-b_2)\mathbf{j}+(z-b_3)\mathbf{k}}{|\mathbf{r}-\mathbf{b}|} \\ &= -\frac{\mathbf{r}-\mathbf{b}}{|\mathbf{r}-\mathbf{b}|^3}.\end{aligned}$$

**6.** For any element $ds$ on the filament $\mathcal{F}$, we have

$$\mathbf{div}\left(d\mathbf{s}\times\frac{\mathbf{r}-\mathbf{s}}{|\mathbf{r}-\mathbf{s}|^3}\right) = 0$$

by Exercise 5, since the divergence is taken with respect to $\mathbf{r}$, and so $\mathbf{s}$ and $ds$ can be regarded as constant. Hence

$$\mathbf{div}\oint_{\mathcal{F}}\frac{d\mathbf{s}\times(\mathbf{r}-\mathbf{s})}{|\mathbf{r}-\mathbf{s}|^3} = \oint_{\mathcal{F}}\mathbf{div}\left(d\mathbf{s}\times\frac{\mathbf{r}-\mathbf{s}}{|\mathbf{r}-\mathbf{s}|^3}\right) = 0.$$

**8.** The first component of $(d\mathbf{s}\bullet\nabla)\mathbf{F}(s)$ is $\nabla F_1(s)\bullet d\mathbf{s}$. Since $\mathcal{F}$ is closed and $\nabla F_1$ is conservative,

$$\mathbf{i}\bullet\oint_{\mathcal{F}}(d\mathbf{s}\bullet\nabla)\mathbf{F}(s) = \oint_{\mathcal{F}}\nabla F_1(s)\bullet d\mathbf{s} = 0.$$

Similarly, the other components have zero line integrals, so

$$\oint_{\mathcal{F}}(d\mathbf{s}\bullet\nabla)\mathbf{F}(s) = \mathbf{0}.$$

**10.** By analogy with the filament case, the current in volume element $dV$ at position $\mathbf{s}$ is $\mathbf{J}(\mathbf{s})\,dV$, which gives rise at position $\mathbf{r}$ to a magnetic field

$$d\mathbf{H}(\mathbf{r}) = \frac{1}{4\pi}\frac{\mathbf{J}(\mathbf{s})\times(\mathbf{r}-\mathbf{s})}{|\mathbf{r}-\mathbf{s}|^3}\,dV.$$

If $R$ is a region of 3-space outside which $\mathbf{J}$ is identically zero, then at any point $\mathbf{r}$ in 3-space, the total magnetic field is

$$\mathbf{H}(\mathbf{r}) = \frac{1}{4\pi}\iiint_R\frac{\mathbf{J}(\mathbf{s})\times(\mathbf{r}-\mathbf{s})}{|\mathbf{r}-\mathbf{s}|^3}\,dV.$$

Now $\mathbf{A}(\mathbf{r})$ was defined to be

$$\mathbf{A}(\mathbf{r}) = \frac{1}{4\pi}\iiint_R\frac{\mathbf{J}(\mathbf{s})}{|\mathbf{r}-\mathbf{s}|}\,dV.$$

We have

$$\begin{aligned}\mathbf{curl}\,\mathbf{A}(\mathbf{r}) &= \frac{1}{4\pi}\iiint_R \nabla_\mathbf{r} \times \left(\frac{1}{|\mathbf{r}-\mathbf{s}|}\mathbf{J}(\mathbf{s})\right) dV\\ &= \frac{1}{4\pi}\iiint_R \nabla_\mathbf{r}\frac{1}{|\mathbf{r}-\mathbf{s}|} \times \mathbf{J}(\mathbf{s})\,dV\\ &\qquad\text{(by Theorem 3(c))}\\ &= -\frac{1}{4\pi}\iiint_R \frac{(\mathbf{r}-\mathbf{s})\times\mathbf{J}(\mathbf{s})}{|\mathbf{r}-\mathbf{s}|^3}\,dV\\ &\qquad\text{(by Exercise 4)}\\ &= \mathbf{H}(\mathbf{r}).\end{aligned}$$

**12.** $\mathbf{A}(\mathbf{r}) = \dfrac{1}{4\pi}\displaystyle\iiint_R \frac{\mathbf{J}(\mathbf{s})\,dV}{|\mathbf{r}-\mathbf{s}|}$, where $R$ is a region of 3-space such that $\mathbf{J}(\mathbf{s}) = \mathbf{0}$ outside $R$. We assume that $\mathbf{J}(\mathbf{s})$ is continuous, so $\mathbf{J}(\mathbf{s}) = \mathbf{0}$ on the surface $\mathcal{S}$ of $R$.
In the following calculations we use subscripts $\mathbf{s}$ and $\mathbf{r}$ to denote the variables with respect to which derivatives are taken. By Theorem 3(b),

$$\begin{aligned}\mathbf{div}_\mathbf{s}\frac{\mathbf{J}(\mathbf{s})}{|\mathbf{r}-\mathbf{s}|} &= \left(\nabla_\mathbf{s}\frac{1}{|\mathbf{r}-\mathbf{s}|}\right)\bullet\mathbf{J}(\mathbf{s}) + \frac{1}{|\mathbf{r}-\mathbf{s}|}\nabla_\mathbf{s}\bullet\mathbf{J}(\mathbf{s})\\ &= -\nabla_\mathbf{r}\left(\frac{1}{|\mathbf{r}-\mathbf{s}|}\right)\bullet\mathbf{J}(\mathbf{s}) + 0\end{aligned}$$

because $\nabla_\mathbf{r}|\mathbf{r}-\mathbf{s}| = -\nabla_\mathbf{s}|\mathbf{r}-\mathbf{s}|$, and because $\nabla\bullet\mathbf{J} = \nabla\bullet(\nabla\times\mathbf{H}) = 0$ by Theorem 3(g). Hence

$$\begin{aligned}\mathbf{div}\,\mathbf{A}(\mathbf{r}) &= \frac{1}{4\pi}\iiint_R\left(\nabla_\mathbf{r}\frac{1}{|\mathbf{r}-\mathbf{s}|}\right)\bullet\mathbf{J}(\mathbf{s})\,dV\\ &= -\frac{1}{4\pi}\iiint_R \nabla_\mathbf{s}\bullet\frac{\mathbf{J}(\mathbf{s})}{|\mathbf{r}-\mathbf{s}|}\,dV\\ &= -\frac{1}{4\pi}\oiint_\mathcal{S}\frac{\mathbf{J}(\mathbf{s})}{|\mathbf{r}-\mathbf{s}|}\bullet\hat{\mathbf{N}}\,dS = 0\end{aligned}$$

since $\mathbf{J}(\mathbf{s}) = \mathbf{0}$ on $\mathcal{S}$.

By Theorem 3(i),

$$\mathbf{J} = \nabla\times\mathbf{H} = \nabla\times(\nabla\times\mathbf{A}) = \nabla(\nabla\bullet\mathbf{A}) - \nabla^2\mathbf{A} = -\nabla^2\mathbf{A}.$$

**13.** If we measure depth in the liquid by $-z$, so that the $z$-axis is vertical and $z = 0$ at the surface, then the pressure at depth $-z$ is $p = -\delta g z$, where $\delta$ is the density of the liquid. Thus

$$\nabla p = -\delta g\mathbf{k} = \delta\mathbf{g},$$

where $\mathbf{g} = -g\mathbf{k}$ is the constant downward vector acceleration of gravity.
The force of the liquid on surface element $dS$ of the solid with outward (from the solid) normal $\hat{\mathbf{N}}$ is

$$d\mathbf{B} = -p\hat{\mathbf{N}}\,dS = -(-\delta g z)\hat{\mathbf{N}}\,dS = \delta g z\hat{\mathbf{N}}\,dS.$$

Thus, the total force of the liquid on the solid (the buoyant force) is

$$\begin{aligned}\mathbf{B} &= \oiint_\mathcal{S}\delta g z\hat{\mathbf{N}}\,dS\\ &= \iiint_R \nabla(\delta g z)\,dV \quad \text{(see Theorem 7)}\\ &= -\iiint_R \delta\mathbf{g}\,dV = -M\mathbf{g},\end{aligned}$$

where $M = \displaystyle\iiint_R \delta\,dV$ is the mass of the liquid which would occupy the same space as the solid. Thus $\mathbf{B} = -\mathbf{F}$, where $\mathbf{F} = M\mathbf{g}$ is the weight of the liquid displaced by the solid.

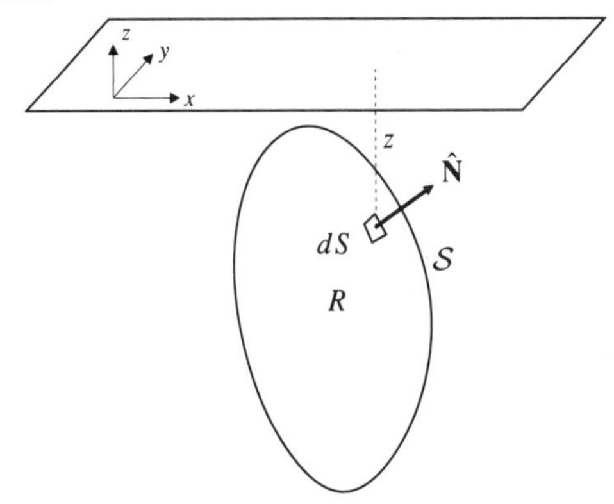


Fig. 9.5.13

**14.** The argument of Exercise 13 extends to the case where the solid is only partly submerged. Let $R^*$ be the part of the region occupied by the solid that is below the surface of the liquid. Let $\mathcal{S}^* = \mathcal{S}_1 \cup \mathcal{S}_2$ be the boundary of $R^*$, with $\mathcal{S}_1 \subset \mathcal{S}$ and $\mathcal{S}_2$ in the plane of the surface of the liquid. Since $p = -\delta g z = 0$ on $\mathcal{S}_2$, we have

$$\iint_{\mathcal{S}_2}\delta g z\hat{\mathbf{N}}\,dS = 0.$$

Therefore the buoyant force on the solid is

$$\begin{aligned}\mathbf{B} &= \iint_{\mathcal{S}_1}\delta g z\hat{\mathbf{N}}\,dS\\ &= \iint_{\mathcal{S}_1}\delta g z\hat{\mathbf{N}}\,dS + \iint_{\mathcal{S}_2}\delta g z\hat{\mathbf{N}}\,dS\\ &= \oiint_{\mathcal{S}^*}\delta g z\hat{\mathbf{N}}\,dS\\ &= -\iiint_{R^*}\delta\mathbf{g}\,dV = -M^*\mathbf{g},\end{aligned}$$

where $M^* = \iiint_{R^*} \delta\, dV$ is the mass of the liquid which would occupy $R^*$. Again we conclude that the buoyant force is the negative of the weight of the liquid displaced.

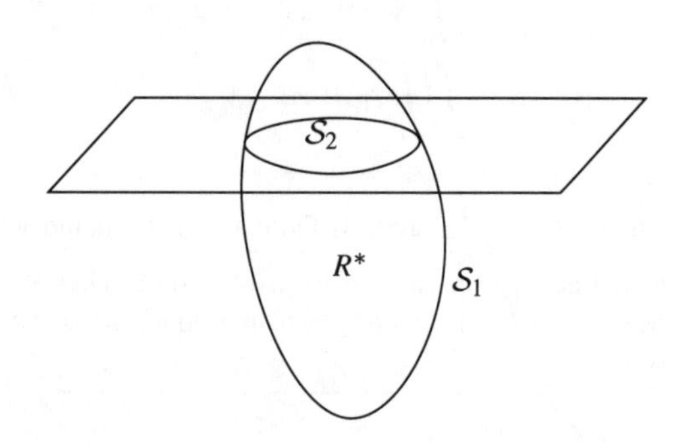


Fig. 9.5.14

## Section 9.6 Orthogonal Curvilinear Coordinates (page 465)

**2.** $f(\rho,\phi,\theta) = \rho\phi\theta$ (spherical coordinates). By Example 10,

$$\begin{aligned}\nabla f &= \frac{\partial f}{\partial \rho}\hat{\boldsymbol{\rho}} + \frac{1}{\rho}\frac{\partial f}{\partial \phi}\hat{\boldsymbol{\phi}} + \frac{1}{\rho\sin\phi}\frac{\partial f}{\partial \theta}\hat{\boldsymbol{\theta}} \\ &= \phi\theta\,\hat{\boldsymbol{\rho}} + \theta\,\hat{\boldsymbol{\phi}} + \frac{\phi}{\sin\phi}\,\hat{\boldsymbol{\theta}}.\end{aligned}$$

**4.** $\mathbf{F}(r,\theta,z) = r\hat{\boldsymbol{\theta}}$

$$\mathbf{div\,F} = \frac{1}{r}\left[\frac{\partial}{\partial\theta}(r)\right] = 0$$

$$\mathbf{curl\,F} = \frac{1}{r}\begin{vmatrix} \hat{\mathbf{r}} & r\hat{\boldsymbol{\theta}} & \mathbf{k} \\ \frac{\partial}{\partial r} & \frac{\partial}{\partial \theta} & \frac{\partial}{\partial z} \\ 0 & r^2 & 0 \end{vmatrix} = 2\mathbf{k}.$$

**6.** $\mathbf{F}(\rho,\phi,\theta) = \rho\hat{\boldsymbol{\phi}}$

$$\mathbf{div\,F} = \frac{1}{\rho^2\sin\phi}\left[\frac{\partial}{\partial\phi}\left(\rho^2\sin\phi\right)\right] = \cot\phi$$

$$\mathbf{curl\,F} = \frac{1}{\rho^2\sin\phi}\begin{vmatrix} \hat{\boldsymbol{\rho}} & \rho\hat{\boldsymbol{\phi}} & \rho\sin\phi\,\hat{\boldsymbol{\theta}} \\ \frac{\partial}{\partial \rho} & \frac{\partial}{\partial \phi} & \frac{\partial}{\partial \theta} \\ 0 & \rho^2 & 0 \end{vmatrix} = 2\hat{\boldsymbol{\theta}}.$$

**8.** $\mathbf{F}(\rho,\phi,\theta) = \rho^2\hat{\boldsymbol{\rho}}$

$$\mathbf{div\,F} = \frac{1}{\rho^2\sin\phi}\left[\frac{\partial}{\partial\rho}\left(\rho^4\sin\phi\right)\right] = 4\rho$$

$$\mathbf{curl\,F} = \frac{1}{\rho^2\sin\phi}\begin{vmatrix} \hat{\boldsymbol{\rho}} & \rho\hat{\boldsymbol{\phi}} & \rho\sin\phi\,\hat{\boldsymbol{\theta}} \\ \frac{\partial}{\partial \rho} & \frac{\partial}{\partial \phi} & \frac{\partial}{\partial \theta} \\ \rho^2 & 0 & 0 \end{vmatrix} = \mathbf{0}.$$

**10.** Since $(u, v, z)$ constitute orthogonal curvilinear coordinates in $\mathbb{R}^3$, with scale factors $h_u$, $h_v$ and $h_z = 1$, we have, for a function $f(u,v)$ independent of $z$,

$$\begin{aligned}\nabla f(u,v) &= \frac{1}{h_u}\frac{\partial f}{\partial u}\hat{\mathbf{u}} + \frac{1}{h_v}\frac{\partial f}{\partial v}\hat{\mathbf{v}} + \frac{1}{1}\frac{\partial f}{\partial z}\mathbf{k} \\ &= \frac{1}{h_u}\frac{\partial f}{\partial u}\hat{\mathbf{u}} + \frac{1}{h_v}\frac{\partial f}{\partial v}\hat{\mathbf{v}}.\end{aligned}$$

For $\mathbf{F}(u,v) = F_u(u,v)\,\hat{\mathbf{u}} + F_v(u,v)\,\hat{\mathbf{v}}$ (independent of $z$ and having no $\mathbf{k}$ component), we have

$$\mathbf{div\,F}(u,v) = \frac{1}{h_u h_v}\left[\frac{\partial}{\partial u}(h_u F_u) + \frac{\partial}{\partial v}(h_v F_v)\right]$$

$$\begin{aligned}\mathbf{curl\,F}(u,v) &= \frac{1}{h_u h_v}\begin{vmatrix} h_u\hat{\mathbf{u}} & h_v\hat{\mathbf{v}} & \mathbf{k} \\ \frac{\partial}{\partial u} & \frac{\partial}{\partial v} & \frac{\partial}{\partial z} \\ h_u F_u & h_v F_v & 0 \end{vmatrix} \\ &= \frac{1}{h_u h_v}\left[\frac{\partial}{\partial u}(h_v F_v) - \frac{\partial}{\partial v}(h_u F_u)\right]\mathbf{k}.\end{aligned}$$

**12.** $x = a\cosh u\cos v$, $\quad y = a\sinh u\sin v$.

a) $u$-curves: If $A = a\cosh u$ and $B = a\sinh u$, then

$$\frac{x^2}{A^2} + \frac{y^2}{B^2} = \cos^2 v + \sin^2 v = 1.$$

Since $A^2 - B^2 = a^2(\cosh^2 u - \sinh^2 u) = a^2$, the $u$-curves are ellipses with foci at $(\pm a, 0)$.

b) $v$-curves: If $A = a\cos v$ and $B = a\sin v$, then

$$\frac{x^2}{A^2} - \frac{y^2}{B^2} = \cosh^2 u - \sinh^2 u = 1.$$

Since $A^2 + B^2 = a^2(\cos^2 v + \sin^2 v) = a^2$, the $v$-curves are hyperbolas with foci at $(\pm a, 0)$.

c) The $u$-curve $u = u_0$ has parametric equations

$$x = a\cosh u_0\cos v, \qquad y = a\sinh u_0\sin v,$$

and therefore has slope at $(u_0, v_0)$ given by

$$m_u = \frac{dy}{dx} = \frac{dy}{dv}\Big/\frac{dx}{dv}\bigg|_{(u_0,v_0)} = \frac{a\sinh u_0\cos v_0}{-a\cosh u_0\sin v_0}.$$

The $v$-curve $v = v_0$ has parametric equations

$$x = a\cosh u\cos v_0, \qquad y = a\sinh u\sin v_0,$$

and therefore has slope at $(u_0, v_0)$ given by

$$m_v = \frac{dy}{dx} = \frac{dy}{du}\Big/\frac{dx}{du}\bigg|_{(u_0,v_0)} = \frac{a\cosh u_0\sin v_0}{a\sinh u_0\cos v_0}.$$

Since the product of these slopes is $m_u m_v = -1$, the curves $u = u_0$ and $v = v_0$ intersect at right angles.

d) $\mathbf{r} = a\cosh u \cos v\,\mathbf{i} + a\sinh u \sin v\,\mathbf{j}$

$$\frac{\partial \mathbf{r}}{\partial u} = a\sinh u \cos v\,\mathbf{i} + a\cosh u \sin v\,\mathbf{j}$$
$$\frac{\partial \mathbf{r}}{\partial v} = -a\cosh u \sin v\,\mathbf{i} + a\sinh u \cos v\,\mathbf{j}.$$

The scale factors are

$$h_u = \left|\frac{\partial \mathbf{r}}{\partial u}\right| = a\sqrt{\sinh^2 u \cos^2 v + \cosh^2 u \sin^2 v}$$
$$h_v = \left|\frac{\partial \mathbf{r}}{\partial v}\right| = a\sqrt{\sinh^2 u \cos^2 v + \cosh^2 u \sin^2 v} = h_u.$$

The area element is

$$\begin{aligned} dA &= h_u h_v\,du\,dv \\ &= a^2\left(\sinh^2 u \cos^2 v + \cosh^2 u \sin^2 v\right) du\,dv. \end{aligned}$$

**14.** $\nabla f(r,\theta,z) = \dfrac{\partial f}{\partial r}\hat{\mathbf{r}} + \dfrac{1}{r}\dfrac{\partial f}{\partial \theta}\hat{\boldsymbol{\theta}} + \dfrac{\partial f}{\partial z}\mathbf{k}$

$$\begin{aligned} \nabla^2 f(r,\theta,z) &= \mathbf{div}\left(\nabla f(r,\theta,z)\right) \\ &= \frac{1}{r}\left[\frac{\partial}{\partial r}\left(r\frac{\partial f}{\partial r}\right) + \frac{\partial}{\partial \theta}\left(\frac{1}{r}\frac{\partial f}{\partial \theta}\right) + \frac{\partial}{\partial z}\left(r\frac{\partial f}{\partial z}\right)\right] \\ &= \frac{\partial^2 f}{\partial r^2} + \frac{1}{r}\frac{\partial f}{\partial r} + \frac{1}{r^2}\frac{\partial^2 f}{\partial \theta^2} + \frac{\partial^2 f}{\partial z^2}. \end{aligned}$$

**16.** $\nabla f(u,v,w) = \dfrac{1}{h_u}\dfrac{\partial f}{\partial u}\hat{\mathbf{u}} + \dfrac{1}{h_v}\dfrac{\partial f}{\partial v}\hat{\mathbf{v}} + \dfrac{1}{h_w}\dfrac{\partial f}{\partial w}\hat{\mathbf{w}}$

$$\begin{aligned} \nabla^2 f(u,v,w) &= \mathbf{div}\left(\nabla f(u,v,w)\right) \\ &= \frac{1}{h_u h_v h_w}\left[\frac{\partial}{\partial u}\left(\frac{h_v h_w}{h_u}\frac{\partial f}{\partial u}\right) + \frac{\partial}{\partial v}\left(\frac{h_u h_w}{h_v}\frac{\partial f}{\partial v}\right)\right. \\ &\qquad \left. + \frac{\partial}{\partial w}\left(\frac{h_u h_v}{h_w}\frac{\partial f}{\partial w}\right)\right] \\ &= \frac{1}{h_u^2}\left[\frac{\partial^2 f}{\partial u^2} + \left(\frac{1}{h_v}\frac{\partial h_v}{\partial u} + \frac{1}{h_w}\frac{\partial h_w}{\partial u} - \frac{1}{h_u}\frac{\partial h_u}{\partial u}\right)\frac{\partial f}{\partial u}\right] \\ &\quad + \frac{1}{h_v^2}\left[\frac{\partial^2 f}{\partial v^2} + \left(\frac{1}{h_u}\frac{\partial h_u}{\partial v} + \frac{1}{h_w}\frac{\partial h_w}{\partial v} - \frac{1}{h_v}\frac{\partial h_v}{\partial v}\right)\frac{\partial f}{\partial v}\right] \\ &\quad + \frac{1}{h_w^2}\left[\frac{\partial^2 f}{\partial w^2} + \left(\frac{1}{h_u}\frac{\partial h_u}{\partial w} + \frac{1}{h_v}\frac{\partial h_v}{\partial w} - \frac{1}{h_w}\frac{\partial h_w}{\partial w}\right)\frac{\partial f}{\partial w}\right]. \end{aligned}$$

## Review Exercises 9 (page 466)

**2.** Let $R$ be the region inside the cylinder $\mathcal{S}$ and between the planes $z=0$ and $z=b$. The oriented boundary of $R$ consists of $\mathcal{S}$ and the disks $D_1$ with normal $\hat{\mathbf{N}}_1 = \mathbf{k}$ and $D_2$ with normal $\hat{\mathbf{N}}_2 = -\mathbf{k}$ as shown in the figure. For $\mathbf{F} = x\mathbf{i} + \cos(z^2)\mathbf{j} + e^z\mathbf{k}$ we have $\mathbf{div}\,\mathbf{F} = 1 + e^z$ and

$$\begin{aligned} \iiint_R \mathbf{div}\,\mathbf{F}\,dV &= \iint_{D_2} dx\,dy \int_0^b (1+e^z)\,dz \\ &= \iint_{D_2} [b + (e^b - 1)]\,dx\,dy \\ &= \pi a^2 b + \pi a^2(e^b - 1). \end{aligned}$$

Also
$$\iint_{D_2} \mathbf{F}\bullet(-\mathbf{k})\,dA = -\iint_{D_2} e^0\,dA = -\pi a^2$$
$$\iint_{D_1} \mathbf{F}\bullet\mathbf{k}\,dA = \iint_{D_1} e^b\,dA = \pi a^2 e^b.$$

By the Divergence Theorem

$$\begin{aligned} &\iint_{\mathcal{S}} \mathbf{F}\bullet\hat{\mathbf{N}}\,dS + \iint_{D_1} \mathbf{F}\bullet\mathbf{k}\,dA + \iint_{D_2} \mathbf{F}\bullet(-\mathbf{k})\,dA \\ &\qquad = \iiint_R \mathbf{div}\,\mathbf{F}\,dV = \pi a^2 b + \pi a^2(e^b - 1). \end{aligned}$$

Therefore, $\displaystyle\iint_{\mathcal{S}} \mathbf{F}\bullet\hat{\mathbf{N}}\,dS = \pi a^2 b.$

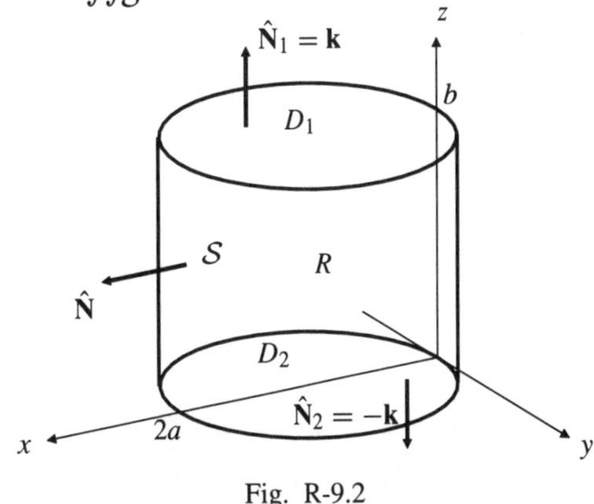


Fig. R-9.2

**4.** If $\mathbf{F} = -z\mathbf{i} + x\mathbf{j} + y\mathbf{k}$, then

$$\mathbf{curl}\,\mathbf{F} = \begin{vmatrix} \mathbf{i} & \mathbf{j} & \mathbf{k} \\ \dfrac{\partial}{\partial x} & \dfrac{\partial}{\partial y} & \dfrac{\partial}{\partial z} \\ -z & x & y \end{vmatrix} = \mathbf{i} - \mathbf{j} + \mathbf{k}.$$

The unit normal $\hat{\mathbf{N}}$ to a region in the plane $2x + y + 2z = 7$ is

$$\hat{\mathbf{N}} = \pm\frac{2\mathbf{i} + \mathbf{j} + 2\mathbf{k}}{3}.$$

If $C$ is the boundary of a disk $D$ of radius $a$ in that plane, then

$$\begin{aligned} \oint_C \mathbf{F}\bullet d\mathbf{r} &= \iint_D \mathbf{curl}\,\mathbf{F}\bullet\hat{\mathbf{N}}\,dS \\ &= \pm\iint_D \frac{2-1+2}{3}\,dS = \pm\pi a^2. \end{aligned}$$

**6.** If $\mathcal{S}$ is any surface with upward normal $\hat{\mathbf{N}}$ and boundary the curve $\mathcal{C}$: $x^2+y^2=1$, $z=2$, then $\mathcal{C}$ is oriented counterclockwise as seen from above, and it has parametrization

$$\mathbf{r} = \cos t\mathbf{i} + \sin t\mathbf{j} + 2\mathbf{k} \quad (0 \le 2 \le 2\pi).$$

Thus $d\mathbf{r} = (-\sin t\mathbf{i} + \cos t\mathbf{j})\,dt$, and if $\mathbf{F} = -y\mathbf{i} + x\cos(1-x^2-y^2)\mathbf{j} + yz\mathbf{k}$, then the flux of $\mathbf{curl\,F}$ upward through $\mathcal{S}$ is

$$\begin{aligned}\iint_{\mathcal{S}} \mathbf{curl\,F} \bullet \hat{\mathbf{N}}\,dS &= \oint_{\mathcal{C}} \mathbf{F} \bullet d\mathbf{r}\\ &= \int_0^{2\pi} (\sin^2 t + \cos^2 t + 0)\,dt = 2\pi.\end{aligned}$$

**8.** If $\mathbf{curl\,F} = \mu\mathbf{F}$ on $\mathbb{R}^3$, where $\mu \ne 0$ is a constant, then

$$\mathbf{div\,F} = \frac{1}{\mu}\mathbf{div\,curl\,F} = 0$$

by Theorem 3(g) of Section 9.2. By part (i) of the same theorem,

$$\begin{aligned}\nabla^2\mathbf{F} &= \nabla(\mathbf{div\,F}) - \mathbf{curl\,curl\,F}\\ &= 0 - \mu\mathbf{curl\,F} = -\mu^2\mathbf{F}.\end{aligned}$$

Thus $\nabla^2\mathbf{F} + \mu^2\mathbf{F} = \mathbf{0}$.

**10.** Let $\mathcal{C}$ be a simple, closed curve in the $xy$-plane bounding a region $R$. If

$$\mathbf{F} = (2y^3 - 3y + xy^2)\mathbf{i} + (x - x^3 + x^2y)\mathbf{j},$$

then by Green's Theorem, the circulation of $\mathbf{F}$ around $\mathcal{C}$ is

$$\begin{aligned}&\oint_{\mathcal{C}} \mathbf{F} \bullet d\mathbf{r}\\ &= \iint_R \left[\frac{\partial}{\partial x}(x - x^3 + x^2y) - \frac{\partial}{\partial y}(2y^3 - 3y + xy^2)\right] dA\\ &= \iint_R (1 - 3x^2 + 2xy - 6y^2 + 3 - 2xy)\,dA\\ &= \iint_R (4 - 3x^2 - 6y^2)\,dx\,dy.\end{aligned}$$

The last integral has a maximum value when the region $R$ is bounded by the ellipse $3x^2 + 6y^2 = 4$, oriented counterclockwise; this is the largest region in the $xy$-plane where the integrand is nonnegative.

**12.** Let $\mathcal{C}$ be a simple, closed curve on the plane $x+y+z=1$, oriented counterclockwise as seen from above, and bounding a plane region $\mathcal{S}$ on $x+y+z=1$. Then $\mathcal{S}$ has normal $\hat{\mathbf{N}} = (\mathbf{i}+\mathbf{j}+\mathbf{k})/\sqrt{3}$. If $\mathbf{F} = xy^2\mathbf{i} + (3z - xy^2)\mathbf{j} + (4y - x^2y)\mathbf{k}$, then

$$\begin{aligned}\mathbf{curl\,F} &= \begin{vmatrix} \mathbf{i} & \mathbf{j} & \mathbf{k}\\ \dfrac{\partial}{\partial x} & \dfrac{\partial}{\partial y} & \dfrac{\partial}{\partial z}\\ xy^2 & 3z - xy^2 & 4y - x^2y \end{vmatrix}\\ &= (1 - x^2)\mathbf{i} + 2xy\mathbf{j} - (y^2 + 2xy)\mathbf{k}.\end{aligned}$$

By Stokes's Theorem we have

$$\oint_{\mathcal{C}} \mathbf{F} \bullet d\mathbf{r} = \iint_{\mathcal{S}} \mathbf{curl\,F} \bullet \hat{\mathbf{N}}\,dS = \iint_{\mathcal{S}} \frac{1 - x^2 - y^2}{\sqrt{3}}\,dS.$$

The last integral will be maximum if the projection of $\mathcal{S}$ onto the $xy$-plane is the disk $x^2 + y^2 \le 1$. This maximum value is

$$\begin{aligned}&\iint_{x^2+y^2\le 1} \frac{1 - x^2 - y^2}{\sqrt{3}}\sqrt{3}\,dx\,dy\\ &= \int_0^{2\pi} d\theta \int_0^1 (1 - r^2)r\,dr = 2\pi\left(\frac{1}{2} - \frac{1}{4}\right) = \frac{\pi}{2}.\end{aligned}$$

## Challenging Problems 9 (page 466)

**2.** a) The steradian measure of a half-cone of semi-vertical angle $\alpha$ is

$$\int_0^{2\pi} d\theta \int_0^{\alpha} \sin\phi\,d\phi = 2\pi(1 - \cos\alpha).$$

b) If $\mathcal{S}$ is the intersection of a smooth surface with the general half-cone $K$, and is oriented with normal field $\hat{\mathbf{N}}$ pointing away from the vertex $P$ of $K$, and if $\mathcal{S}_a$ is the intersection with $K$ of a sphere of radius $a$ centred at $P$, with $a$ chosen so that $\mathcal{S}$ and $\mathcal{S}_a$ do not intersect in $K$, then $\mathcal{S}$, $\mathcal{S}_a$, and the walls of $K$ bound a solid region $R$ that does not contain the origin. If $\mathbf{F} = \mathbf{r}/|\mathbf{r}|^3$, then $\mathbf{div\,F} = 0$ in $R$ (see Example 3 in Section 9.1), and $\mathbf{F} \bullet \hat{\mathbf{N}} = 0$ on the walls of $K$. It follows from the Divergence Theorem applied to $\mathbf{F}$ over $R$ that

$$\begin{aligned}\iint_{\mathcal{S}} \mathbf{F} \bullet \hat{\mathbf{N}}\,dS &= \iint_{\mathcal{S}_a} \mathbf{F} \bullet \frac{\mathbf{r}}{|\mathbf{r}|}\,dS\\ &= \frac{a^2}{a^4}\iint_{\mathcal{S}_a} dS = \frac{1}{a^2}(\text{area of } \mathcal{S}_a)\\ &= \text{area of } \mathcal{S}_1.\end{aligned}$$

The area of $\mathcal{S}_1$ (the part of the sphere of radius 1 in $K$) is the measure (in steradians) of the solid angle subtended by $K$ at its vertex $P$. Hence this measure is given by

$$\iint_{\mathcal{S}} \frac{\mathbf{r}}{|\mathbf{r}|^3} \bullet \hat{\mathbf{N}}\,dS.$$

**4.** a) Verification of the identity

$$\frac{\partial}{\partial t}\left(\mathbf{G}\bullet\left[\frac{\partial \mathbf{r}}{\partial u}\times\frac{\partial \mathbf{r}}{\partial v}\right]\right)-\frac{\partial}{\partial u}\left(\mathbf{G}\bullet\left[\frac{\partial \mathbf{r}}{\partial t}\times\frac{\partial \mathbf{r}}{\partial v}\right]\right)$$
$$-\frac{\partial}{\partial v}\left(\mathbf{G}\bullet\left[\frac{\partial \mathbf{r}}{\partial u}\times\frac{\partial \mathbf{r}}{\partial t}\right]\right)$$
$$=\frac{\partial \mathbf{F}}{\partial t}\bullet\left[\frac{\partial \mathbf{r}}{\partial u}\times\frac{\partial \mathbf{r}}{\partial v}\right]+(\boldsymbol{\nabla}\bullet\mathbf{F})\frac{\partial \mathbf{r}}{\partial t}\bullet\left[\frac{\partial \mathbf{r}}{\partial u}\times\frac{\partial \mathbf{r}}{\partial v}\right]$$

can be carried out using the following MapleV commands:

```
with(linalg):
F:=(x,y,z,t)->
[F1(x,y,z,t),F2(x,y,z,t),F3(x,y,z,t)];
r:=(u,v,t)
->[x(u,v,t),y(u,v,t),z(u,v,t)];
ru:=(u,v,t)->diff(r(u,v,t),u);
rv:=(u,v,t)->diff(r(u,v,t),v);
rt:=(u,v,t)->diff(r(u,v,t),t);
G:=(u,v,t)
->F(x(u,v,t),y(u,v,t),z(u,v,t),t);
ruxv:=(u,v,t)
->crossprod(ru(u,v,t),rv(u,v,t));
rtxv:=(u,v,t)
->crossprod(rt(u,v,t),rv(u,v,t));
ruxt:=(u,v,t)
->crossprod(ru(u,v,t),rt(u,v,t));
LH1:=
diff(dotprod(G(u,v,t),ruxv(u,v,t)),t);
LH2:=
diff(dotprod(G(u,v,t),rtxv(u,v,t)),u);
LH3:=
diff(dotprod(G(u,v,t),ruxt(u,v,t)),v);
LHS:=simplify(LH1-LH2-LH3);
RH1:=dotprod(subs(x=x(u,v,t),
y=y(u,v,t),z=z(u,v,t),
diff(F(x,y,z,t),t)),ruxv(u,v,t));
RH2:=(divf(u,v,t))*
(dotprod(rt(u,v,t),ruxv(u,v,t)));
RHS:=simplify(RH1+RH2);
simplify(LHS-RHS);
```

Again the final output is 0, indicating that the identity is valid.

b) If $\mathcal{C}_t$ is the oriented boundary of $\mathcal{S}_t$ and $L_t$ is the corresponding counterclockwise boundary of the parameter region $R$ in the $uv$-plane, then

$$\oint_{\mathcal{C}_t}\left(\mathbf{F}\times\frac{\partial \mathbf{r}}{\partial t}\right)\bullet d\mathbf{r}$$
$$=\oint_{L_t}\left(\mathbf{G}\times\frac{\partial \mathbf{r}}{\partial t}\right)\bullet\left(\frac{\partial \mathbf{r}}{\partial u}\,du+\frac{\partial \mathbf{r}}{\partial v}\,dv\right)$$
$$=\oint_{L_t}\left[-\mathbf{G}\bullet\left(\frac{\partial \mathbf{r}}{\partial u}\times\frac{\partial \mathbf{r}}{\partial t}\right)+\mathbf{G}\bullet\left(\frac{\partial \mathbf{r}}{\partial t}\times\frac{\partial \mathbf{r}}{\partial v}\right)\right]dt$$
$$=\iint_R\left[\frac{\partial}{\partial u}\left(\mathbf{G}\bullet\left(\frac{\partial \mathbf{r}}{\partial t}\times\frac{\partial \mathbf{r}}{\partial v}\right)\right)\right.$$
$$\left.+\frac{\partial}{\partial v}\left(\mathbf{G}\bullet\left(\frac{\partial \mathbf{r}}{\partial u}\times\frac{\partial \mathbf{r}}{\partial t}\right)\right)\right]du\,dv,$$

by Green's Theorem.

c) Using the results of (a) and (b), we calculate

$$\frac{d}{dt}\iint_{\mathcal{S}_t}\mathbf{F}\bullet\hat{\mathbf{N}}\,dS=\iint_R\frac{\partial}{\partial t}\left[\mathbf{G}\bullet\left(\frac{\partial \mathbf{r}}{\partial u}\times\frac{\partial \mathbf{r}}{\partial v}\right)\right]du\,dv$$
$$=\iint_R\frac{\partial \mathbf{F}}{\partial t}\bullet\left(\frac{\partial \mathbf{r}}{\partial u}\times\frac{\partial \mathbf{r}}{\partial v}\right)du\,dv$$
$$+\iint_R(\mathbf{div}\,\mathbf{F})\frac{\partial \mathbf{r}}{\partial t}\bullet\left(\frac{\partial \mathbf{r}}{\partial u}\times\frac{\partial \mathbf{r}}{\partial v}\right)du\,dv$$
$$+\iint_R\left[\frac{\partial}{\partial u}\left(\mathbf{G}\bullet\left(\frac{\partial \mathbf{r}}{\partial t}\times\frac{\partial \mathbf{r}}{\partial v}\right)\right)\right.$$
$$\left.+\frac{\partial}{\partial v}\left(\mathbf{G}\bullet\left(\frac{\partial \mathbf{r}}{\partial u}\times\frac{\partial \mathbf{r}}{\partial t}\right)\right)\right]du\,dv$$
$$=\iint_{\mathcal{S}_t}\frac{\partial \mathbf{F}}{\partial t}\bullet\hat{\mathbf{N}}\,dS+\iint_{\mathcal{S}_t}(\mathbf{div}\,\mathbf{F})\mathbf{v}_S\bullet\hat{\mathbf{N}}\,dS$$
$$+\oint\!\!\!\int \mathcal{C}_t(\mathbf{F}\times\mathbf{v}_C)\bullet d\mathbf{r}.$$

# CHAPTER 10. ORDINARY DIFFERENTIAL EQUATIONS

## Section 10.1 Classifying Differential Equations (page 472)

**2.** $\dfrac{d^2y}{dx^2}+x=y$: 2nd order, linear, nonhomogeneous.

**4.** $y'''+xy'=x\sin x$: 3rd order, linear, nonhomogeneous.

**6.** $y''+4y'-3y=2y^2$: 2nd order, nonlinear.

**8.** $\cos x\,\dfrac{dx}{dt}+x\sin t=0$: 1st order, nonlinear, homogeneous.

**10.** $x^2y''+e^xy'=\dfrac{1}{y}$: 2nd order, nonlinear.

**12.** If $y=e^x$, then $y''-y=e^x-e^x=0$; if $y=e^{-x}$, then $y''-y=e^{-x}-e^{-x}=0$. Thus $e^x$ and $e^{-x}$ are both solutions of $y''-y=0$. Since $y''-y=0$ is linear and homogeneous, any function of the form

$$y=Ae^x+Be^{-x}$$

is also a solution. Thus $\cosh x=\frac{1}{2}(e^x+e^{-x})$ is a solution, but neither $\cos x$ nor $x^e$ is a solution.

**14.** Given that $y_1=e^{kx}$ is a solution of $y''-k^2y=0$, we suspect that $y_2=e^{-kx}$ is also a solution. This is easily verified since

$$y_2''-k^2y_2=k^2e^{-kx}-k^2e^{-kx}=0.$$

Since the DE is linear and homogeneous,

$$y=Ay_1+By_2=Ae^{kx}+Be^{-kx}$$

is a solution for any constants $A$ and $B$. It will satisfy

$$\begin{aligned}0&=y(1)=Ae^k+Be^{-k}\\2&=y'(1)=Ake^k-Bke^{-k},\end{aligned}$$

provided $A=e^{-k}/k$ and $B=-e^k/k$. The required solution is

$$y=\frac{1}{k}e^{k(x-1)}-\frac{1}{k}e^{-k(x-1)}.$$

**16.** $y=e^{rx}$ is a solution of the equation $y''-y'-2y=0$ if $r^2e^{rx}-re^{rx}-2e^{rx}=0$, that is, if $r^2-r-2=0$. This quadratic has two roots, $r=2$, and $r=-1$. Since the DE is linear and homogeneous, the function $y=Ae^{2x}+Be^{-x}$ is a solution for any constants $A$ and $B$. This solution satisfies

$$1=y(0)=A+B,\quad 2=y'(0)=2A-B,$$

provided $A=1$ and $B=0$. Thus, the required solution is $y=e^{2x}$.

**18.** If $y=y_1(x)=-e$, then $y_1'=0$ and $y_1''=0$. Thus $y_1''-y_1=0+e=e$. By Exercise 12 we know that $y_2=Ae^x+Be^{-x}$ satisfies the homogeneous DE $y''-y=0$. Therefore, by Theorem 2,

$$y=y_1(x)+y_2(x)=-e+Ae^x+Be^{-x}$$

is a solution of $y''-y=e$. This solution satisfies

$$0=y(1)=Ae+\frac{B}{e}-e,\quad 1=y'(1)=Ae-\frac{B}{e},$$

provided $A=(e+1)/(2e)$ and $B=e(e-1)/2$. Thus the required solution is $y=-e+\frac{1}{2}(e+1)e^{x-1}+\frac{1}{2}(e-1)e^{1-x}$.

## Section 10.2 First-Order Separable and Homogeneous Equations (page 476)

**2.**
$$\begin{aligned}&\frac{dy}{dx}=\frac{3y-1}{x}\\&\int\frac{dy}{3y-1}=\int\frac{dx}{x}\\&\frac{1}{3}\ln|3y-1|=\ln|x|+\frac{1}{3}\ln C\\&\frac{3y-1}{x^3}=C\\\Rightarrow\quad &y=\frac{1}{3}(1+Cx^3).\end{aligned}$$

**4.**
$$\begin{aligned}&\frac{dy}{dx}=x^2y^2\\&\int\frac{dy}{y^2}=\int x^2\,dx\\&-\frac{1}{y}=\frac{1}{3}x^3+\frac{1}{3}C\\\Rightarrow\quad &y=-\frac{3}{x^3+C}.\end{aligned}$$

**6.**
$$\begin{aligned}&\frac{dy}{dx}=x^2y^3\\&\int\frac{dy}{y^3}=\int x^2\,dx\\&-\frac{1}{2y^2}=\frac{x^3}{3}+C_1\\&y^2=\frac{1}{C-(2/3)x^3}.\end{aligned}$$

**8.**
$$\begin{aligned}&\frac{dx}{dt}=e^x\sin t\\&\int e^{-x}\,dx=\int\sin t\,dt\\&-e^{-x}=-\cos t-C\\\Rightarrow\quad &x=-\ln(\cos t+C).\end{aligned}$$

**10.** $$\frac{dy}{dx} = 1 + y^2$$
$$\int \frac{dy}{1+y^2} = \int dx$$
$$\tan^{-1} y = x + C$$
$$\Rightarrow \quad y = \tan(x + C).$$

**12.** We have
$$\frac{dy}{dx} = y^2(1-y)$$
$$\int \frac{dy}{y^2(1-y)} = \int dx = x + K.$$

Expand the left side in partial fractions
$$\frac{1}{y^2(1-y)} = \frac{A}{y} + \frac{B}{y^2} + \frac{C}{1-y}$$
$$= \frac{A(y-y^2) + B(1-y) + Cy^2}{y^2(1-y)}$$
$$\Rightarrow \begin{cases} -A + C = 0; \\ A - B = 0; \\ B = 1. \end{cases} \Rightarrow A = B = C = 1.$$

Hence,
$$\int \frac{dy}{y^2(1-y)} = \int \left(\frac{1}{y} + \frac{1}{y^2} + \frac{1}{1-y}\right) dy$$
$$= \ln|y| - \frac{1}{y} - \ln|1-y|.$$

Therefore,
$$\ln\left|\frac{y}{1-y}\right| - \frac{1}{y} = x + K.$$

**14.** $$x\frac{dy}{dx} = y \ln x$$
$$\int \frac{dy}{y} = \int \frac{\ln x}{x}\, dx$$
$$\ln|y| = \frac{1}{2}(\ln x)^2 + C_1$$
$$y = Ce^{(\ln x)^2/2}, \qquad (x > 0).$$

**16.** $$\frac{dy}{dx} = \frac{xy}{x^2 + 2y^2} \qquad \text{Let } y = vx$$
$$v + x\frac{dv}{dx} = \frac{vx^2}{(1+2v^2)x^2}$$
$$x\frac{dv}{dx} = \frac{v}{1+2v^2} - v = -\frac{2v^3}{1+2v^2}$$
$$\int \frac{1+2v^2}{v^3}\, dv = -2\int \frac{dx}{x}$$
$$-\frac{1}{2v^2} + 2\ln|v| = -2\ln|x| + C_1$$
$$-\frac{x^2}{2y^2} + 2\ln|y| = C_1$$
$$x^2 - 4y^2 \ln|y| = Cy^2.$$

**18.** $$\frac{dy}{dx} = \frac{x^3 + 3xy^2}{3x^2y + y^3} \qquad \text{Let } y = vx$$
$$v + x\frac{dv}{dx} = \frac{x^3(1+3v^2)}{x^3(3v+v^3)}$$
$$x\frac{dv}{dx} = \frac{1+3v^2}{3v+v^3} - v = \frac{1-v^4}{v(3+v^2)}$$
$$\int \frac{(3+v^2)v\, dv}{1-v^4} = \int \frac{dx}{x} \qquad \begin{array}{l}\text{Let } u = v^2 \\ du = 2v\, dv\end{array}$$
$$\frac{1}{2}\int \frac{3+u}{1-u^2}\, du = \ln|x| + C_1$$
$$\frac{3}{4}\ln\left|\frac{u+1}{u-1}\right| - \frac{1}{4}\ln|1-u^2| = \ln|x| + C_1$$
$$3\ln\left|\frac{y^2+x^2}{y^2-x^2}\right| - \ln\left|\frac{x^4-y^4}{x^4}\right| = 4\ln|x| + C_2$$
$$\ln\left|\left(\frac{x^2+y^2}{x^2-y^2}\right)^3 \frac{1}{x^4-y^4}\right| = C_2$$
$$\ln\left|\frac{(x^2+y^2)^2}{(x^2-y^2)^4}\right| = C_2$$
$$x^2 + y^2 = C(x^2-y^2)^2.$$

**20.** $$\frac{dy}{dx} = \frac{y}{x} - e^{-y/x} \quad \text{(let y=vx)}$$
$$v + x\frac{dv}{dx} = v - e^{-v}$$
$$e^v\, dv = -\frac{dx}{x}$$
$$e^v = -\ln|x| + \ln|C|$$
$$e^{y/x} = \ln\left|\frac{C}{x}\right|$$
$$y = x \ln\ln\left|\frac{C}{x}\right|.$$

**22.** $$\frac{dy}{dx} = 1 + \frac{2y}{x} \qquad \text{Let } y = vx$$
$$v + x\frac{dv}{dx} = 1 + 2v$$
$$x\frac{dv}{dx} = 1 + v$$
$$\int \frac{dv}{1+v} = \int \frac{dx}{x}$$
$$\ln|1+v| = \ln|x| + C_1$$
$$1 + \frac{y}{x} = Cx \quad \Rightarrow \quad x + y = Cx^2.$$

Since $(1, 3)$ lies on the curve, $4 = C$. Thus the curve has equation $x + y = 4x^2$.

**24.** The system $x_0+2y_0-4=0$, $2x_0-y_0-3=0$ has solution $x_0=2$, $y_0=1$. Thus, if $\xi=x-2$ and $\eta=y-1$, where

$$\frac{dy}{dx}=\frac{x+2y-4}{2x-y-3},$$

then

$$\frac{d\eta}{d\xi}=\frac{\xi+2\eta}{2\xi-\eta} \qquad \text{Let } \eta=v\xi$$

$$v+\xi\frac{dv}{d\xi}=\frac{1+2v}{2-v}$$

$$\xi\frac{dv}{d\xi}=\frac{1+2v}{2-v}-v=\frac{1+v^2}{2-v}$$

$$\int\left(\frac{2-v}{1+v^2}\right)dv=\int\frac{d\xi}{\xi}$$

$$2\tan^{-1}v-\frac{1}{2}\ln(1+v^2)=\ln|\xi|+C_1$$

$$4\tan^{-1}\frac{\eta}{\xi}-\ln(\xi^2+\eta^2)=C.$$

Hence the solution of the original equation is

$$4\tan^{-1}\frac{y-1}{x-2}-\ln\big((x-2)^2+(y-1)^2\big)=C.$$

## Section 10.3 Exact Equations and Integrating Factors (page 479)

**2.** $(e^x\sin y+2x)\,dx+(e^x\cos y+2y)\,dy=0$

$d(e^x\sin y+x^2+y^2)=0$

$e^x\sin y+x^2+y^2=C.$

**4.**
$$\left(2x+1-\frac{y^2}{x^2}\right)dx+\frac{2y}{x}\,dy=0$$

$$d\left(x^2+x+\frac{y^2}{x}\right)=0$$

$$x^2+x+\frac{y^2}{x}=C.$$

**6.**
$$(xe^x+x\ln y+y)\,dx+\left(\frac{x^2}{y}+x\ln x+x\sin y\right)dy=0$$

$$M=xe^x+x\ln y+y, \qquad N=\frac{x^2}{y}+x\ln x+x\sin y$$

$$\frac{\partial M}{\partial y}=\frac{x}{y}+1, \qquad \frac{\partial N}{\partial x}=\frac{2x}{y}+\ln x+1+\sin y$$

$$\frac{1}{N}\left(\frac{\partial M}{\partial y}-\frac{\partial N}{\partial x}\right)=\frac{1}{N}\left(-\frac{x}{y}-\ln x-\sin y\right)=-\frac{1}{x}$$

$$\frac{d\mu}{\mu}=-\frac{1}{x}\,dx \quad\Rightarrow\quad \mu=\frac{1}{x}$$

$$\left(e^x+\ln y+\frac{y}{x}\right)dx+\left(\frac{x}{y}+\ln x+\sin y\right)dy$$

$$d\left(e^x+x\ln y+y\ln x-\cos y\right)=0$$

$$e^x+x\ln y+y\ln x-\cos y=C.$$

**8.**
$$2y^2(x+y^2)\,dx+xy(x+6y^2)\,dy=0$$

$$(2xy^2+2y^4)\mu(y)\,dx+(x^2y+6xy^3)\mu(y)\,dy=0$$

$$\frac{\partial M}{\partial y}=(4xy+8y^3)\mu(y)+(2xy^2+2y^4)\mu'(y)$$

$$\frac{\partial N}{\partial x}=(2xy+6y^3)\mu(y).$$

For exactness we require

$$(2xy^2+2y^4)\mu'(y)=[(2xy+6y^3)-(4xy+8y^3)]\mu(y)$$

$$y(2xy+2y^3)\mu'(y)=-(2xy+2y^3)\mu(y)$$

$$y\mu'(y)=-\mu(y) \quad\Rightarrow\quad \mu(y)=\frac{1}{y}$$

$$(2xy+2y^3)\,dx+(x^2+6xy^2)\,dy=0$$

$$d(x^2y+2xy^3)=0 \quad\Rightarrow\quad x^2y+2xy^3=C.$$

**10.** If $\mu(xy)$ is an integrating factor for $M\,dx+N\,dy=0$, then

$$\frac{\partial}{\partial y}(\mu M)=\frac{\partial}{\partial x}(\mu N), \qquad \text{or}$$

$$x\mu'(xy)M+\mu(xy)\frac{\partial M}{\partial y}=y\mu'(xy)N+\mu(xy)\frac{\partial N}{\partial x}.$$

Thus $M$ and $N$ will have to be such that the right-hand side of the equation

$$\frac{\mu'(xy)}{\mu(xy)}=\frac{1}{xM-yN}\left(\frac{\partial N}{\partial x}-\frac{\partial M}{\partial y}\right)$$

depends only on the product $xy$.

## Section 10.4 First-Order Linear Equations (page 482)

**2.** We have $\dfrac{dy}{dx}+\dfrac{2y}{x}=\dfrac{1}{x^2}$. Let $\mu=\displaystyle\int\frac{2}{x}\,dx=2\ln x=\ln x^2$, then $e^\mu=x^2$, and

$$\frac{d}{dx}(x^2y)=x^2\frac{dy}{dx}+2xy$$

$$=x^2\left(\frac{dy}{dx}+\frac{2y}{x}\right)=x^2\left(\frac{1}{x^2}\right)=1$$

$$\Rightarrow\quad x^2y=\int dx=x+C \quad\Rightarrow\quad y=\frac{1}{x}+\frac{C}{x^2}.$$

**4.** We have $\dfrac{dy}{dx}+y=e^x$. Let $\mu=\int dx=x$, then $e^\mu=e^x$, and

$$\frac{d}{dx}(e^xy)=e^x\frac{dy}{dx}+e^xy=e^x\left(\frac{dy}{dx}+y\right)=e^{2x}$$

$$\Rightarrow\quad e^xy=\int e^{2x}\,dx=\frac{1}{2}e^{2x}+C.$$

Hence, $y = \frac{1}{2}e^x + Ce^{-x}$.

**6.** We have $\frac{dy}{dx} + 2e^x y = e^x$. Let $\mu = \int 2e^x\,dx = 2e^x$; then

$$\begin{aligned}\frac{d}{dx}\left(e^{2e^x}y\right) &= e^{2e^x}\frac{dy}{dx} + 2e^x e^{2e^x}y\\ &= e^{2e^x}\left(\frac{dy}{dx} + 2e^x y\right) = e^{2e^x}e^x.\end{aligned}$$

Therefore

$$\begin{aligned}e^{2e^x}y &= \int e^{2e^x}e^x\,dx \quad \text{Let } u = 2e^x\\ &\qquad\qquad\qquad\quad du = 2e^x\,dx\\ &= \frac{1}{2}\int e^u\,du = \frac{1}{2}e^{2e^x} + C.\end{aligned}$$

Hence, $y = \frac{1}{2} + Ce^{-2e^x}$.

**8.** $\frac{dy}{dx} + 3x^2y = x^2, \qquad y(0) = 1$

$$\mu = \int 3x^2\,dx = x^3$$

$$\frac{d}{dx}(e^{x^3}y) = e^{x^3}\frac{dy}{dx} + 3x^2e^{x^3}y = x^2e^{x^3}$$

$$e^{x^3}y = \int x^2e^{x^3}\,dx = \frac{1}{3}e^{x^3} + C$$

$$y(0) = 1 \Rightarrow 1 = \frac{1}{3} + C \Rightarrow C = \frac{2}{3}$$

$$y = \frac{1}{3} + \frac{2}{3}e^{-x^3}.$$

**10.** $x^2y' + y = x^2e^{1/x}, \qquad y(1) = 3e$

$$y' + \frac{1}{x^2}y = e^{1/x}$$

$$\mu = \int \frac{1}{x^2}\,dx = -\frac{1}{x}$$

$$\frac{d}{dx}\left(e^{-1/x}y\right) = e^{-1/x}\left(y' + \frac{1}{x^2}y\right) = 1$$

$$e^{-1/x}y = \int 1\,dx = x + C$$

$$y(1) = 3e \Rightarrow 3 = 1 + C \Rightarrow C = 2$$

$$y = (x+2)e^{1/x}.$$

**12.** We proceed by separation of variables:

$$\begin{aligned}m\frac{dv}{dt} &= mg - kv^2\\ \frac{dv}{dt} &= g - \frac{k}{m}v^2\\ \frac{dv}{g - \frac{k}{m}v^2} &= dt\\ \int \frac{dv}{\frac{mg}{k} - v^2} &= \frac{k}{m}\int dt = \frac{kt}{m} + C.\end{aligned}$$

Let $a^2 = mg/k$, where $a > 0$. Thus, we have

$$\begin{aligned}\int \frac{dv}{a^2 - v^2} &= \frac{kt}{m} + C\\ \frac{1}{2a}\ln\left|\frac{a+v}{a-v}\right| &= \frac{kt}{m} + C\\ \ln\left|\frac{a+v}{a-v}\right| &= \frac{2akt}{m} + C_1 = 2\sqrt{\frac{kg}{m}}\,t + C_1\\ \frac{a+v}{a-v} &= C_2e^{2t\sqrt{kg/m}}\end{aligned}$$

Assuming $v(0) = 0$, we get $C_2 = 1$.

$$\begin{aligned}a + v &= e^{2t\sqrt{kg/m}}(a - v)\\ v\left(1 + e^{2t\sqrt{kg/m}}\right) &= a\left(e^{2t\sqrt{kg/m}} - 1\right)\\ &= \sqrt{\frac{mg}{k}}\left(e^{2t\sqrt{kg/m}} - 1\right)\\ v &= \sqrt{\frac{mg}{k}}\,\frac{e^{2t\sqrt{kg/m}} - 1}{e^{2t\sqrt{kg/m}} + 1}.\end{aligned}$$

Clearly $v \to \sqrt{\frac{mg}{k}}$ as $t \to \infty$. This also follows from setting $\frac{dv}{dt} = 0$ in the given differential equation.

## Section 10.5 Existence, Uniqueness, and Numerical Methods (page 490)

**A computer spreadsheet was used in Exercises 1–12. The intermediate results appearing in the spreadsheet are not shown in these solutions.**

**2.** We start with $x_0 = 1$, $y_0 = 0$, and calculate

$$\begin{aligned}x_{n+1} &= x_n + h, \qquad u_{n+1} = y_n + h(x_n + y_n)\\ y_{n+1} &= y_n + \frac{h}{2}(x_n + y_n + x_{n+1} + u_{n+1}).\end{aligned}$$

a) For $h = 0.2$ we get $x_5 = 2$, $y_5 = 2.405416$.

b) For $h = 0.1$ we get $x_{10} = 2$, $y_{10} = 2.428162$.

c) For $h = 0.05$ we get $x_{20} = 2$, $y_{20} = 2.434382$.

**4.** We start with $x_0 = 0$, $y_0 = 0$, and calculate

$$x_{n+1} = x_n + h, \qquad y_{n+1} = hx_n\,e^{-y_n}.$$

a) For $h = 0.2$ we get $x_{10} = 2$, $y_{10} = 1.074160$.

b) For $h = 0.1$ we get $x_{20} = 2$, $y_{20} = 1.086635$.

**6.** We start with $x_0 = 0$, $y_0 = 0$, and calculate

$$\begin{aligned} x_{n+1} &= x_n + h \\ p_n &= x_n e^{-y_n} \\ q_n &= \left(x_n + \frac{h}{2}\right) e^{-(y_n+(h/2)p_n} \\ r_n &= \left(x_n + \frac{h}{2}\right) e^{-(y_n+(h/2)q_n} \\ s_n &= (x_n + h)e^{-(y_n+hr_n)} \\ y_{n+1} &= y_n + \frac{h}{6}(p_n + 2q_n + 2r_n + s_n). \end{aligned}$$

a) For $h = 0.2$ we get $x_{10} = 2$, $y_{10} = 1.098614$.

b) For $h = 0.1$ we get $x_{20} = 2$, $y_{20} = 1.098612$.

**8.** We start with $x_0 = 0$, $y_0 = 0$, and calculate

$$x_{n+1} = x_n + h, \qquad u_{n+1} = y_n + h\cos y_n$$
$$y_{n+1} = y_n + \frac{h}{2}(\cos y_n + \cos u_{n+1}).$$

a) For $h = 0.2$ we get $x_5 = 1$, $y_5 = 0.862812$.

b) For $h = 0.1$ we get $x_{10} = 1$, $y_{10} = 0.865065$.

c) For $h = 0.05$ we get $x_{20} = 1$, $y_{20} = 0.865598$.

**10.** We start with $x_0 = 0$, $y_0 = 0$, and calculate

$$x_{n+1} = x_n + h, \qquad y_{n+1} = y_n + h\cos(x_n^2).$$

a) For $h = 0.2$ we get $x_5 = 1$, $y_5 = 0.944884$.

b) For $h = 0.1$ we get $x_{10} = 1$, $y_{10} = 0.926107$.

c) For $h = 0.05$ we get $x_{20} = 1$, $y_{20} = 0.915666$.

**12.** We start with $x_0 = 0$, $y_0 = 0$, and calculate

$$\begin{aligned} x_{n+1} &= x_n + h \\ p_n &= \cos(x_n^2) \\ q_n &= \cos((x_n + (h/2))^2) \\ r_n &= \cos((x_n + (h/2))^2) \\ q_n &= \cos((x_n + h)^2) \\ y_{n+1} &= y_n + \frac{h}{6}(p_n + 2q_n + 2r_n + s_n). \end{aligned}$$

a) For $h = 0.2$ we get $x_5 = 1$, $y_5 = 0.904524$.

b) For $h = 0.1$ we get $x_{10} = 1$, $y_{10} = 0.904524$.

c) For $h = 0.05$ we get $x_{20} = 1$, $y_{20} = 0.904524$.

**14.** $u(x) = 1 + 3\int_2^x t^2 u(t)\,dt$

$$\begin{aligned} &\frac{du}{dx} = 3x^2 u(x), \qquad u(2) = 1 + 0 = 1 \\ &\frac{du}{u} = 3x^2\,dx \ \Rightarrow\ \ln u = x^3 + C \\ &0 = \ln 1 = \ln u(2) = 2^3 + C \ \Rightarrow\ C = -8 \\ &u = e^{x^3-8}. \end{aligned}$$

**16.** If $\phi(0) = A \ge 0$ and $\phi'(x) \ge k\phi(x)$ on an interval $[0, X]$, where $k > 0$ and $X > 0$, then

$$\frac{d}{dx}\left(\frac{\phi(x)}{e^{kx}}\right) = \frac{e^{kx}\phi'(x) - ke^{kx}\phi(x)}{e^{2kx}} \ge 0.$$

Thus $\phi(x)/e^{kx}$ is increasing on $[0, X]$. Since its value at $x = 0$ is $\phi(0) = A \ge 0$, therefore $\phi(x)/e^{kx} \ge A$ on $[0, X]$, and $\phi(x) \ge Ae^{kx}$ there.

## Section 10.6 Differential Equations of Second Order (page 494)

**2.** If $y_1 = e^{-2x}$, then $y_1'' - y_1' - 6y_1 = e^{-2x}(4 + 2 - 6) = 0$, so $y_1$ is a solution of the DE $y'' - y' - 6y = 0$. Let $y = e^{-2x}v$. Then

$$\begin{aligned} y' &= e^{-2x}(v' - 2v), \qquad y'' = e^{-2x}(v'' - 4v' + 4v) \\ y'' - y' - 6y &= e^{-2x}(v'' - 4v' + 4v - v' + 2v - 6v) \\ &= e^x(v'' - 5v'). \end{aligned}$$

$y$ satisfies $y'' - y' - 6y = 0$ provided $w = v'$ satisfies $w' - 5w = 0$. This equation has solution $v' = w = (C_1/5)e^{5x}$, so $v = C_1e^{5x} + C_2$. Thus the given DE has solution $y = e^{-2x}v = C_1e^{3x} + C_2e^{-2x}$.

**4.** If $y_1 = x^2$ on $(0, \infty)$, then

$$x^2y_1'' - 3xy_1' + 4y_1 = 2x^2 - 6x^2 + 4x^2 = 0,$$

so $y_1$ is a solution of the DE $x^2y'' - 3xy' + 4y = 0$. Let $y = x^2v(x)$. Then

$$\begin{aligned} y' &= x^2v' + 2xv, \qquad y'' = x^2v'' + 4xv' + 2v \\ x^2y'' - 3xy' + 4y &= x^4v'' + 4x^3v' + 2x^2v \\ &\quad - 3x^3v' - 6x^2v + 4x^2v \\ &= x^3(xv'' + v'). \end{aligned}$$

$y$ satisfies $x^2y'' - 3xy' + 4y = 0$ provided $w = v'$ satisfies $xw' + w = 0$. This equation has solution $v' = w = C_1/x$ (obtained by separation of variables), so $v = C_1\ln x + C_2$. Thus the given DE has solution $y = x^2v = C_1x^2\ln x + C_2x^2$.

**6.** If $y = x^{-1/2}\cos x$, then

$$\begin{aligned} y' &= -\frac{1}{2}x^{-3/2}\cos x - x^{-1/2}\sin x \\ y'' &= \frac{3}{4}x^{-5/2}\cos x + x^{-3/2}\sin x - x^{-1/2}\cos x. \end{aligned}$$

Thus

$$x^2y'' + xy' + \left(x^2 - \frac{1}{4}\right)y$$
$$= \frac{3}{4}x^{-1/2}\cos x + x^{1/2}\sin x - x^{3/2}\cos x$$
$$- \frac{1}{2}x^{-1/2}\cos x - x^{1/2}\sin x + x^{3/2}\cos x - \frac{1}{4}x^{-1/2}\cos x$$
$$= 0.$$

Therefore $y = x^{-1/2}\cos x$ is a solution of the Bessel equation

$$x^2y'' + xy' + \left(x^2 - \frac{1}{4}\right)y = 0. \qquad (*)$$

Now let $y = x^{-1/2}(\cos x)v(x)$. Then

$$y' = -\frac{1}{2}x^{-3/2}(\cos x)v - x^{-1/2}(\sin x)v + x^{-1/2}(\cos x)v'$$
$$y'' = \frac{3}{4}x^{-5/2}(\cos x)v + x^{-3/2}(\sin x)v - x^{-3/2}(\cos x)v'$$
$$- x^{-1/2}(\cos x)v - 2x^{-1/2}(\sin x)v' + x^{-1/2}(\cos x)v''.$$

If we substitute these expressions into the equation $(*)$, many terms cancel out and we are left with the equation

$$(\cos x)v'' - 2(\sin x)v' = 0.$$

Substituting $u = v'$, we rewrite this equation in the form

$$(\cos x)\frac{du}{dx} = 2(\sin x)u$$
$$\int \frac{du}{u} = 2\int \tan x\,dx \Rightarrow \ln|u| = 2\ln|\sec x| + C_0.$$

Thus $v' = u = C_1\sec^2 x$, from which we obtain

$$v = C_1\tan x + C_2.$$

Thus the general solution of the Bessel equation $(*)$ is

$$y = x^{-1/2}(\cos x)v = C_1x^{-1/2}\sin x + C_2x^{-1/2}\cos x.$$

**8.** If $y$ satisfies

$$y^{(n)} + a_{n-1}(x)y^{(n-1)} + \cdots + a_1(x)y' + a_0(x)y = f(x),$$

then let

$$y_1 = y, \quad y_2 = y', \quad y_3 = y'', \quad \ldots \quad y_n = y^{(n-1)}.$$

Therefore

$$y_1' = y_2, \quad y_2' = y_3, \quad \ldots \quad y_{n-2}' = y_{n-1}, \quad \text{and}$$
$$y_n' = -a_0y_1 - a_1y_2 - a_2y_3 - \cdots - a_{n-1}y_n + f,$$

and we have

$$\frac{d}{dx}\begin{pmatrix} y_1 \\ y_2 \\ \vdots \\ y_n \end{pmatrix} = \begin{pmatrix} 0 & 1 & 0 & \ldots & 0 \\ 0 & 0 & 1 & \ldots & 0 \\ \vdots & \vdots & \vdots & & \vdots \\ 0 & 0 & 0 & \ldots & 1 \\ -a_0 & -a_1 & -a_2 & \ldots & -a_n \end{pmatrix}\begin{pmatrix} y_1 \\ y_2 \\ \vdots \\ y_n \end{pmatrix} + \begin{pmatrix} 0 \\ 0 \\ \vdots \\ 0 \\ f \end{pmatrix}.$$

**10.** $$\begin{vmatrix} 2-\lambda & 1 \\ 2 & 3-\lambda \end{vmatrix} = 6 - 5\lambda + \lambda^2 - 2$$
$$= \lambda^2 - 5\lambda + 4$$
$$= (\lambda - 1)(\lambda - 4) = 0$$

if $\lambda = 1$ or $\lambda = 4$.

Let $\mathcal{A} = \begin{pmatrix} 2 & 1 \\ 2 & 3 \end{pmatrix}$.

If $\lambda = 1$ and $\mathcal{A}\mathbf{v} = \mathbf{v}$, then

$$\mathcal{A} = \begin{pmatrix} 2 & 1 \\ 2 & 3 \end{pmatrix}\begin{pmatrix} v_1 \\ v_2 \end{pmatrix} = \begin{pmatrix} v_1 \\ v_2 \end{pmatrix} \Leftrightarrow v_1 + v_2 = 0.$$

Thus we may take $\mathbf{v} = \mathbf{v}_1 = \begin{pmatrix} 1 \\ -1 \end{pmatrix}$.

If $\lambda = 4$ and $\mathcal{A}\mathbf{v} = 4\mathbf{v}$, then

$$\mathcal{A} = \begin{pmatrix} 2 & 1 \\ 2 & 3 \end{pmatrix}\begin{pmatrix} v_1 \\ v_2 \end{pmatrix} = 4\begin{pmatrix} v_1 \\ v_2 \end{pmatrix} \Leftrightarrow 2v_1 - v_2 = 0.$$

Thus we may take $\mathbf{v} = \mathbf{v}_2 = \begin{pmatrix} 1 \\ 2 \end{pmatrix}$.

By the result of Exercise 9, $\mathbf{y} = e^x\mathbf{v}_1$ and $\mathbf{y} = e^{4x}\mathbf{v}_2$ are solutions of the homogeneous linear system $\mathbf{y}' = \mathcal{A}\mathbf{y}$. Therefore the general solution of the system is

$$\mathbf{y} = C_1e^x\mathbf{v}_1 + C_2e^{4x}\mathbf{v}_2,$$

that is

$$\begin{pmatrix} y_1 \\ y_2 \end{pmatrix} = C_1e^x\begin{pmatrix} 1 \\ -1 \end{pmatrix} + C_2e^{4x}\begin{pmatrix} 1 \\ 2 \end{pmatrix}, \quad \text{or}$$
$$y_1 = C_1e^x + C_2e^{4x}$$
$$y_2 = -C_1e^x + 2C_2e^{4x}.$$

## Section 10.7 Linear Differential Equations with Constant Coefficients (page 502)

**2.** $$y'' - 2y' - 3y = 0$$
$$\text{auxiliary eqn} \quad r^2 - 2r - 3 = 0 \quad \Rightarrow r = -1, \ r = 3$$
$$y = Ae^{-t} + Be^{3t}.$$

**4.** $4y'' - 4y' - 3y = 0$
$4r^2 - 4r - 3 = 0 \Rightarrow (2r+1)(2r-3) = 0$
Thus, $r_1 = -\frac{1}{2}$, $r_2 = \frac{3}{2}$, and $y = Ae^{-(1/2)t} + Be^{(3/2)t}$.

**6.** $y'' - 2y' + y = 0$
$r^2 - 2r + 1 = 0 \Rightarrow (r-1)^2 = 0$
Thus, $r = 1, 1$, and $y = Ae^t + Bte^t$.

**8.** $9y'' + 6y' + y = 0$
$9r^2 + 6r + 1 = 0 \Rightarrow (3r+1)^2 = 0$
Thus, $r = -\frac{1}{3}, -\frac{1}{3}$, and $y = Ae^{-(1/3)t} + Bte^{-(1/3)t}$.

**10.** For $y'' - 4y' + 5y = 0$ the auxiliary equation is $r^2 - 4r + 5 = 0$, which has roots $r = 2 \pm i$. Thus, the general solution of the DE is $y = Ae^{2t}\cos t + Be^{2t}\sin t$.

**12.** Given that $y'' + y' + y = 0$, hence $r^2 + r + 1 = 0$. Since $a = 1$, $b = 1$ and $c = 1$, the discriminant is $D = b^2 - 4ac = -3 < 0$ and $-(b/2a) = -\frac{1}{2}$ and $\omega = \sqrt{3}/2$. Thus, the general solution is
$$y = Ae^{-(1/2)t}\cos\left(\frac{\sqrt{3}}{2}t\right) + Be^{-(1/2)t}\sin\left(\frac{\sqrt{3}}{2}t\right).$$

**14.** Given that $y'' + 10y' + 25y = 0$, hence $r^2 + 10r + 25 = 0 \Rightarrow (r+5)^2 = 0 \Rightarrow r = -5$. Thus,
$$y = Ae^{-5t} + Bte^{-5t}$$
$$y' = -5e^{-5t}(A + Bt) + Be^{-5t}.$$
Since
$$0 = y(1) = Ae^{-5} + Be^{-5}$$
$$2 = y'(1) = -5e^{-5}(A + B) + Be^{-5},$$
we have $A = -2e^5$ and $B = 2e^5$.
Thus, $y = -2e^5e^{-5t} + 2te^5e^{-5t} = 2(t-1)e^{-5(t-1)}$.

**16.** $\begin{cases} y'' + y' + y = 0 \\ y(2\pi/\sqrt{3}) = 0 \\ y'(2\pi/\sqrt{3}) = 1 \end{cases}$
The auxiliary equation for the DE is $r^2 + r + 1 = 0$, which has roots $r = \frac{1}{2}(-1 \pm i\sqrt{3})$. Thus
$$y = Ae^{-t/2}\cos\frac{\sqrt{3}t}{2} + Be^{-t/2}\sin\frac{\sqrt{3}t}{2}$$
$$y' = \frac{-A + \sqrt{3}B}{2}e^{-t/2}\cos\frac{\sqrt{3}t}{2} - \frac{\sqrt{3}A + B}{2}e^{-t/2}\sin\frac{\sqrt{3}t}{2}.$$
Now $0 = y(2\pi/\sqrt{3}) = -e^{-\pi/\sqrt{3}}A \Rightarrow A = 0$, and
$$1 = y'(2\pi/\sqrt{3}) = \frac{A - \sqrt{3}B}{2}e^{-\pi/\sqrt{3}} \Rightarrow B = \frac{-2e^{\pi/\sqrt{3}}}{\sqrt{3}}.$$
Therefore $y = -\dfrac{2}{\sqrt{3}}e^{(\pi/\sqrt{3})-(t/2)}\sin\dfrac{\sqrt{3}t}{2}$.

**18.** The auxiliary equation $ar^2 + br + c = 0$ has roots
$$r_1 = \frac{-b - \sqrt{D}}{2a}, \quad r_2 = \frac{-b + \sqrt{D}}{2a},$$
where $D = b^2 - 4ac$. Note that $a(r_2 - r_1) = \sqrt{D} = -(2ar_1 + b)$. If $y = e^{r_1 t}u$, then $y' = e^{r_1 t}(u' + r_1 u)$, and $y'' = e^{r_1 t}(u'' + 2r_1 u' + r_1^2 u)$. Substituting these expressions into the DE $ay'' + by' + cy = 0$, and simplifying, we obtain
$$e^{r_1 t}(au'' + 2ar_1 u' + bu') = 0,$$
or, more simply, $u'' - (r_2 - r_1)u' = 0$. Putting $v = u'$ reduces this equation to first order:
$$v' = (r_2 - r_1)v,$$
which has general solution $v = Ce^{(r_2 - r_1)t}$. Hence
$$u = \int Ce^{(r_2 - r_1)t}\,dt = Be^{(r_2 - r_1)t} + A,$$
and $y = e^{r_1 t}u = Ae^{r_1 t} + Be^{r_2 t}$.

**20.** $y^{(4)} - 2y'' + y = 0$
Auxiliary: $r^4 - 2r^2 + 1 = 0$
$(r^2 - 1)^2 = 0 \Rightarrow r = -1, -1, 1, 1$
General solution: $y = C_1e^{-t} + C_2te^{-t} + C_3e^t + C_4te^t$.

**22.** $y^{(4)} + 4y^{(3)} + 6y'' + 4y' + y = 0$
Auxiliary: $r^4 + 4r^3 + 6r^2 + 4r + 1 = 0$
$(r+1)^4 = 0 \Rightarrow r = -1, -1, -1, -1$
General solution: $y = e^{-t}(C_1 + C_2t + C_3t^2 + C_4t^3)$.

**24.** Aux. eqn: $(r^2 - r - 2)^2(r^2 - 4)^2 = 0$
$(r+1)^2(r-2)^2(r-2)^2(r+2)^2 = 0$
$r = 2, 2, 2, 2, -1, -1, -2, -2.$
The general solution is
$$y = e^{2t}(C_1 + C_2t + C_3t^2 + C_4t^3) + e^{-t}(C_5 + C_6t) + e^{-2t}(C_7 + C_8t).$$

**26.** $t^2y'' - ty' - 3y = 0$
$r(r-1) - r - 3 = 0 \Rightarrow r^2 - 2r - 3 = 0$
$\Rightarrow (r-3)(r+1) = 0 \Rightarrow r_1 = -1$ and $r_2 = 3$
Thus, $y = At^{-1} + Bt^3$.

**28.** Consider $t^2y'' - ty' + 5y = 0$. Since $a = 1$, $b = -1$, and $c = 5$, therefore $(b-a)^2 < 4ac$. Then $k = (a-b)/2a = 1$ and $\omega^2 = 4$. Thus, the general solution is $y = At\cos(2\ln t) + Bt\sin(2\ln t)$.

**30.** Given that $t^2y'' + ty' + y = 0$. Since $a = 1$, $b = 1$, $c = 1$ therefore $(b-a)^2 < 4ac$. Then $k = (a-b)/2a = 0$ and $\omega^2 = 1$. Thus, the general solution is $y = A\cos(\ln t) + B\sin(\ln t)$.

## Section 10.8 Nonhomogeneous Linear Equations (page 507)

**2.** $y'' + y' - 2y = x$.
The complementary function is $y_h = C_1e^{-2x} + C_2e^x$, as shown in Exercise 1. For a particular solution try $y = Ax + B$. Then $y' = A$ and $y'' = 0$, so $y$ satisfies the given equation if

$$x = A - 2(Ax + B) = A - 2B - 2Ax.$$

We require $A - 2B = 0$ and $-2A = 1$, so $A = -1/2$ and $B = -1/4$. The general solution of the given equation is

$$y = -\frac{2x+1}{4} + C_1e^{-2x} + C_2e^x.$$

**4.** $y'' + y' - 2y = e^x$.
The complementary function is $y_h = C_1e^{-2x} + C_2e^x$, as shown in Exercise 1. For a particular solution try $y = Axe^x$. Then

$$y' = Ae^x(1+x), \qquad y'' = Ae^x(2+x),$$

so $y$ satisfies the given equation if

$$e^x = Ae^x(2 + x + 1 + x - 2x) = 3Ae^x.$$

We require $A = 1/3$. The general solution of the given equation is

$$y = \frac{1}{3}xe^x + C_1e^{-2x} + C_3e^x.$$

**6.** $y'' + 4y = x^2$. The complementary function is $y = C_1\cos(2x) + C_2\sin(2x)$. For the given equation, try $y = Ax^2 + Bx + C$. Then

$$x^2 = y'' + 4y = 2A + 4Ax^2 + 4Bx + 4C$$

Thus $2A + 4C = 0$, $4A = 1$, $4B = 0$, and we have $A = \frac{1}{4}$, $B = 0$, and $C = -\frac{1}{8}$. The given equation has general solution

$$y = \frac{1}{4}x^2 - \frac{1}{8} + C_1\cos(2x) + C_2\sin(2x).$$

**8.** $y'' + 4y' + 4y = e^{-2x}$.
The homogeneous equation has auxiliary equation $r^2 + 4r + 4 = 0$ with roots $r = -2, -2$. Thus the complementary function is

$$y_h = C_1e^{-2x} + C_2xe^{-2x}.$$

For a particular solution, try $y = Ax^2e^{-2x}$. Then $y' = e^{-2x}(2Ax - 2Ax^2)$ and $y'' = e^{-2x}(2A - 8Ax + 4Ax^2)$. We have

$$\begin{aligned} e^{-2x} &= y'' + 4y' + 4y \\ &= e^{-2x}(2A - 8Ax + 4Ax^2 + 8Ax - 8Ax^2 + 4Ax^2) \\ &= 2Ae^{-2x}. \end{aligned}$$

Thus we require $A = 1/2$. The given equation has general solution

$$y = e^{-2x}\left(\frac{x^2}{2} + C_1 + C_2x\right).$$

**10.** $y'' + 2y' + 2y = e^{-x}\sin x$.
The complementary function is the same as in Exercise 9, but for a particular solution we try

$$\begin{aligned} y &= Axe^{-x}\cos x + Bxe^{-x}\sin x \\ y' &= e^{-x}\cos x(A - Ax + Bx) + e^{-x}\sin x(B - Bx - Ax) \\ y'' &= e^{-x}\cos x(2B - 2Bx - 2A) \\ &\quad + e^{-x}\sin x(2Ax - 2A - 2B). \end{aligned}$$

This satisfies the nonhomogeneous DE if

$$\begin{aligned} e^{-x}\sin x &= y'' + 2y' + 2y \\ &= 2Be^{-x}\cos x - 2Ae^{-x}\sin x. \end{aligned}$$

Thus we require $B = 0$ and $A = -1/2$. The given equation has general solution

$$y = -\frac{1}{2}xe^{-x}\cos x + e^{-x}(C_1\cos x + C_2\sin x).$$

**12.** $y'' + 2y' + y = xe^{-x}$.
The homogeneous equation has auxiliary equation $r^2 + 2r + 1 = 0$ with roots $r = -1$ and $r = -1$. Thus the complementary function is $y_h = C_1e^{-x} + C_2xe^{-x}$. For a particular solution, try $y = e^{-x}(Ax^2 + Bx^3)$. Then

$$\begin{aligned} y' &= e^{-x}(2Ax + (3B - A)x^2 - Bx^3) \\ y'' &= e^{-x}(2A + (6B - 4A)x - (6B - A)x^2 + Bx^3). \end{aligned}$$

This satisfies the nonhomogeneous DE if

$$\begin{aligned} xe^{-x} &= y'' + 2y' + y \\ &= e^{-x}(2A + 6Bx). \end{aligned}$$

Thus we require $A = 0$ and $B = 1/6$. The given equation has general solution

$$y = \frac{1}{6}x^3e^{-x} + C_1e^{-x} + C_2xe^{-x}.$$

**14.** $y'' + y' - 2y = e^x$.
The complementary function is $y_h = C_1e^{-2x} + C_2e^x$. For a particular solution use

$$y_p = e^{-2x}u_1(x) + e^x u_2(x),$$

where the coefficients $u_1$ and $u_2$ satisfy

$$\begin{aligned} -2e^{-2x}u_1' + e^x u_2' &= e^x \\ e^{-2x}u_1' + e^x u_2' &= 0. \end{aligned}$$

Thus

$$\begin{aligned} u_1' &= -\frac{1}{3}e^{3x} & u_2' &= \frac{1}{3} \\ u_1 &= -\frac{1}{9}e^{3x} & u_2 &= \frac{1}{3}x. \end{aligned}$$

Thus $y_p = -\frac{1}{9}e^x + \frac{1}{3}xe^x$. The general solution of the given equation is

$$\begin{aligned} y &= -\frac{1}{9}e^x + \frac{1}{3}xe^x + C_1e^{-2x} + C_2e^x \\ &= \frac{1}{3}xe^x + C_1e^{-2x} + C_3e^x. \end{aligned}$$

**16.** $x^2y'' + xy' - y = x^r$ has a solution of the form $y = Ax^r$ provided $r \neq \pm 1$. If this is the case, then

$$x^r = Ax^r\big(r(r-1) + r - 1\big) = Ax^r(r^2 - 1).$$

Thus $A = 1/(r^2 - 1)$ and a particular solution of the DE is

$$y = \frac{1}{r^2 - 1}x^r.$$

**18.** $x^2y'' + xy' - y = x$.
Try $y = xu_1(x) + \frac{1}{x}u_2(x)$, where $u_1$ and $u_2$ satisfy

$$xu_1' + \frac{u_2'}{x} = 0, \quad u_1' - \frac{u_2'}{x^2} = \frac{1}{x}.$$

Solving these equations for $u_1'$ and $u_2'$, we get

$$u_2' = -\frac{x}{2}, \qquad u_1' = \frac{1}{2x}.$$

Thus $u_1 = \frac{1}{2}\ln x$ and $u_2 = -\frac{x^2}{4}$. A particular solution is

$$y = \frac{1}{2}x\ln x - \frac{x}{4}.$$

The term $-x/4$ can be absorbed into the term $C_1x$ in the complementary function, so the general solution is

$$y = \frac{1}{2}x\ln x + C_1x + \frac{C_2}{x}.$$

**20.** $x^2y'' + xy' + \left(x^2 - \frac{1}{4}\right)y = x^{3/2}$.
A particular solution can be obtained in the form

$$y = x^{-1/2}(\cos x)u_1(x) + x^{-1/2}(\sin x)u_2(x),$$

where $u_1$ and $u_2$ satisfy

$$\begin{aligned} &x^{-1/2}(\cos x)u_1' + x^{-1/2}(\sin x)u_2' = 0 \\ &\left(-\frac{1}{2}x^{-3/2}\cos x - x^{-1/2}\sin x\right)u_1' \\ &\qquad - \left(\frac{1}{2}x^{-3/2}\sin x - x^{-1/2}\cos x\right)u_2' = x^{-1/2}. \end{aligned}$$

We can simplify these equations by dividing the first by $x^{-1/2}$, and adding the first to $2x$ times the second, then dividing the result by $2x^{1/2}$. The resulting equations are

$$\begin{aligned} (\cos x)u_1' + (\sin x)u_2' &= 0 \\ -(\sin x)u_1' + (\cos x)u_2' &= 1, \end{aligned}$$

which have solutions $u_1' = -\sin x$, $u_2' = \cos x$, so that $u_1 = \cos x$ and $u_2 = \sin x$. Thus a particular solution of the given equation is

$$y = x^{-1/2}\cos^2 x + x^{-1/2}\sin^2 x = x^{-1/2}.$$

The general solution is

$$y = x^{-1/2}\big(1 + C_2\cos x + C_2\sin x\big).$$

## Section 10.9 Series Solutions (page 512)

**2.** $y'' = xy$. Try $\sum_{n=0}^{\infty} a_nx^n$. Then

$$\begin{aligned} y' &= \sum_{n=0}^{\infty} na_nx^{n-1} = \sum_{n=1}^{\infty} na_nx^{n-1} \\ y'' &= \sum_{n=2}^{\infty} n(n-1)a_nx^{n-2} = \sum_{n=0}^{\infty}(n+2)(n+1)a_{n+2}x^n. \end{aligned}$$

Thus we have

$$\begin{aligned}
0 &= y'' - xy \\
&= \sum_{n=0}^{\infty}(n+2)(n+1)a_{n+2}x^n - \sum_{n=0}^{\infty}a_n x^{n+1} \\
&= \sum_{n=0}^{\infty}(n+2)(n+1)a_{n+2}x^n - \sum_{n=1}^{\infty}a_{n-1}x^n \\
&= 2a_2 + \sum_{n=1}^{\infty}\big[(n+2)(n+1)a_{n+2} - a_{n-1}\big]x^n.
\end{aligned}$$

Thus $a_2 = 0$ and $a_{n+2} = \dfrac{a_{n-1}}{(n+2)(n+1)}$ for $n \ge 1$. Given $a_0$ and $a_1$, we have

$$\begin{aligned}
a_3 &= \frac{a_0}{2\times 3} \\
a_6 &= \frac{a_3}{5\times 6} = \frac{a_0}{2\times 3\times 5\times 6} = \frac{1\times 4\times a_0}{6!} \\
a_9 &= \frac{a_6}{8\times 9} = \frac{1\times 4\times 7\times a_0}{9!} \\
&\vdots \\
a_{3n} &= \frac{1\times 4\times\cdots\times(3n-2)a_0}{(3n)!} \\
a_4 &= \frac{a_1}{3\times 4} = \frac{2\times a_1}{4!} \\
a_7 &= \frac{a_4}{6\times 7} = \frac{2\times 5\times a_1}{7!} \\
&\vdots \\
a_{3n+1} &= \frac{2\times 5\times\cdots\times(3n-1)a_1}{(3n+1)!} \\
0 &= a_2 = a_5 = a_8 = \cdots = a_{3n+2}.
\end{aligned}$$

Thus the general solution of the given equation is

$$\begin{aligned}
y &= a_0\left(1 + \sum_{n=1}^{\infty}\frac{1\times 4\times\cdots\times(3n-2)}{(3n)!}x^{3n}\right) \\
&\quad + a_1\sum_{n=1}^{\infty}\frac{2\times 5\times\cdots\times(3n-1)}{(3n+1)!}x^{3n+1}.
\end{aligned}$$

**4.** $(1-x^2)y'' - xy' + 9y = 0$, $y(0) = 0$, $y'(0) = 1$. Try

$$y = \sum_{n=0}^{\infty}a_n x^n.$$

Then $a_0 = 0$ and $a_1 = 1$. We have

$$\begin{aligned}
y' &= \sum_{n=1}^{\infty}na_n x^{n-1} \\
y'' &= \sum_{n=2}^{\infty}n(n-1)a_n x^{n-2} \\
0 &= (1-x^2)y'' - xy' + 9y \\
&= \sum_{n=0}^{\infty}(n+2)(n+1)a_{n+2}x^n - \sum_{n=2}^{\infty}n(n-1)a_n x^n \\
&\quad - \sum_{n=1}^{\infty}na_n x^n + 9\sum_{n=0}^{\infty}a_n x^n \\
&= 2a_2 + 9a_0 + (6a_3 + 8a_1)x \\
&\quad + \sum_{n=2}^{\infty}\big[(n+2)(n+1)a_{n+2} - (n^2-9)a_n\big]x^n.
\end{aligned}$$

Thus $2a_2 + 9a_0 = 0$, $6a_3 + 8a_1 = 0$, and

$$a_{n+2} = \frac{(n^2-9)a_n}{(n+1)(n+2)}.$$

Therefore we have

$$\begin{aligned}
a_2 &= a_4 = a_6 = \cdots = 0 \\
a_3 &= -\frac{4}{3}, \quad a_5 = 0 = a_7 = a_9 = \cdots.
\end{aligned}$$

The initial-value problem has solution

$$y = x - \frac{4}{3}x^3.$$

**6.** $xy'' + y' + xy = 0$.
Since $x = 0$ is a regular singular point of this equation, try

$$\begin{aligned}
y &= \sum_{n=0}^{\infty}a_n x^{n+\mu} \qquad (a_0 = 1) \\
y' &= \sum_{n=0}^{\infty}(n+\mu)a_n x^{n+\mu-1} \\
y'' &= \sum_{n=0}^{\infty}(n+\mu)(n+\mu-1)a_n x^{n+\mu-2}.
\end{aligned}$$

Then we have

$$\begin{aligned}
0 &= xy'' + y' + xy \\
&= \sum_{n=0}^{\infty}\big[(n+\mu)(n+\mu-1) + (n+\mu)\big]a_n x^{n+\mu-1} \\
&\quad + \sum_{n=0}^{\infty}a_n x^{n+\mu+1} \\
&= \sum_{n=0}^{\infty}(n+\mu)^2 a_n x^{n+\mu-1} + \sum_{n=2}^{\infty}a_{n-2}x^{n+\mu-1} \\
&= \mu^2 x^{\mu-1} + (1+\mu)^2 a_1 x^{\mu} \\
&\quad + \sum_{n=2}^{\infty}\big[(n+\mu)^2 a_n + a_{n-2}\big]x^{n+\mu-1}.
\end{aligned}$$

Thus $\mu = 0$, $a_1 = 0$, and $a_n = -\dfrac{a_{n-2}}{n^2}$ for $n \ge 2$.
It follows that $0 = a_1 = a_3 = a_5 = \cdots$, and, since $a_0 = 1$,

$$a_2 = -\frac{1}{2^2}, \qquad a_4 = \frac{1}{2^2 4^2}, \quad \ldots$$
$$a_{2n} = \frac{(-1)^n}{2^2 4^2 \cdots (2n)^2} = \frac{(-1)^n}{2^{2n}(n!)^2}.$$

One series solution is

$$y = 1 + \sum_{n=1}^{\infty} \frac{(-1)^n x^{2n}}{2^{2n}(n!)^2}.$$

## Review Exercises 10 (page 512)

**2.** $\dfrac{dy}{dx} = e^{-y}\sin x$
$e^y\,dy = \sin x\,dx \Rightarrow e^y = -\cos x + C$
$y = \ln(C - \cos x)$

**4.** $\dfrac{dy}{dx} = \dfrac{x^2 + y^2}{2xy}$ (let $y = xv(x)$)
$v + x\dfrac{dv}{dx} = \dfrac{1 + v^2}{2v}$
$x\dfrac{dv}{dx} = \dfrac{1 + v^2}{2v} - v = \dfrac{1 - v^2}{2v}$
$\dfrac{2v\,dv}{v^2 - 1} = -\dfrac{dx}{x}$
$\ln(v^2 - 1) = \ln\dfrac{1}{x} + \ln C = \ln\dfrac{C}{x}$
$\dfrac{y^2}{x^2} - 1 = \dfrac{C}{x} \Rightarrow y^2 - x^2 = Cx$

**6.** $\dfrac{dy}{dx} = -\dfrac{y + e^x}{x + e^y}$
$(y + e^x)\,dx + (x + e^y)\,dy = 0$ (exact)
$d\left(xy + e^x + e^y\right) = 0$
$xy + e^x + e^y = C$

**8.** $2\dfrac{d^2y}{dt^2} + 5\dfrac{dy}{dt} + 2y = 0$
Aux: $2r^2 + 5r + 2 = 0 \Rightarrow r = -1/2,\ -2$
$y = C_1 e^{-t/2} + C_2 e^{-2t}$

**10.** $2x^2 y'' + y = 0$
Aux: $2r(r - 1) + 1 = 0$
$2r^2 - 2r + 1 = 0 \Rightarrow r = \dfrac{1}{2}(1 \pm i)$
$y = C_1|x|^{1/2}\cos\left(\frac{1}{2}\ln|x|\right) + C_2|x|^{1/2}\sin\left(\frac{1}{2}\ln|x|\right)$

**12.** $\dfrac{d^3y}{dt^3} + 8\dfrac{d^2y}{dt^2} + 16\dfrac{dy}{dt} = 0$
Aux: $r^3 + 8r^2 + 16r = 0$
$r(r + 4)^2 = 0 \Rightarrow r = 0, -4, -4$
$y = C_1 + C_2 e^{-4t} + C_3 t e^{-4t}$

**14.** $\dfrac{d^2y}{dx^2} - 5\dfrac{dy}{dx} + 6y = xe^{2x}$
Same complementary function as in Exercise 13. For a particular solution we try $y = (Ax^2 + Bx)e^{2x}$. Substituting this into the given DE leads to

$$xe^{2x} = (2A - B)e^{2x} - 2Axe^{2x},$$

so that we need $A = -1/2$ and $B = 2A = -1$. The general solution is

$$y = -\left(\frac{1}{2}x^2 + x\right)e^{2x} + C_1 e^{2x} + C_2 e^{3x}.$$

**16.** $x^2\dfrac{d^2y}{dx^2} - 2y = x^3.$
The corresponding homogeneous equation has auxiliary equation $r(r - 1) - 2 = 0$, with roots $r = 2$ and $r = -1$, so the complementary function is
$y = C_1 x^2 + C_2/x$. A particular solution of the nonhomogeneous equation can have the form $y = Ax^3$. Substituting this into the DE gives

$$6Ax^3 - 2Ax^3 = x^3,$$

so that $A = 1/4$. The general solution is

$$y = \frac{1}{4}x^3 + C_1 x^2 + \frac{C_2}{x}.$$

**18.** $\dfrac{dy}{dx} = \dfrac{y^2}{x^2}$, $y(2) = 1$
$\dfrac{dy}{y^2} = \dfrac{dx}{x^2} \Rightarrow -\dfrac{1}{y} = -\dfrac{1}{x} - C$
$1 = \dfrac{1}{2} + C \Rightarrow C = \dfrac{1}{2}$
$y = \left(\dfrac{1}{x} + \dfrac{1}{2}\right)^{-1} = \dfrac{2x}{x + 2}$

**20.** $\dfrac{dy}{dx} + (\cos x)y = 2\cos x, \qquad y(\pi) = 1$
$\dfrac{d}{dx}\left(e^{\sin x}y\right) = e^{\sin x}\left(\dfrac{dy}{dx} + (\cos x)y\right) = 2\cos x\, e^{\sin x}$
$e^{\sin x}y = 2e^{\sin x} + C$
$y = 2 + Ce^{-\sin x}$
$1 = 2 + Ce^0 \Rightarrow C = -1$
$y = 2 - e^{-\sin x}$

**22.** $y'' + 2y' + (1 + \pi^2)y = 0$, $y(1) = 0$, $y'(1) = \pi$
Aux: $r^2 + 2r + 1 + \pi^2 = 0 \Rightarrow r = -1 \pm \pi i$.

$$\begin{aligned} y &= Ae^{-x}\cos(\pi x) + Be^{-x}\sin(\pi x) \\ y' &= e^{-x}\cos(\pi x)(-A + B\pi) + e^{-x}\sin(\pi x)(-B - A\pi). \end{aligned}$$

Thus $-Ae^{-1} = 0$ and $(A - B\pi)e^{-1} = \pi$, so that $A = 0$ and $B = -e$. The solution is $y = -e^{1-x}\sin(\pi x)$.

**24.** $x^2y'' - 3xy' + 4y = 0$, $y(e) = e^2$, $y'(e) = 0$
Aux: $r(r-1) - 3r + 4 = 0$, or $(r-2)^2 = 0$, so that $r = 2, 2$.

$$\begin{aligned} y &= Ax^2 + Bx^2\ln x \\ y' &= 2Ax + 2Bx\ln x + Bx. \end{aligned}$$

We require $e^2 = Ae^2 + Be^2$ and $0 = 2Ae + 3Be$. Thus $A + B = 1$ and $2A = -3B$, so that $A = 3$ and $B = -2$. The solution is $y = 3x^2 - 2x^2\ln x$, valid for $x > 0$.

**26.** $2\dfrac{d^2y}{dx^2} + 5\dfrac{dy}{dx} - 3y = 6 + 7e^{x/2}$, $y(0) = 0$, $y'(0) = 1$
Aux: $2r^2 + 5r - 3 = 0 \Rightarrow r = 1/2, -3$.
Complementary function: $y = C_1e^{x/2} + C_2e^{-3x}$.
Particular solution: $y = A + Bxe^{x/2}$

$$\begin{aligned} y' &= Be^{x/2}\left(1 + \frac{x}{2}\right) \\ y'' &= Be^{x/2}\left(1 + \frac{x}{4}\right). \end{aligned}$$

We need

$$Be^{x/2}\left(2 + \frac{x}{2} + 5 + \frac{5x}{2} - 3x\right) - 3A = 6 + 7e^{x/2}.$$

This is satisfied if $A = -2$ and $B = 1$. The general solution of the DE is

$$y = -2 + xe^{x/2} + C_1e^{x/2} + C_2e^{-3x}.$$

Now the initial conditions imply that

$$\begin{aligned} 0 &= y(0) = -2 + C_1 + C_2 \\ 1 &= y'(0) = 1 + \frac{C_1}{2} - 3C_2, \end{aligned}$$

which give $C_1 = 12/7$, $C_2 = 2/7$. Thus the IVP has solution

$$y = -2 + xe^{x/2} + \frac{1}{7}(12e^{x/2} + 2e^{-3x}).$$

**28.** $(x^2 + 3y^2)\,dx + xy\,dy = 0$. Multiply by $x^n$:

$$x^n(x^2 + 3y^2)\,dx + x^{n+1}y\,dy = 0$$

is exact provided $6x^ny = (n+1)x^ny$, that is, provided $n = 5$. In this case the left side is $d\phi$, where

$$\phi(x, y) = \frac{1}{2}x^6y^2 + \frac{1}{8}x^8.$$

The general solution of the given DE is

$$4x^6y^2 + x^8 = C.$$

**30.** $x^2y'' - x(2 + x\cot x)y' + (2 + x\cot x)y = x^3\sin x$
Look for a particular solution of the form $y = xu_1(x) + x\cos xu_2(x)$, where

$$\begin{aligned} xu_1' + x\cos xu_2' &= 0 \\ u_1' + (\cos x - x\sin x)u_2' &= x\sin x. \end{aligned}$$

Divide the first equation by $x$ and subtract from the second equation to get

$$-x\sin xu_2' = x\sin x.$$

Thus $u_2' = -1$ and $u_2 = -x$. The first equation now gives $u_1' = \cos x$, so that $u_1 = \sin x$. The general solution of the DE is

$$y = x\sin x - x^2\cos x + C_1x + C_2x\cos x.$$

# APPENDICES

## Appendix I. Complex Numbers (page A-10)

**2.** $z = 4 - i$, $\mathrm{Re}(z) = 4$, $\mathrm{Im}(z) = -1$

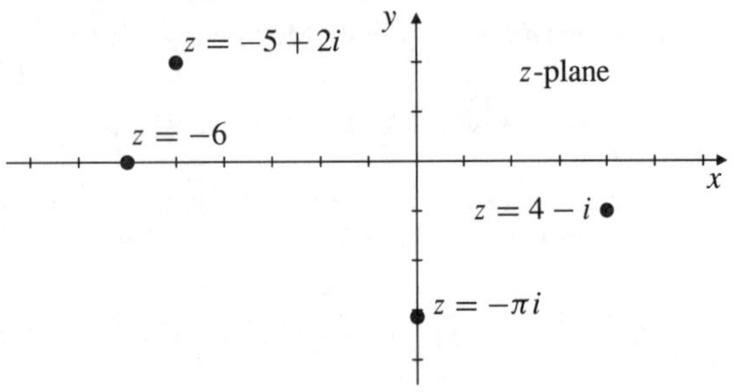

$z = -5 + 2i$ $y$ $z$-plane $z = -6$ $x$ $z = 4 - i$ $z = -\pi i$

Fig. A.2

**4.** $z = -6$, $\mathrm{Re}(z) = -6$, $\mathrm{Im}(z) = 0$

**6.** $z = -2$, $|z| = 2$, $\mathrm{Arg}\,(z) = \pi$
$z = 2(\cos\pi + i\sin\pi)$

**8.** $z = -5i$, $|z| = 5$, $\mathrm{Arg}\,(z) = 3\pi/2$
$z = 5(\cos(3\pi/2) + i\sin(3\pi/2))$

**10.** $z = -2 + i$, $|z| = \sqrt{5}$, $\theta = \mathrm{Arg}\,(z) = \pi - \tan^{-1}(1/2)$
$z = \sqrt{5}(\cos\theta + i\sin\theta)$

**12.** $z = 3 - 4i$, $|z| = 5$, $\theta = \mathrm{Arg}\,(z) = 2\pi - \tan^{-1}(4/3)$
$z = 5(\cos\theta + i\sin\theta)$

**14.** $z = -\sqrt{3} - 3i$, $|z| = 2\sqrt{3}$, $\mathrm{Arg}\,(z) = 4\pi/3$
$z = 2\sqrt{3}(\cos(4\pi/3) + i\sin(4\pi/3))$

**16.** If $\mathrm{Arg}\,(z) = \dfrac{7\pi}{4}$ and $\mathrm{Arg}\,(w) = \dfrac{\pi}{2}$, then
$\arg(zw) = \dfrac{7\pi}{4} + \dfrac{\pi}{2} = \dfrac{9\pi}{4}$, so $\mathrm{Arg}\,(zw) = \dfrac{9\pi}{4} - 2\pi = \dfrac{\pi}{4}$.

**18.** $|z| = 2$, $\arg(z) = \pi \Rightarrow z = 2(\cos\pi + i\sin\pi) = -2$

**20.** $|z| = 1$, $\arg(z) = \dfrac{3\pi}{4} \Rightarrow z = \left(\cos\dfrac{3\pi}{4} + i\sin\dfrac{3\pi}{4}\right)$

$\Rightarrow z = -\dfrac{1}{\sqrt{2}} + \dfrac{1}{\sqrt{2}}i$

**22.** $|z| = 0 \Rightarrow z = 0$ for any value of $\arg(z)$

**24.** $\overline{5 + 3i} = 5 - 3i$

**26.** $\overline{4i} = -4i$

**28.** $|z| = 2$ represents all points on the circle of radius 2 centred at the origin.

**30.** $|z - 2i| \le 3$ represents all points in the closed disk of radius 3 centred at the point $2i$.

**32.** $\arg(z) = \pi/3$ represents all points on the ray from the origin in the first quadrant, making angle $60^\circ$ with the positive direction of the real axis.

**34.** $(2 + 5i) + (3 - i) = 5 + 4i$

**36.** $(4 + i)(4 - i) = 16 - i^2 = 17$

**38.** $(a + bi)(\overline{2a - bi}) = (a + bi)(2a + bi) = 2a^2 - b^2 + 3abi$

**40.** $\dfrac{2 - i}{2 + i} = \dfrac{(2 - i)^2}{4 - i^2} = \dfrac{3 - 4i}{5}$

**42.** $\dfrac{1 + i}{i(2 + 3i)} = \dfrac{1 + i}{-3 + 2i} = \dfrac{(1 + i)(-3 - 2i)}{9 + 4} = \dfrac{-1 - 5i}{13}$

**44.** If $z = x + yi$ and $w = u + vi$, where $x$, $y$, $u$, and $v$ are real, then

$$\begin{aligned}\overline{z + w} &= \overline{x + u + (y + v)i}\\ &= x + u - (y + v)i = x - yi + u - vi = \bar{z} + \bar{w}.\end{aligned}$$

**46.**
$$z = 3 + i\sqrt{3} = 2\sqrt{3}\left(\cos\frac{\pi}{6} + i\sin\frac{\pi}{6}\right)$$
$$w = -1 + i\sqrt{3} = 2\left(\cos\frac{2\pi}{3} + i\sin\frac{2\pi}{3}\right)$$
$$zw = 4\sqrt{3}\left(\cos\frac{5\pi}{6} + i\sin\frac{5\pi}{6}\right)$$
$$\frac{z}{w} = \sqrt{3}\left(\cos\frac{-\pi}{2} + i\sin\frac{-\pi}{2}\right) = -i\sqrt{3}$$

**48.** $\cos(3\theta) + i\sin(3\theta) = (\cos\theta + i\sin\theta)^3$
$= \cos^3\theta + 3i\cos^2\theta\sin\theta - 3\cos\theta\sin^2\theta - i\sin^3\theta$
Thus

$$\cos(3\theta) = \cos^3\theta - 3\cos\theta\sin^2\theta = 4\cos^3\theta - 3\cos\theta$$
$$\sin(3\theta) = 3\cos^2\theta\sin\theta - \sin^3\theta = 3\sin\theta - 4\sin^3\theta.$$

**50.** If $z = w = -1$, then $zw = 1$, so $\sqrt{zw} = 1$. But if we use $\sqrt{z} = \sqrt{-1} = i$ and the same value for $\sqrt{w}$, then $\sqrt{z}\sqrt{w} = i^2 = -1 \ne \sqrt{zw}$.

**52.** The three cube roots of $-8i = 8\left(\cos\dfrac{3\pi}{2} + i\sin\dfrac{3\pi}{2}\right)$ are of the form $2(\cos\theta + i\sin\theta)$ where $\theta = \pi/2$, $\theta = 7\pi/6$, and
$\theta = 11\pi/6$. Thus they are

$$2i, \quad -\sqrt{3} - i, \quad \sqrt{3} - i.$$

**54.** The four fourth roots of $4 = 4(\cos 0 + i\sin 0)$ are of the form $\sqrt{2}(\cos\theta + i\sin\theta)$ where $\theta = 0$, $\theta = \pi/2$, $\pi$, and $\theta = 3\pi/2$. Thus they are $\sqrt{2}$, $i\sqrt{2}$, $-\sqrt{2}$, and $-i\sqrt{2}$.

**56.** The equation $z^5 + a^5 = 0$ $(a > 0)$ has solutions that are the five fifth roots of $-a^5 = a(\cos\pi + i\sin\pi)$; they are of the form $a(\cos\theta + i\sin\theta)$, where $\theta = \pi/5$, $3\pi/5$, $\pi$, $7\pi/5$, and $9\pi/5$.

## Appendix II. Complex Functions (page A-20)

**In Solutions 1–12, $z = x + yi$ and $w = u + vi$, where $x$, $y$, $u$, and $v$ are real.**

**2.** The function $w = \bar{z}$ transforms the line $x + y = 1$ to the line $u - v = 1$.

**4.** The function $w = z^3$ transforms the closed quarter-circular disk $0 \le |z| \le 2$, $0 \le \arg(z) \le \pi/2$ to the closed three-quarter disk $0 \le |w| \le 8$, $0 \le \arg(w) \le 3\pi/2$.

**6.** The function $w = -iz$ rotates the $z$-plane $-90°$, so transforms the wedge $\pi/4 \le \arg(z) \le \pi/3$ to the wedge $-\pi/4 \le \arg(z) \le -\pi/6$.

**8.** The function $w = z^2 = x^2 - y^2 + 2xyi$ transforms the line $x = 1$ to $u = 1 - y^2$, $v = 2y$, which is the parabola $v^2 = 4 - 4u$ with vertex at $w = 1$, opening to the left.

**10.** The function $w = 1/z = (x - yi)/(x^2 + y^2)$ transforms the line $x = 1$ to the curve given parametrically by

$$u = \frac{1}{1 + y^2}, \qquad v = \frac{-y}{1 + y^2}.$$

This curve is, in fact, a circle,

$$u^2 + v^2 = \frac{1 + y^2}{(1 + y^2)^2} = u,$$

with centre $w = 1/2$ and radius $1/2$.

**12.** The function $w = e^{iz} = e^{-y}(\cos x + i \sin x)$ transforms the vertical half-strip $0 < x < \pi/2$, $0 < y < \infty$ to the first-quadrant part of the unit open disk $|w| = e^{-y} < 1$, $0 < \arg(w) = x < \pi/2$, that is $u > 0$, $v > 0$, $u^2 + v^2 < 1$.

**14.**
$$f(z) = z^3 = (x + yi)^3 = x^3 - 3xy^2 + (3x^2y - y^3)i$$
$$u = x^3 - 3xy^2, \qquad v = 3x^2y - y^3$$
$$\frac{\partial u}{\partial x} = 3(x^2 - y^2) = \frac{\partial v}{\partial y}, \quad \frac{\partial u}{\partial y} = -6xy = -\frac{\partial v}{\partial x}$$
$$f'(z) = \frac{\partial u}{\partial x} + i\frac{\partial v}{\partial x} = 3(x^2 - y^2 + 2xyi) = 3z^2.$$

**16.**
$$f(z) = e^{z^2} = e^{x^2-y^2}(\cos(2xy) + i\sin(2xy))$$
$$u = e^{x^2-y^2}\cos(2xy), \qquad v = e^{x^2-y^2}\sin(2xy)$$
$$\frac{\partial u}{\partial x} = e^{x^2-y^2}(2x\cos(2xy) - 2y\sin(2xy)) = \frac{\partial v}{\partial y}$$
$$\frac{\partial u}{\partial y} = -e^{x^2-y^2}(2y\cos(2xy) + 2x\sin(2xy)) = -\frac{\partial v}{\partial x}$$
$$\begin{aligned} f'(z) &= \frac{\partial u}{\partial x} + i\frac{\partial v}{\partial x} \\ &= e^{x^2-y^2}[2x\cos(2xy) - 2y\sin(2xy) \\ &\quad + i(2y\cos(2xy) + 2x\sin(2xy))] \\ &= (2x + 2yi)e^{x^2-y^2}(\cos(2xy) + i\sin(2xy)) = 2ze^{z^2}. \end{aligned}$$

**18.**
$$\begin{aligned} e^{z+2\pi i} &= e^x(\cos(y + 2\pi) + i\sin(y + 2\pi)) \\ &= e^x(\cos y + i\sin y) = e^z. \end{aligned}$$

Thus $e^z$ is periodic with period $2\pi i$. So is $e^{-z} = 1/e^z$. Since $e^{i(z+2\pi)} = e^{zi+2\pi i} = e^{zi}$, therefore $e^{zi}$ and also $e^{-zi}$ are periodic with period $2\pi$. Hence

$$\cos z = \frac{e^{zi} + e^{-zi}}{2} \text{ and } \sin z = \frac{e^{zi} - e^{-zi}}{2i}$$

are periodic with period $2\pi$, and

$$\cosh z = \frac{e^z + e^{-z}}{2} \text{ and } \sinh z = \frac{e^z - e^{-z}}{2}$$

are periodic with period $2\pi i$.

**20.**
$$\begin{aligned} \cosh(iz) &= \frac{e^{iz} + e^{-iz}}{2} = \cosh z \\ -i\sinh(iz) &= \frac{1}{i}\,\frac{e^{iz} - e^{-iz}}{2} = \sin z \\ \cos(iz) &= \frac{e^{-z} + e^z}{2} = \cosh z \\ \sin(iz) &= \frac{e^{-z} - e^z}{2i} = i\,\frac{-e^{-z} + e^z}{2} = i\sinh z \end{aligned}$$

**22.**
$$\begin{aligned} \sin z = 0 &\Leftrightarrow e^{zi} = e^{-zi} \Leftrightarrow e^{2zi} = 1 \\ &\Leftrightarrow e^{-2y}[\cos(2x) + i\sin(2x)] = 1 \\ &\Leftrightarrow \sin(2x) = 0, \quad e^{-2y}\cos(2x) = 1 \\ &\Leftrightarrow y = 0, \ \cos(2x) = 1 \\ &= \Leftrightarrow y = 0, \ x = 0, \pm\pi, \ \pm 2\pi, \ldots \end{aligned}$$

Thus the only complex zeros of $\sin z$ are its real zeros at $z = n\pi$ for integers $n$.

**24.**
$$\begin{aligned} e^z &= e^{x+yi} = e^x\cos y + ie^x\sin y \\ e^{-z} &= e^{-x-yi} = e^{-x}\cos y - e^{-x}\sin y \\ \cosh z &= \frac{e^z + e^{-z}}{2} = \frac{e^x + e^{-x}}{2}\cos y + i\frac{e^x - e^{-x}}{2}\sin y \\ &= \cosh x\cos y + i\sinh x\sin y \end{aligned}$$
$$\mathrm{Re}(\cosh z) = \cosh x\cos y, \quad \mathrm{Im}(\cosh z) = \sinh x\sin y.$$

**26.**
$$\begin{aligned} e^{iz} &= e^{-y+xi} = e^{-y}\cos x + ie^{-y}\sin x \\ e^{-iz} &= e^{y-xi} = e^y\cos x - ie^y\sin x \\ \cos z &= \frac{e^{iz} + e^{-iz}}{2} = \frac{e^{-y} + e^y}{2}\cos x + i\frac{e^{-y} - e^y}{2}\sin x \\ &= \cos x\cosh y - i\sin x\sinh y \end{aligned}$$
$$\mathrm{Re}(\cos z) = \cos x\cosh y, \quad \mathrm{Im}(\cos z) = -\sin x\sinh y$$
$$\begin{aligned} \sin z &= \frac{e^{iz} - e^{-iz}}{2i} = \frac{e^{-y} - e^y}{2i}\cos x + i\frac{e^{-y} + e^y}{2i}\sin x \\ &= \sin x\cosh y + i\cos x\sinh y \end{aligned}$$
$$\mathrm{Re}(\sin z) = \sin x\cosh y, \quad \mathrm{Im}(\sin z) = \cos x\sinh y.$$

**28.** $z^2 - 2z + i = 0 \Rightarrow (z-1)^2 = 1 - i$

$$= \sqrt{2}\left(\cos\frac{7\pi}{4} + i\sin\frac{7\pi}{4}\right)$$

$$\Rightarrow z = 1 \pm 2^{1/4}\left(\cos\frac{7\pi}{8} + i\sin\frac{7\pi}{8}\right)$$

**30.** $z^2 - 2iz - 1 = 0 \Rightarrow (z-i)^2 = 0$

$$\Rightarrow z = i \quad \text{(double root)}$$

**32.** $z^4 - 2z^2 + 4 = 0 \;\Rightarrow\; (z^2-1)^2 = -3$

$z^2 = 1 - i\sqrt{3}$ or $z^2 = 1 + i\sqrt{3}$

$$z^2 = 2\left(\cos\frac{5\pi}{3} + i\sin\frac{5\pi}{3}\right),\quad z^2 = 2\left(\cos\frac{\pi}{3} + i\sin\frac{\pi}{3}\right)$$

$$z = \pm\sqrt{2}\left(\cos\frac{5\pi}{6} + i\sin\frac{5\pi}{6}\right),\quad \text{or}$$

$$z = \pm\sqrt{2}\left(\cos\frac{\pi}{6} + i\sin\frac{\pi}{6}\right)$$

$$z = \pm\left(\sqrt{\frac{3}{2}} - \frac{i}{\sqrt{2}}\right),\quad z = \pm\left(\sqrt{\frac{3}{2}} + \frac{i}{\sqrt{2}}\right)$$

**34.** Since $P(z) = z^4 - 4z^3 + 12z^2 - 16z + 16$ has real coefficients, if $z_1 = 1 - \sqrt{3}i$ is a zero of $P(z)$, then so is $\overline{z_1}$. Now

$$(z - z_1)(z - \overline{z_1}) = (z-1)^2 + 3 = z^2 - 2z + 4.$$

By long division (details omitted) we discover that

$$\frac{z^4 - 4z^3 + 12z^2 - 16z + 16}{z^2 - 2z + 4} = z^2 - 2z + 4.$$

Thus $z_1$ and $\overline{z_1}$ are both *double zeros* of $P(z)$. These are the only zeros.

**36.** Since $P(z) = z^5 - 2z^4 - 8z^3 + 8z^2 + 31z - 30$ has real coefficients, if $z_1 = -2 + i$ is a zero of $P(z)$, then so is $z_2 = -2 - i$. Now

$$(z - z_1)(z - z_2) = z^2 + 4z + 5.$$

By long division (details omitted) we discover that

$$\frac{z^5 - 2z^4 - 8z^3 + 8z^2 + 31z - 30}{z^2 + 4z + 5} = z^3 - 6z^2 + 11z - 6.$$

Observe that $z_3 = 1$ is a zero of $z^3 - 6z^2 + 11z - 6$. By long division again:

$$\frac{z^3 - 6z^2 + 11z - 6}{z - 1} = z^2 - 5z + 6 = (z-2)(z-3).$$

Hence $P(z)$ has the five zeros $-2 + i$, $-2 - i$, 1, 2, and 3.